Studies in Computational Intelligence

Volume 914

Series Editor

Janusz Kacprzyk, Polish Academy of Sciences, Warsaw, Poland

The series "Studies in Computational Intelligence" (SCI) publishes new developments and advances in the various areas of computational intelligence—quickly and with a high quality. The intent is to cover the theory, applications, and design methods of computational intelligence, as embedded in the fields of engineering, computer science, physics and life sciences, as well as the methodologies behind them. The series contains monographs, lecture notes and edited volumes in computational intelligence spanning the areas of neural networks, connectionist systems, genetic algorithms, evolutionary computation, artificial intelligence, cellular automata, self-organizing systems, soft computing, fuzzy systems, and hybrid intelligent systems. Of particular value to both the contributors and the readership are the short publication timeframe and the world-wide distribution, which enable both wide and rapid dissemination of research output.

The books of this series are submitted to indexing to Web of Science, EI-Compendex, DBLP, SCOPUS, Google Scholar and Springerlink.

More information about this series at http://www.springer.com/series/7092

Arash Shaban-Nejad · Martin Michalowski ·
David L. Buckeridge
Editors

Explainable AI in Healthcare and Medicine

Building a Culture of Transparency and Accountability

Editors
Arash Shaban-Nejad
Department of Pediatrics,
College of Medicine
The University of Tennessee
Health Science Center (UTHSC),
Oak-Ridge National Lab (ORNL)
Memphis, TN, USA

Martin Michalowski
School of Nursing
University of Minnesota
Minneapolis, MN, USA

David L. Buckeridge
McGill Clinical & Health Informatics
Montreal, QC, Canada

ISSN 1860-949X ISSN 1860-9503 (electronic)
Studies in Computational Intelligence
ISBN 978-3-030-53354-0 ISBN 978-3-030-53352-6 (eBook)
https://doi.org/10.1007/978-3-030-53352-6

© The Editor(s) (if applicable) and The Author(s), under exclusive license
to Springer Nature Switzerland AG 2021
This work is subject to copyright. All rights are solely and exclusively licensed by the Publisher, whether the whole or part of the material is concerned, specifically the rights of translation, reprinting, reuse of illustrations, recitation, broadcasting, reproduction on microfilms or in any other physical way, and transmission or information storage and retrieval, electronic adaptation, computer software, or by similar or dissimilar methodology now known or hereafter developed.
The use of general descriptive names, registered names, trademarks, service marks, etc. in this publication does not imply, even in the absence of a specific statement, that such names are exempt from the relevant protective laws and regulations and therefore free for general use.
The publisher, the authors and the editors are safe to assume that the advice and information in this book are believed to be true and accurate at the date of publication. Neither the publisher nor the authors or the editors give a warranty, expressed or implied, with respect to the material contained herein or for any errors or omissions that may have been made. The publisher remains neutral with regard to jurisdictional claims in published maps and institutional affiliations.

This Springer imprint is published by the registered company Springer Nature Switzerland AG
The registered company address is: Gewerbestrasse 11, 6330 Cham, Switzerland

Preface

Artificial intelligence is revolutionizing healthcare and medicine. Advances in technology come with many responsibilities, restrictions, and concerns related to transparency, accountability, and ethics. One of the main challenges in the application of AI to healthcare and medicine is the lack of explanation that is understandable by patients and caregivers regarding diagnostic and therapeutic solutions, prescriptions, and interventions.

This book highlights the latest achievements in the application of artificial intelligence to healthcare and medicine. The edited volume contains selected papers presented at the 2020 International Workshop on Health Intelligence, co-located with the Association for the Advancement of Artificial Intelligence (AAAI) annual conference. The papers present an overview of the issues, challenges, and potential in the field, along with new research results. The book makes the emerging topics of digital health and explainable AI in health and medicine accessible to a broad readership with a wide range of practical applications. It provides information for scientists, researchers, students, industry professionals, national and international public health agencies, and NGOs interested in the theory and practice of digital and precision medicine and health, with an emphasis on individual risk factors for disease prevention, diagnosis, and intervention.

Memphis, USA	Arash Shaban-Nejad
Minneapolis, USA	Martin Michalowski
Montreal, Canada	David L. Buckeridge

Contents

List of Contributors

Vered Aharonson University of the Witwatersrand, Johannesburg, South Africa

Nariman Ammar Department of Pediatrics, College of Medicine, The University of Tennessee Health Science Center - Oak-Ridge National Lab (UTHSC-ORNL), Center for Biomedical Informatics, Memphis, TN, USA

Wangpeng An Tsinghua University, Beijing, China

Corey W. Arnold Computational Diagnostics Lab, Department of Radiological Sciences, Los Angeles, CA, USA;
Department of Bioengineering, Los Angeles, CA, USA;
Department of Pathology, Los Angeles, CA, USA

James E. Bailey Center for Health System Improvement, College of Medicine, University of Tennessee Health Science Center, Memphis, TN, USA

Sivaraman Balakrishnan Carnegie Mellon University, Pittsburgh, PA, USA

Tathagata Banerjee BlackThorn Therapeutics, San Francisco, CA, USA

Dennis Barbour Washington University in St. Louis, St. Louis, MO, USA

Megan Barnes Department of Linguistics, University of Washington, Seattle, WA, USA

Hande Batan University of Colorado Boulder, Boulder, CO, USA

Jeffrey P. Bigham Carnegie Mellon University, Pittsburgh, USA

Joschka Boedecker Neurorobotics Lab, Department of Computer Science, University of Freiburg, Freiburg, Germany

David L. Buckeridge McGill Clinical and Health Informatics, School of Population and Global Health, McGill University, Montreal, QC, Canada

Jonathon Byrd Carnegie Mellon University, Pittsburgh, PA, USA

Sophie Mai Chau IBM Research–Zurich, Rüschlikon, Switzerland

Hugh Chen Paul G. Allen School of CSE, University of Washington, Seattle, USA

Lin Chen Cambia Health Solutions, Seattle, WA, USA

Trevor Cohen Department of Biomedical Informatics and Medical Education, University of Washington, Seattle, WA, USA

Yuan Cui Oberlin College, Oberlin, USA

Gari D. Clifford Georgia Institute of Technology, Atlanta, GA, USA;
Emory University, Atlanta, GA, USA

Tirthankar Dasgupta TCS Research and Innovation, Kolkata, India

Robert L. Davis Department of Pediatrics, College of Medicine, The University of Tennessee Health Science Center - Oak-Ridge National Lab (UTHSC-ORNL), Center for Biomedical Informatics, Memphis, TN, USA

Lipika Dey TCS Research and Innovation, Kolkata, India

John P. Dickerson University of Maryland, College Park, USA

Samuel Dooley University of Maryland, College Park, USA

Ruth A. Engelberg Cambia Palliative Care Center of Excellence, University of Washington, Seattle, WA, USA;
Division of Pulmonary, Critical Care, and Sleep Medicine, Department of Medicine, University of Washington, Seattle, WA, USA

Libby Ferland Department of Computer Science and Engineering, University of Minnesota, Minneapolis, MN, USA

Qingzhu Gao BlackThorn Therapeutics, San Francisco, CA, USA

Pantelis Georgiou Department of Electrical and Electronic Engineering, Imperial College London, London, UK

Marzyeh Ghassemi Department of Computer Science, University of Toronto, Toronto, Canada

Maria Gini Department of Computer Science and Engineering, University of Minnesota, Minneapolis, MN, USA

Lucas Glass IQVIA, Plymouth Meeting, PA, USA

Stephen Gou Department of Computer Science, University of Toronto, Toronto, Canada

Kendra Griesman Haverford College, Haverford, USA

Yang Gu Suning USA, Palo Alto, CA, USA

Yang Guo Institute for Infocomm Research Agency for Science, Technology and Research (A*STAR), Singapore, Singapore

Martin Horn Cambia Health Solutions, Seattle, WA, USA

Kristen Howell Department of Linguistics, University of Washington, Seattle, WA, USA

Yanke Hu Humana, Irving, TX, USA

Ming Huang Department of Health Sciences Research, Mayo Clinic, Rochester, MN, USA

Songfang Huang IBM Research-China, Beijing, China

Ziming Huang IBM Research-China, Beijing, China

Maria Hügle Neurorobotics Lab, Department of Computer Science, University of Freiburg, Freiburg, Germany

Thomas Hügle Department of Rheumatology, University Hospital Lausanne, CHUV, Lausanne, Switzerland

Taiki Inoue The University of Tokyo, Tokyo, Japan

Yusuke Iwasawa The University of Tokyo, Tokyo, Japan

Robert Jenssen Department of Physics and Technology, UiT The Arctic University of Norway, Tromsø, Norway

Hawoong Jeong Department of Physics, Korea Advanced Institute of Science and Technology, Daejeon, South Korea

Xiaoqian Jiang University of Texas Health Science Center at Houston, Houston, TX, USA

Sabin Kafle Cambia Health Solutions, Seattle, WA, USA

Gabriel Kalweit Neurorobotics Lab, Department of Computer Science, University of Freiburg, Freiburg, Germany

Hamid R. Karimian Michigan State University, East Lansing, MI, USA

Michie Kawashima Kyoto Sangyo University, Kyoto, Japan

Yoshimasa Kawazoe The University of Tokyo, Tokyo, Japan

Ariane Kehlbacher University of Reading, Reading, Berkshire, UK

Malik Khadar Department of Computer Science and Engineering, University of Minnesota, Minneapolis, MN, USA

Nisei Kimura The University of Tokyo, Tokyo, Japan

Judith Klein-Seetharaman Colorado School of Mines, Golden, CO, USA

Matthew Kollada BlackThorn Therapeutics, San Francisco, CA, USA

Sandeep Konam Abridge AI Inc., Pittsburgh, USA

Parisa Kordjamshidi Michigan State University, East Lansing, MI, USA

Kundan Krishna Carnegie Mellon University, Pittsburgh, USA

Pavitra Krishnaswamy Institute for Infocomm Research Agency for Science, Technology and Research (A*STAR), Singapore, Singapore

Satomi Kuroshima Tamagawa University, Tokyo, Japan

Trevor Larsen Washington University in St. Louis, St. Louis, MO, USA

Byunghwee Lee Department of Physics, Korea Advanced Institute of Science and Technology, Daejeon, South Korea

Robert Y. Lee Cambia Palliative Care Center of Excellence, University of Washingt, Seattle, WA, USA;
Critical Care, and Sleep Medicine, Department of Medicine, University of Washingt, Seattle, WA, USA

Su-In Lee Paul G. Allen School of CSE, University of Washington, Seattle, USA

Brenna Li Department of Computer Science, University of Toronto, Toronto, Canada

Kezhi Li Institute of Health Informatics, University College London, London, UK

Shaochun Li IBM Research-China, Beijing, China

Xiang Li Cambia Health Solutions, Seattle, WA, USA

Zachary C. Lipton Carnegie Mellon University, Pittsburgh, PA, USA

Yuyang Liu Department of Computer Science, University of Toronto, Toronto, Canada

Zhengyuan Liu Institute for Infocomm Research Agency for Science, Technology and Research (A*STAR), Singapore, Singapore

William B. Lober Cambia Palliative Care Center of Excellence, University of Washington, Seattle, WA, USA;
Department of Biobehavioral Nursing and Health Informatics, University of Washington, Seattle, WA, USA;
Department of Biomedical Informatics and Medical Education, University of Washington, Seattle, WA, USA

Michael Lucke Department of Computer Science and Engineering, University of Minnesota, Minneapolis, MN, USA

Scott Lundberg Microsoft Research, Albuquerque, USA

Gustavo Malkomes SigOpt, San Francisco, CA, USA

Muhammad Mamdani Unity Health Toronto, Toronto, Canada

William J. Martin BlackThorn Therapeutics, San Francisco, CA, USA

María Rodríguez Martínez IBM Research–Zurich, Rüschlikon, Switzerland

Yutaka Matsuo The University of Tokyo, Tokyo, Japan

Duncan C. McElfresh University of Maryland, College Park, USA

Jing Mei IBM Research-China, Beijing, China

Monika S. Mellem BlackThorn Therapeutics, San Francisco, CA, USA

Yiwen Meng Computational Diagnostics Lab, Department of Radiological Sciences, Los Angeles, CA, USA;
Department of Bioengineering, Los Angeles, CA, USA

Martin Michalowski School of Nursing, University of Minnesota, Minneapolis, MN, USA

Bruno Michel IBM Research–Zurich, Rüschlikon, Switzerland

Karl Øyvind Mikalsen University Hospital of North-Norway, Tromsø, Norway; Department of Physics and Technology, UiT The Arctic University of Norway, Tromsø, Norway

Ishani Mondal IIT Kharagpur, Kharagpur, India

Michael J. Moore Tulane University, New Orleans, LA, USA;
Tulane Brain Institute; and AxoSim Inc., New Orleans, LA, USA

Joshua Murray Unity Health Toronto, Toronto, Canada

Kotaro Nakayama The University of Tokyo, Tokyo, Japan;
NABLAS Inc, Bunkyo City, Japan

Abir Naskar TCS Research and Innovation, Kolkata, India

Runpeng Nie Department of Computer Science and Engineering, University of Minnesota, Minneapolis, MN, USA

Jorge J. Nieva USC Norris Comprehensive Cancer Center, Los Angeles, CA, USA

Tanachat Nilanon University of Southern California, Los Angeles, CA, USA

Luciano P. Nocera University of Southern California, Los Angeles, CA, USA

Kazuhiko Ohe The University of Tokyo, Tokyo, Japan

Michael Ong Department of Medicine, Los Angeles, CA, USA

Masaki Onishi National Institute of Advanced Industrial Science and Technology, Ibaraki, Japan

Ali Oskooei IBM Research–Zurich, Rüschlikon, Switzerland

Serguei Pakhomov Department of Pharmaceutical Care and Health Systems, University of Minnesota, Minneapolis, MN, USA

Michael J. Paul University of Colorado Boulder, Boulder, CO, USA

Amy Pavel Carnegie Mellon University, Pittsburgh, USA

Kevin Pollard Tulane University, New Orleans, LA, USA

Chloe Pou-Prom Unity Health Toronto, Toronto, Canada

Dianna Radpour University of Colorado Boulder, Boulder, CO, USA

Savitha Ramasamy Institute for Infocomm Research Agency for Science, Technology and Research (A*STAR), Singapore, Singapore

J. Randall Curtis Cambia Palliative Care Center of Excellence, University of Washington, Seattle WA, USA;
Division of Pulmonary, Critical Care, and Sleep Medicine, Department of Medicine, University of Washington, Seattle WA, USA;
Department of Biobehavioral Nursing and Health Informatics, University of Washington, Seattle WA, USA

Zahra Sabet AIG, London, UK

Kenya Sakka The University of Tokyo, Tokyo, Japan

Kei Sato University of Tsukuba, Ibaraki, Japan

Jude Sauve Department of Computer Science and Engineering, University of Minnesota, Minneapolis, MN, USA

Benjamin Schloss Abridge AI Inc., Pittsburgh, USA

Nabeel Seedat University of the Witwatersrand, Johannesburg, South Africa;
Shutterstock, New York, USA

Neil Sehgal University of Maryland, College Park, USA

Sai P. Selvaraj Abridge AI Inc., Pittsburgh, USA

Arash Shaban-Nejad Oak-Ridge National Lab (UTHSC-ORNL), Center for Biomedical Informatics, Department of Pediatrics, College of Medicine, The University of Tennessee Health Science Center, Memphis, TN, USA

Nilay D. Shah Department of Health Sciences Research, Mayo Clinic, Rochester, MN, USA

Cyrus Shahabi University of Southern California, Los Angeles, CA, USA

John Shawe-Taylor University College London, London, UK

Eun Kyong Shin Department of Sociology, Korea University, Seoul, South Korea

James Sibley Cambia Palliative Care Center of Excellence, University of Washington, Seattle WA, USA;
Department of Biobehavioral Nursing and Health Informatics, University of Washington, Seattle WA, USA;
Department of Biomedical Informatics and Medical Education, University of Washington, Seattle WA, USA

Gaurav Singh University College London, London, UK

Cristina Soguero-Ruiz Department of Physics and Technology, UiT The Arctic University of Norway, Tromsø, Norway;
Rey Juan Carlos University, Móstoles, Spain

William Speier Computational Diagnostics Lab, Department of Radiological Sciences, Los Angeles, CA, USA

Arvind Sridhar IBM Research–Zurich, Rüschlikon, Switzerland

Raj Subramanian Humana, Irving, TX, USA

Pradyumna Byappanahalli Suresha Georgia Institute of Technology, Atlanta, GA, USA

James Thomas University College London, London, UK

Kotaro Uchida Tokyo Medical University, Tokyo, Japan

Amol Verma Unity Health Toronto, Toronto, Canada

Ke Wang IBM Research-China, Beijing, China

Weiqin Wang Pennsylvania State University, State College, USA

Yunlong Wang IQVIA, Plymouth Meeting, PA, USA

Jonas Weiss IBM Research–Zurich, Rüschlikon, Switzerland

Tyler Will Michigan State University, East Lansing, USA

Weili Wu Department of Computer Science, The University of Texas at Dallas, Richardson, TX, USA

Eryu Xia IBM Research-China, Beijing, China

Cao Xiao IQVIA, Boston, MA, USA

Ryohei Yamaguchi The University of Tokyo, Tokyo, Japan

Jienan Yao Department of Computer Science, University of Toronto, Toronto, Canada

Lixia Yao Department of Health Sciences Research, Mayo Clinic, Rochester, MN, USA

Ikushi Yoda National Institute of Advanced Industrial Science and Technology, Ibaraki, Japan

Yilian Yuan IQVIA, Plymouth Meeting, PA, USA

Na Zhao Peking University School and Hospital of Stomatology, Beijing, China

Shiwan Zhao IBM Research-China, Beijing, China

Taiyu Zhu Department of Electrical and Electronic Engineering, Imperial College London, London, UK

Abbreviations

ACS	Acute Coronary Syndrome
ACP	Advance Care Planning
AD	Advance Directive
AF	Atrial Fibrillation
AL	Anastomosis Leakage
AI	Artificial Intelligence
AMED	Agency for Medical Research and Development
ANS	Autonomic Nervous System
ART	Adaptive Semi-Supervised Recursive Tree Partitioning
ASR	Automatic Speech Recognition
ATOM-HP	Analytical Tools to Objectively Measure Human Performance
AUROC	Area Under the ROC Curve
BBB	Bayes by Backprop
BLSTM	Bayesian Long Short-Term Memory
C2AE	Canonical Correlated Auto Encoder
CA	Conversational Agent
CAE	Convolutional Autoencoder
CCA	Canonical Component Analysis
CCE	Categorical Cross Entropy
CDSM	Chronic Disease Self-Management
CEWS	Clinical Early Warnings System
CGM	Continuous Glucose Monitoring
CNN	Convolutional Neural Networks
CRF	Conditional Random Fields
CTW	Canonical Time Warping
GCTW	Generalized Canonical Time Warping
GNNs	Graph Neural Networks
GRAS	Generally Recognized as Safe
CRC	Colon Rectal Cancer
CTs	Clinical Trials

DCTW	Deep Canonical Time Warping
DL	Deep Learning
DNR	Do Not Resuscitate
DQN	Deep Q-Networks
DRNN	Dilated Recurrent Neural Networks
DTW	Dynamic Time Warping
ECG	Electro Cardiography
ECOG	Eastern Cooperative Oncology Group
EHR	Electronic Health Records
EMD	Empirical Mode Decomposition
EMRs	Emergency Medicine Residents
EMRs	Electronic Medical Records
EOL	End of Life
EpiNet	Epidemic Networks
FAQ	Frequently Asked Question
FCNs	Fully Connected Networks
FDA	Food and Drug Administration
FOAF	Friend of A Friend
FT	FastText
GAC	Global Alignment Kernel
GIM	General Internal Medicine
GLoVe	Global Vectors for Word Representation
GRU	Gated Recurrent Unit
HCC	Hepatocellular Carcinoma
HF	High Frequency
HITECH	Health Information Technology for Economic and Clinical Health
HR	Heart Rate
HRV	Heart-Rate Variability
ICD-9	International Classification of Disease, ninth revision
ICU	Intensive Care Units
INR	International Normalized Ratio
IOB	Inside-Outside-Beginning
IRB	Institutional Review Board
ISCR	Insulin Suspension and Carbohydrate Recommendation
IW-DTW	Invariant Weighted Dynamic Time Warping
JSPS	Japan Society for the Promotion of Science
JSRT	Japanese Social of Radiological Technology
JWT	JSON Web Token
KELSA	Knowledge-Enriched Local Sequence Alignment
LAE	LSTM Autoencoder
LDA	Latent Dirichlet Allocation
LF	Low Frequency
LPS	Learned Pattern Similarity
LSH	Locally Sensitive Hashing
LST	Life-Sustaining Treatments

LSTM	Long Short-Term Memory
LW	Living Well
MACE	Major Adverse Cardiovascular Events
MAE	Mean Absolute Error
MAR	Missing at Random
MaxEnt	Maximum Entropy
MCAR	Missing Completely at Random
MCV	Medical Code Vectors
MDP	Markov Decision Process
MEL	Medical Embedding Layer
MELD	Model for End-stage Liver Disease
MERS	Middle East Respiratory Syndrome
MeSH	Medical Subject Heading
MIMIC-III	Medical Information Mart for Intensive Care
ML	Machine Learning
MLET	Multi-Level Embedding with Topic modeling
MLP	Multi-Layer Perceptron
MM	Multiple Medicine
MN	Multiple Numbers
MNAR	Missing Not at Random
MPGO	Massachusetts General Physicians Organization
MPOA	Medical Power of Attorney
MR	Medication Regimen
MSE	Mean Squared Error
MTS	Multivariate Time Series
MW	Manifold Warping
NAFLD	Nonalcoholic Fatty Liver Disease
NASH	Nonalcoholic Steatohepatitis
NCBO	National Center for Biomedical Ontology
NLP	Natural Language Processing
NLU	Natural Language Understanding
NIOSH	National Institute for Occupational Safety and Health
OPOM	Optimized Predictor of Mortality
OPTN	Organ Procurement and Transplantation Network
PCA	Principal Components Analysis
PDE	Patient Demographics Embedding
PEL	Patient Embedding Layer
PD	Parkinson's Disease
PHI	Protected Health Information
PHL	Personal Health Library
PICO	Population, Intervention/Control, Outcome
PODs	Personal Online Data stores
POLST	Physician Orders for Life-Sustaining Treatment
PS	Patient Performance Status
QA	Question Answering

RA Rheumatoid Arthritis
REF Reference Alignments
ReLU Rectified Linear Units
REP Rochester Epidemiology Project
RL Reinforcement Learning
RMSSD Root Mean Square of Successive Differences
RNN Recurrent Neural Network
RoS Review of Systems
PRAUC Precision Recall Area Under the Curve
RS Random Forest
SCQM Swiss Clinical Quality Management
SOLID Social Linked Data
SOTA State-of-the-art
SSI Surgical Site Infection
STAR Standard Transplant Analysis and Research
SWA Smith–Waterman Algorithm
T1D Type 1 Diabetes
t-SNE t-distributed Stochastic Neighbor Embedding
TCK Time Series Cluster Kernel
TCR Transplant Candidate Registration
TIPS Transjugular Intrahepatic Portosystemic Shunt
TweetNet Networks on Twitter
UI User Interface
UMLS Unified Medical Language System
UTS Univariate Time Series
WHO World Health Organization
WOZ Wizard of Oz Protocol
XAI Explainable AI
XGB XGBoost

Explainability and Interpretability: Keys to Deep Medicine

Arash Shaban-Nejad, Martin Michalowski, and David L. Buckeridge

Abstract Deep medicine, which aims to push the boundaries of artificial intelligence to reshape the health and medical intelligence and decision making, is a promising concept that is gaining attention over traditional EMR-based medical information management systems. The success of intelligent solutions in health and medicine depends on the degree to which they support interoperability, to allow consistent integration of different systems and data sources, and explainability, to make their decisions understandable, interpretable, and justifiable by humans.

Keywords Deep medicine · Precision medicine · Explainable AI · Interpretability

1 Introduction

Artificial intelligence (AI) is revolutionizing [1, 2] healthcare and medicine, empowering precision medicine with its seven dimensions [3]: precision observation and assessment, health promotion, engagement, early detection, prevention, treatments, and equity. Eric Topol [4], who coined the term Deep medicine, suggests that AI promises "to provide composite, panoramic views of individuals' medical data; to improve decision-making; to avoid errors such as misdiagnosis and unnecessary

A. Shaban-Nejad (✉)
Oak-Ridge National Lab (UTHSC-ORNL), Center for Biomedical Informatics, Department of Pediatrics, College of Medicine, The University of Tennessee Health Science Center, Memphis, TN, USA
e-mail: ashabann@uthsc.edu

M. Michalowski
School of Nursing, University of Minnesota, Minneapolis, MN, USA
e-mail: martinm@umn.edu

D. L. Buckeridge
McGill Clinical and Health Informatics, School of Population and Global Health, McGill University, Montreal, QC H3A 1A3, Canada
e-mail: david.buckeridge@mcgill.ca

© The Editor(s) (if applicable) and The Author(s), under exclusive license to Springer Nature Switzerland AG 2021

A. Shaban-Nejad et al. (eds.), *Explainable AI in Healthcare and Medicine*, Studies in Computational Intelligence 914,
https://doi.org/10.1007/978-3-030-53352-6_1

procedures; to help in the ordering and interpretation of appropriate tests, and to recommend treatment". The concept of Deep medicine rests on three major processes namely "deep phenotyping", "deep learning" and "deep empathy and human connection" [4]. There are many ways to empower patients about their health and well-being. Creating personalized treatment and interventions (e.g. personalized diet regimens) often requires aggregating data in the problem domain with genomics data and data from wearable devices [5, 6]. However, many of these data are scattered across siloed data sources and run on multiple platforms and operating systems, lacking the interoperability needed for integration, analysis, and interpretation [7]. Also, in medicine the notions of transparency, accountability, and ethics are crucial and the need for explainability and interpretability of the predicated outcomes and their underlying models and algorithms is evident. Systems are interpretable "if their operations can be understood by a human, either through introspection or through a produced explanation" [8].

2 Explainability

One of the main challenges in today's clinical and public health decision support systems is the lack of sufficient explanation understandable by stakeholders (e.g. physicians, patients, researchers, public) regarding their diagnostic and therapeutic solutions and interventions [9]. Unlike classic analytical models built with traditional statistical analysis, complex models built using Machine Learning (ML) methods can be more difficult to explain and justify to a human user or decision-maker. As such, there is a growing area of research concerning the explainability of ML systems [8, 10] referred to as Explainable Artificial Intelligence (XAI). XAI aims to provide justification, transparency, and traceability of black-box machine learning methods as well as testability of causal assumptions [5, 6]. The black-box mechanism in ML methods often makes it difficult (or legally impossible due to confidentiality concerns) for outsiders to fully understand the underlying algorithm and identify potential biases. Also, there are other issues besides confidentiality, such as complexity, unreasonableness, and injustice (when decisions are justifiable but are discriminatory or unfair) [11]. Explainable models support more trustable inference to answer "how" and "why" questions, in addition to "what", "when" "where", and "who" questions [12]. The quality of an XAI system can be assessed using various metrics to examine the quality of explanations, users' satisfaction, understandability, trustability, and examine the "human-XAI" system performance [13]. An integrative review of methods in Human-XAI systems can be found in [14, 15].

XAI is especially instrumental in medicine and healthcare to ensure that the recommendations made by intelligent systems are sound, correct, and justifiable and enable the care providers to make better decisions and infer new knowledge, insights, or discovery.

Also, explanation is needed to justify why a decision is made, a treatment is prescribed, a health intervention is implemented, and how they are preferred over other different options available. XAI has been used in a wide range of biomedical and health applications such as neuroscience and behavioral neurostimulation [16], genomics [17], breast cancer [18], and public health intervention evaluation [12].

To have a more explainable AI, one may consider incorporating prior knowledge into the AI algorithms or using ontologies [19] that capture a domain knowledge within concepts, properties, individuals, and axioms. Ontologies have been widely used to support public health intelligence [20–22], clinical intelligence [23–25], global health surveillance [26, 27]. Ontologies can provide semantic explainability for causal links and relations within a complex medical and non-medical data sets and generate more human-understandable recommendations and evaluate existing treatments or health interventions [28, 29].

3 Interoperability

Interoperability is another prerequisite for deep medicine [4]. According to the Healthcare Information and Management Systems Society (HIMMS) [30] interoperability in healthcare is defined as "the ability of different information systems, devices and applications ('systems') to access, exchange, integrate and cooperatively use data in a coordinated manner, within and across organizational, regional and national boundaries, to provide timely and seamless portability of information and optimize the health of individuals and populations globally." Interoperability has gained increasing attention in the health and medical domain [7, 31]. The concept of explainability is tightly connected to interoperability, especially when studying cause and effect within an integrated system. For example, when evaluating an intervention [28], interoperability helps you to predict the outcome, given a change in other elements or input parameters.

Four levels of interoperability [30] can be specified: Foundational, the interconnectivity between two systems to securely exchange data; Structural, compatibility of format, and syntax for data exchange; Semantics, standardization of common models and underlying shared meaning for data integration and exchange often through the use of ontologies and controlled vocabularies; and Organizational, to maintain data governance, policy, social and legal processes, and workflows.

Maintaining interoperability is a key not only for success in clinical data exchange but also for multi-disciplinary health surveillance at local [20] and global [26, 27] levels. Without a proper level of interoperability, the integration of multi-dimensional socio-economic [32], environmental [33, 34], and health data into a consistent integrated repository is hard to imagine.

4 Artificial Intelligence Tools and Methods

Advances in Artificial Intelligence tools and techniques facilitate knowledge modeling, analysis, and discovery to improve our understanding of underlying causes and risk factors for many adverse health conditions and their interactions. This, in turn, can improve the explainability of decisions made (or recommended) by intelligent health information systems and result in a more transparent healthcare system. Some examples of such AI-based methods are as follows.

In an attempt to accurately and efficiently retrieve similar patients from large scale healthcare data, Wang et al. [35] presented an approach for similar patient retrieval using deep learning-based binary hash codes and embedding vectors derived from an artificial neural network. Mikalsen et al. [36] designed a kernel, using an ensemble learning strategy based on novel mixed-mode Bayesian mixture models, for exploiting both the information from the observed values and the information hidden in the missing patterns in multivariate time series (MTS) originating from electronic health records. Seedat et al. [37] applied machine learning to discriminate six stages of Parkinson's Disease using data from walker-mounted sensors at a movement disorders clinic. Zhu et al. [38] showed a dual-hormone control algorithm to assist people with Type 1 Diabetes (T1D) using deep reinforcement learning (RL). Kollada et al. [39] presented a machine learning method for automating the quality control (QC) of fMRI scanned images. Guo et al. [40] proposed a Bayesian Long Short-term Memory (LSTM) framework to characterize both modeling (epistemic) and data related (aleatoric) uncertainties in prediction tasks for clinical time-series data. The method has been evaluated for mortality prediction using ICU data sets.

EMRs, registries, and trials provide a large source of data to be used by machine learning methods. However, to take full advantage of deep learning methods applied to clinical data, architectures must be robust concerning missing and wrong values, and be able to deal with highly variable-sized lists and long-term dependencies of individual diagnosis, procedures, measurements, and medication prescriptions. Hügle et al. [41] elaborate on limitations of fully connected neural networks and classical machine learning methods in this context and propose Adaptive Net, a novel recurrent neural network architecture that can deal with multiple lists of different events, alleviating the aforementioned limitations. Oskooei et al. [42] presented a platform, DeStress, that employs deep learning for unsupervised identification of mental stress in firefighters from heart-rate variability (HRV) data. After exploring and comparing different methods they demonstrate that the clusters produced by the convolutional autoencoders consistently and successfully stratify stressed versus normal samples, as validated by several established physiological stress markers. Suresha et al. [43] presented a deep learning approach for discriminating between patients with nonalcoholic steatohepatitis and those with nonalcoholic fatty liver disease using electronic medical records. Yao et al. [44] discussed predicting patient's health outcomes through the visualization of deep models based on convolutional neural networks (CNN) applied to in-house hospital nursing notes and physiological data with temporally segmented sliding windows.

Text mining and Natural Language processing tools and techniques [45] have been widely used in biomedical domain. Singh et al. [46] proposed a method for biomedical text annotation using pre-trained bidirectional encoders for mapping clinical texts to output concepts. Sakka et al. [47] proposed an attention-based model that generates Japanese findings at the character-level from chest radiographs. This method aims to overcome some challenges in Japanese language processing where the boundaries of words are not clear and there are numerous orthographic variants. Krishna et al. [48] explored the possibility of using data related to the conversations between physicians and patients at the time of care to extract structured information that might assist physicians with post-visit documentation in electronic health records, and potentially lighten the clerical burden. Horn et al. [49] described the implementation of a secure, multi-task healthcare chatbot to provide easy, fast, on-demand conversational access to a variety of health-related resources. Batan et al. [50] presented a study to investigate the use of social media, more specifically Twitter, data to analyze consumer behaviors regarding natural and artificial sweeteners. They performed a topic model analysis and characterize tweet volumes over time, representing a variety of sweetener-related content. To examine the relationship between the epidemic networks (EpiNet) and the corresponding discourse networks on Twitter (TweetNet) and to explain and unpack the epidemic diffusion process and the epidemic-related social media discourse diffusion Lee et al. [51] linked data from patients with confirmed MERS to social media data mentioning MERS and used network analyses and simulations. Selvaraj and Konam [52] explored the Medication Regimen (MR) extraction task to find out about the dosage and frequency for the medications mentioned in a doctor-patient conversation transcript. Sato et al. [53] developed a quantitative method for evaluating non-technical skills of Emergency Medicine Residents (EMRs) and created a workflow event database based on the trajectories of and conversations among the medical personnel and scores an EMR's non-technical skills based on that database.

Electronic health records and personal health records [54] continue to play a central role in intelligent clinical and public health systems. Ammar et al. [55] discussed the implementation of an integrated Personal Health Library (PHL) for Chronic disease self-management (CDSM) using the Social Linked Data framework through a decentralized linked open data platform. The PHL grants patients true ownership over their data while giving them fine-grained access control mechanisms to exchange and share their data while supporting data interoperability and portability. Huang et al. [56] proposed KELSA, a knowledge-enriched local sequence alignment algorithm for comparing patient medical records, which incorporates meaningful medical knowledge during sequence alignment. The algorithm has been evaluated by comparing it to the Smith-Waterman algorithm (SWA) on synthetic EHR data where the reference alignments are known. To predict depression using different time windows before a clinical diagnosis Meng et al. [57] presented a model that utilizes heterogeneous data and attempts to address sparsity using temporal Multi-Level Embeddings of diagnoses, procedures, and medication codes with demographic information and Topic modeling (MLET). MLET aggregates various categories of

EHR data and learns the inherent structure based on hospital visits in a patient's medical history.

Other research include Hu et al. [58] that provided a filter-based feature engineering method and a two-phase auto hyper parameter optimization approach, combined with tree boosting methods for faster clinical time series classification. Accurate predictions with complex models (e.g., neural networks, ensembles/stacks of different models) impacts patients' well-being. To make these models explainable, Chen et al. [59] presented DeepSHAP for mixed model types, a framework for layer-wise propagation of Shapley values that build upon DeepLIFT (an existing approach for explaining neural networks). This framework enables attributions for stacks of mixed models (e.g., neural network feature extractor into a tree model) as well as attributions of the loss. Using a clinical dataset with a great deal of variation in terms of author and setting, Howell et al. [60] proposed a method for controlling confounding variables through a backdoor adjustment approach. Karimian et al. [61] presented an approach for learning representations to augment the statistical analysis of drug effects on nerve tissues. They aimed to understand the influence of distinct neuropathic conditions on the properties of electrophysiological waveforms produced by such tissues treated with known neurotoxic compounds and healthy controls. Dasgupta et al. [62] presented a method for automatic segregation and classification of inclusion and exclusion criteria of clinical trials to improve patient eligibility matching.

Ferland et al. [63] described the first phases of an ongoing project to create a system designed to detect stress in users and deliver a multimodal touch- and speech-based intervention when users experience. Larsen et al. [64] research aimed at accelerating psychometric screening tests with prior information. They developed a rapid screening algorithm for a change in the psychometric function estimation of patients, then used Bayesian active model selection to perform an automated pure-tone audiometry test to quickly identify if the current estimation will be different from the previous one. Modern AI-based tools are considered to help increase Advance Care Planning (ACP) participation rates and therefore assist patients in future end of life (EOL) care decision making. McElfresh et al. [65] discussed two AI-based applications and their potential implications to improve ACP. Byrd et al. [66] discussed computational methods to improve predicting mortality in liver transplant candidates. Patient Performance Status (PS) is used in cancer medicine to predict prognosis and prescribe treatment. Motion tracking gadgets are useful devices to improve PS assessments. Nilanon et al. [67] applied temporal alignment algorithms to the extracted motion skeleton time series and evaluate their performance in aligning the key frames that separate corresponding motions.

References

1. Shaban-Nejad, A., Michalowski, M., Buckeridge, D.L.: Health intelligence: how artificial intelligence transforms population and personalized health. NPJ Digit. Med. **1** (2018). Article

number: 53
2. Shaban-Nejad, A., Michalowski, M.: Precision Health and Medicine - A Digital Revolution in Healthcare. Studies in Computational Intelligence, vol. 843. Springer, Cham (2020). ISBN 978-3-030-24408-8
3. Shaban-Nejad, A., Michalowski, M., Peek, N., Brownstein, J.S., Buckeridge, D.L.: Seven pillars of precision digital health and medicine. Artif. Intell. Med. **103**, 101793 (2020)
4. Topol, E. Deep Medicine: How Artificial Intelligence Can Make Healthcare Human Again, 1 edn. Basic Books, 11 July 2019
5. Holzinger, A., Langs, G., Denk, H., Zatloukal, K., Müller, H.: Causability and explainability of artificial intelligence in medicine. Wiley Interdiscip. Rev. Data Min. Knowl. Discov. **9**(4), e1312 (2019). https://doi.org/10.1002/widm.1312
6. Pearl, J.: Interpretability and explainability from a causal lens. IPAM Workshop, 16 October 2019. www.helper.ipam.ucla.edu/publications/mlpws2/mlpws2_15879.pdf. Accessed 05 Apr 2020
7. Lehne, M., Sass, J., Essenwanger, A., et al.: Why digital medicine depends on interoperability. NPJ Digit. Med. **2**, 79 (2019). https://doi.org/10.1038/s41746-019-0158-1
8. Biran, O., Cotton, C.: Explanation and justification in machine learning: a survey. In: Proceedings of IJCAI 2017 Workshop on Explainable Artificial Intelligence (XAI). Melbourne, Australia 2017
9. Matuchansky, C.: Deep medicine, artificial intelligence, and the practicing clinician. Lancet **394**(10200), 736 (2019)
10. Adadi, A., Berrada, M.: Peeking inside the black-box: a survey on explainable artificial intelligence (XAI). IEEE Access **6**, 52138–52160 (2018)
11. Felton, E.: What does it mean to ask for an "explainable" algorithm? 31 May 2017. https://freedom-to-tinker.com/2017/05/31/what-does-it-mean-to-ask-for-an-explainable-algorithm/. Accessed 5 May 2020
12. Gunning, D.: Explainable artificial intelligence (XAI). Defense Advanced Research Projects Agency (DARPA) (2017)
13. Hoffman, R.R., Mueller, S.T., Klein, G., Litman, J.: Metrics for explainable AI: Challenges and Prospects. arXiv:1812.04608 [cs.AI], February 2019
14. Miller, T.: Explanation in artificial intelligence: Insights from the social sciences. Artif. Intell. **267**, 1–38 (2019)
15. Mueller, S.T., Hoffman, R.R., Clancey, W., Emrey, Klein, G.A.: Explanation in h-AI systems: a literature meta-review synopsis of key ideas and publications and bibliography for explainable AI. DARPA XAI Literature Review. DARPA XAI Program February 2019. arXiv:1902.01876 [cs.AI]
16. Fellous, J.M., Sapiro, G., Rossi, A., Mayberg, H., Ferrante, M.: Explainable artificial intelligence for neuroscience: behavioral neurostimulation. Front Neurosci. **13**, 1346 (2019). https://doi.org/10.3389/fnins.2019.01346. eCollection 2019
17. Anguita-Ruiz, A., Segura-Delgado, A., Alcalá, R., Aguilera, C.M., Alcalá-Fdez, J.: eXplainable Artificial Intelligence (XAI) for the identification of biologically relevant gene expression patterns in longitudinal human studies, insights from obesity research. PLoS Comput Biol. **16**(4), e1007792 (2020). https://doi.org/10.1371/journal.pcbi.1007792. eCollection 2020 Apr. PMID: 32275707
18. Lamy, J.B., Sekar, B., Guezennec, G., Bouaud, J., Séroussi, B.: Explainable artificial intelligence for breast cancer: a visual case-based reasoning approach. Artif. Intell. Med. **94**, 42–53 (2019). https://doi.org/10.1016/j.artmed.2019.01.001. Epub 2019 Jan 14 PMID: 30871682
19. Landgrebe, J., Smith, B.: Making AI meaningful again. Synthese (2019). https://doi.org/10.1007/s11229-019-02192-y
20. Shaban-Nejad, A., Lavigne, M., Okhmatovskaia, A., Buckeridge, D.L.: PopHR: a knowledge-based platform to support integration, analysis, and visualization of population health data. Ann. N. Y. Acad. Sci. **1387**(1), 44–53 (2017)
21. Buckeridge, D.L., Izadi, M.T., Shaban-Nejad, A., Mondor, L., Jauvin, C., Dubé, L., Jang, Y., Tamblyn, R.: An infrastructure for real-time population health assessment and monitoring. IBM J. Res. Dev. **56**(5), 2 (2012)

22. Shaban-Nejad, A., Buckeridge, D.L., Dubé, L.: COPE: Childhood obesity prevention [Knowledge] enterprise. In: Peleg, M., Lavrač, N., Combi, C. (eds) Artificial Intelligence in Medicine. AIME 2011. Lecture Notes in Computer Science, vol. 6747, pp. 225–229, Springer, Heidelberg (2011)
23. Brenas, J.H., Shin, E.K., Shaban-Nejad, A.: Adverse childhood experiences ontology for mental health surveillance, research, and evaluation: advanced knowledge representation and semantic web techniques. JMIR Ment. Health **6**(5), e13498 (2019). https://doi.org/10.2196/13498
24. Shaban-Nejad, A., Mamiya, H., Riazanov, A., Forster, A.J., Baker, C.J., Tamblyn, R., Buckeridge, D.L.: From cues to nudge: a knowledge-based framework for surveillance of healthcare-associated infections. J. Med. Syst. **40**(1), 23 (2016). https://doi.org/10.1007/s10916-015-0364-6
25. Riazanov, A., Rose, G.W., Klein, A., Forster, A.J., Baker, C.J.O., Shaban-Nejad, A., Buckeridge, D.L.: Towards clinical intelligence with SADI semantic web services: a case study with hospital-acquired infections data. SWAT4LS 2011, pp. 106–113 (2011)
26. Brenas, J.H., Al Manir, M.S., Baker, C.J.O., Shaban-Nejad, A.: A malaria analytics framework to support evolution and interoperability of global health surveillance systems. IEEE Access **5**, 21605–21619 (2017)
27. Al Manir, M.S., Brenas, J.H., Baker, C.J., Shaban-Nejad, A.: A surveillance infrastructure for malaria analytics: provisioning data access and preservation of interoperability MIR public health surveill **4**(2), e10218 (2018). https://doi.org/10.2196/10218
28. Brenas, J.H., Shaban-Nejad, A.: Health intervention evaluation using semantic explainability and causal reasoning. IEEE Access **8**, 9942–9952 (2020)
29. Shaban-Nejad, A., Okhmatovskaia, A., Shin, E.K., Davis, R.L., Franklin, B.E., Buckeridge, D.L.: A semantic framework for logical cross-validation, evaluation and impact analyses of population health interventions. Stud. Health Technol. Inform. **235**, 481–485 (2017)
30. What is Interoperability in Healthcare? https://www.himss.org/what-interoperability. Accessed 5 May 2020
31. Perlin, J.B.: Health information technology interoperability and use for better care and evidence. JAMA **316**, 1667–1668 (2016)
32. Shin, E.K., Mahajan, R., Akbilgic, O., Shaban-Nejad, A.: Sociomarkers and biomarkers: predictive modeling in identifying pediatric asthma patients at risk of hospital revisits. NPJ Digit. Med. **1**, 50, https://doi.org/10.1038/s41746-018-0056-y (2018)
33. Shin, E.K., Shaban-Nejad, A.: Urban decay and pediatric asthma prevalence in Memphis, tennessee: urban data integration for efficient population health surveillance. IEEE Access **6**, 46281–46289 (2018). https://doi.org/10.1109/ACCESS.2018.2866069
34. Shin, E.K., Kwon, Y., Shaban-Nejad, A.: Geo-clustered chronic affinity: pathways from socio-economic disadvantages to health disparities. JAMIA Open **2**(3), 317–322 (2019)
35. Wang, K., Xia, E., Zhao, S., Huang, Z., Huang, S., Mei, J., Li, S.: Fast Similar Patient Retrieval from Large Scale Healthcare Data: A Deep Learning-based Binary Hashing Approach. Explainable AI in Healthcare and Medicine: Building a Culture of Transparency and Accountability. Studies in Computational Intelligence. Springer (2020)
36. Mikalsen, K.Ø., Soguero-Ruiz, C., Jenssen, R.: A kernel to exploit informative missingness in multivariate time series from EHRs. In: Explainable AI in Healthcare and Medicine: Building a Culture of Transparency and Accountability. Studies in Computational Intelligence. Springer, 2020
37. Seedat, N., Aharonson, V.A.: Machine learning discrimination of Parkinson's Disease stages from walker-mounted sensors data. In: Explainable AI in Healthcare and Medicine: Building a Culture of Transparency and Accountability. Studies in Computational Intelligence. Springer (2020)
38. Zhu, T., Li, K., Georgiou, P.: Personalized dual-hormone control for type 1 diabetes using deep reinforcement learning. In: Explainable AI in Healthcare and Medicine: Building a Culture of Transparency and Accountability. Studies in Computational Intelligence. Springer (2020)
39. Kollada, M., Gao, Q., Mellem, M.S., Banerjee, T., Martin, W.J.: A generalizable method for automated quality control of functional neuroimaging datasets. In: Explainable AI in

Healthcare and Medicine: Building a Culture of Transparency and Accountability. Studies in Computational Intelligence. Springer (2020)
40. Guo, Y., Liu, Z., Ramasamy, S., Krishnaswamy, P.: Uncertainty characterization for predictive analytics with clinical time series data. In: Explainable AI in Healthcare and Medicine: Building a Culture of Transparency and Accountability. Studies in Computational Intelligence. Springer (2020)
41. Hügle, M., Kalweit, G., Hügle, T., Boedecker, J.: A dynamic deep neural network for multimodal clinical data analysis. In: Explainable AI in Healthcare and Medicine: Building a Culture of Transparency and Accountability. Studies in Computational Intelligence. Springer (2020)
42. Oskooei, A., Chau, S.M., Weiss, J., Sridhar, A., Martínez, M.R., Michel, B.: DeStress: deep learning for unsupervised identification of mental stress in firefighters from Heart-rate Variability (HRV) data. Explainable AI in Healthcare and Medicine: Building a Culture of Transparency and Accountability. Studies in Computational Intelligence. Springer (2020)
43. Suresha, P.B., Wang, Y., Xiao, C., Glass, L., Yuan, Y., Clifford, G.D.: A deep learning approach for classifying nonalcoholic steatohepatitis patients from nonalcoholic fatty liver disease patients using electronic medical records. In: Explainable AI in Healthcare and Medicine: Building a Culture of Transparency and Accountability. Studies in Computational Intelligence. Springer (2020)
44. Yao, J., Liu, Y., Li, B., Gou, S., Pou-Prom, C., Murray, J., Verma, A., Mamdani, M., Ghassemi, M.: Visualization of deep models on nursing notes and physiological data for predicting health outcomes through temporal sliding windows. In: Explainable AI in Healthcare and Medicine: Building a Culture of Transparency and Accountability. Studies in Computational Intelligence. Springer (2020)
45. Nadkarni, P.M., Ohno-Machado, L., Chapman, W.W.: Natural language processing: an introduction. J. Am. Med. Inform. Assoc. **18**(5), 544–51 (2011). https://doi.org/10.1136/amiajnl-2011-000464
46. Singh, G., Sabet, Z., Shawe-Taylor, J., Thomas, J.: Constructing artificial data for fine-tuning for low-resource biomedical text tagging with applications in PICO annotation. In: Explainable AI in Healthcare and Medicine: Building a Culture of Transparency and Accountability. Studies in Computational Intelligence. Springer (2020)
47. Sakka, K., Nakayama, K., Kimura, N., Inoue, T., Iwasawa, Y., Yamaguchi, R., Kawazoe, Y., Ohe, K., Matsuo, Y.: Character-level Japanese text generation with attention mechanism for chest radiography diagnosis. Explainable AI in Healthcare and Medicine: Building a Culture of Transparency and Accountability. Studies in Computational Intelligence. Springer (2020)
48. Krishna, K., Pavel, A., Schloss, B., Bigham, J.P., Lipton, Z.C.: Extracting structured data from physician-patient conversations by predicting noteworthy utterances. In: Explainable AI in Healthcare and Medicine: Building a Culture of Transparency and Accountability. Studies in Computational Intelligence. Springer (2020)
49. Horn, M., Li, X., Chen, L., Kae, S.: A multi-talent healthcare AI bot platform. In: Explainable AI in Healthcare and Medicine: Building a Culture of Transparency and Accountability. Studies in Computational Intelligence. Springer (2020)
50. Batan, H., Radpour, D., Kehlbacher, A., Klein-Seetharaman, J., Paul, M.J.: Natural vs. artificially sweet tweets: characterizing discussions of non-nutritive sweeteners on Twitter. In: Explainable AI in Healthcare and Medicine: Building a Culture of Transparency and Accountability. Studies in Computational Intelligence. Springer (2020)
51. Lee, B., Jeong, H., Shin, E.K.: On-line (TweetNet) and Off-line (EpiNet): The Distinctive Structures of the Infectious. Explainable AI in Healthcare and Medicine: Building a Culture of Transparency and Accountability. Studies in Computational Intelligence – Springer, 2020
52. Selvaraj, S.P., and Konam, S. Medication Regimen Extraction From Medical Conversations. Explainable AI in Healthcare and Medicine: Building a Culture of Transparency and Accountability. Studies in Computational Intelligence. Springer (2020)
53. Sato, K., Onishi, M., Yoda, I., Uchida, K., Kuroshima, S., Kawashima, M.: Quantitative evaluation of emergency medicine resident's non-technical skills based on trajectory and conversation analysis. In: Explainable AI in Healthcare and Medicine: Building a Culture of Transparency and Accountability. Studies in Computational Intelligence. Springer (2020)

54. Caligtan, C.A., Dykes, P.C.: Electronic health records and personal health records. Semin. Oncol. Nurs. **27**(3), 218–28 (2011). https://doi.org/10.1016/j.soncn.2011.04.007
55. Ammar, N., Bailey, J.E., Davis, R.L., Shaban-Nejad, A.: Implementation of a Personal Health Library (PHL) to support chronic disease self-management. In: Explainable AI in Healthcare and Medicine: Building a Culture of Transparency and Accountability. Studies in Computational Intelligence. Springer, (2020)
56. Huang, M., Shah, N.D., Yao, L.: KELSA: A Knowledge-Enriched Local Sequence Alignment Algorithm for Comparing Patient Medical Records. Explainable AI in Healthcare and Medicine: Building a Culture of Transparency and Accountability. Studies in Computational Intelligence. Springer (2020)
57. Meng, Y., Speier, W., Ong, M., Arnold, C.W.: Multi-level embedding with topic modeling on electronic health records for predicting depression. In: Explainable AI in Healthcare and Medicine: Building a Culture of Transparency and Accountability. Studies in Computational Intelligence. Springer (2020)
58. Hu, Y., An, Y., Subramanian, R., Zhao, N., Gu, Y., Wu, W.: Faster clinical time series classification with filter based feature engineering tree boosting methods. In: Explainable AI in Healthcare and Medicine: Building a Culture of Transparency and Accountability. Studies in Computational Intelligence. Springer (2020)
59. Chen, H., Lundberg, S., and Lee, S.I. Explaining models by propagating shapley values of local components. In: Explainable AI in Healthcare and Medicine: Building a Culture of Transparency and Accountability. Studies in Computational Intelligence. Springer (2020)
60. Howell, K. Barnes, M., Curtis, J.R., Engelberg, R.A., Lee, R.Y., Lober, W.B., Sibley, J., Cohen, T.: Controlling for confounding variables: accounting for dataset bias in classifying patient-provider interactions. In: Explainable AI in Healthcare and Medicine: Building a Culture of Transparency and Accountability. Studies in Computational Intelligence. Springer (2020)
61. Karimian, H.R., Pollard, K., Moore, M.J., Kordjamshidi, P.: Learning representations to augment statistical analysis of drug effects on nerve tissues? In: Explainable AI in Healthcare and Medicine: Building a Culture of Transparency and Accountability. Studies in Computational Intelligence. Springer (2020)
62. Dasgupta, T., Mondal, I., Naskar, A., Dey, L.: Automatic Segregation and Classification of Inclusion and Exclusion Criteria of Clinical Trials to Improve Patient Eligibility Matching. Explainable AI in Healthcare and Medicine: Building a Culture of Transparency and Accountability. Studies in Computational Intelligence. Springer (2020)
63. Ferland, L., Sauve, J., Lucke, M., Nie, R., Khadar, M., Pakhomov, S., Gini, M.: Tell me about your day: designing a conversational agent for time and stress management. In: Explainable AI in Healthcare and Medicine: Building a Culture of Transparency and Accountability. Studies in Computational Intelligence. Springer (2020)
64. Larsen, T., Malkomes, G., Barbour, D.: Accelerating psychometric screening tests with prior information. In: Explainable AI in Healthcare and Medicine: Building a Culture of Transparency and Accountability. Studies in Computational Intelligence. Springer (2020)
65. McElfresh, D.C., Dooley, S., Cui, Y., Griesman, K., Wang, W., Will, T., Sehgal, N., Dickerson, J.P.: Can an algorithm be my healthcare proxy? Explainable AI in Healthcare and Medicine: Building a Culture of Transparency and Accountability. Studies in Computational Intelligence. Springer (2020)
66. Byrd, J., Balakrishnan, S., Jiang, X., Lipton, Z.C.: Predicting mortality in liver transplant candidates. In: Explainable AI in Healthcare and Medicine: Building a Culture of Transparency and Accountability. Studies in Computational Intelligence. Springer (2020)
67. Nilanon, T., Nocera, L.P., Nieva, J.J., Shahabi, C.: Towards automated performance status assessment: temporal alignment of motion skeleton time series. In: Explainable AI in Healthcare and Medicine: Building a Culture of Transparency and Accountability. Studies in Computational Intelligence. Springer (2020)

Fast Similar Patient Retrieval from Large Scale Healthcare Data: A Deep Learning-Based Binary Hashing Approach

Ke Wang, Eryu Xia, Shiwan Zhao, Ziming Huang, Songfang Huang, Jing Mei, and Shaochun Li

Abstract Patient similarity plays an important role in precision evidence-based medicine. While great efforts have been made to derive clinically meaningful similarity measures, how to accurately and efficiently retrieve similar patients from large scale healthcare data remains less explored. Similar patient retrieval has become increasingly important and challenging as the volume of healthcare data grows rapidly. To address the challenge, we propose a coarse-to-fine approach using binary hash codes and embedding vectors derived from an artificial neural network. Experimental results demonstrated that this approach can reduce the time for retrieval by up to over 50.6% without sacrificing the retrieval accuracy. The time reduction became more evident as the data size increased. The retrieval efficiency increased as the number of bits in binary hash codes increased. Descriptive analysis revealed distinct profiles between similar patients and the overall patient cohort.

K. Wang (✉) · E. Xia · S. Zhao · Z. Huang · S. Huang · J. Mei · S. Li
IBM Research-China, Beijing, China
e-mail: wkebj@cn.ibm.com

E. Xia
e-mail: xerbj@cn.ibm.com

S. Zhao
e-mail: zhaosw@cn.ibm.com

Z. Huang
e-mail: hzmzi@cn.ibm.com

S. Huang
e-mail: huangsf@cn.ibm.com

J. Mei
e-mail: meijing@cn.ibm.com

S. Li
e-mail: lishaoc@cn.ibm.com

© The Editor(s) (if applicable) and The Author(s), under exclusive license to Springer Nature Switzerland AG 2021

A. Shaban-Nejad et al. (eds.), *Explainable AI in Healthcare and Medicine*, Studies in Computational Intelligence 914,
https://doi.org/10.1007/978-3-030-53352-6_2

Keywords Patient retrieval · Hashing · Patient similarity

1 Introduction

Precision medicine is an emerging approach for effective disease treatment and prevention that takes individual traits into account [4]. Patient similarity has become a sound step toward precision medicine, which, in individualized patient care, provides data-driven insights from a similar patient cohort instead of the overall patient cohort [11, 15]. It has been proven useful in a number of scenarios including constructing precision cohorts [12], enabling personalized predictive modeling and risk factor identification [10], reducing healthcare costs [13], and improving healthcare systems [6]. Previous studies in patient similarity are focused on deriving clinically meaningful similarity measures [3, 7, 14]. While for patient similarity to be used for precision medicine, defining a clinically meaningful similarity measure is a prerequisite, its wide application relies on fast and accurate similar patient retrieval. Similar patient retrieval refers to the process of retrieving the most similar patients (quantitatively defined either by the top K most similar patients or by patients with a similarity value larger than a threshold) for an individual patient from a patient database based on a defined similarity measure. To retrieve simlilar patient fast, we present a deep learning-based binary hashing approach, which was demonstrated in an outcome-driven cluster analysis using multi-task neural network with attention [16].

2 Related Work

Similar patient retrieval is a topic relatively less studied in the medical informatics domain. ART (Adaptive Semi-Supervised Recursive Tree Partitioning) was proposed as a framework to correctly and efficiently retrieve similar patients for large scale data [15] but suffers from the problem of the dependence on expertise knowledge, which is in the form of pairwise constraints on patients. Fast locality sensitive hashing [5] is an effective technique that hashes similar data items into the same hashing code with high probability. FALCONN [1] is an approximate nearest neighbor search approach for high-dimensional data. A similar problem to similarity retrieval is image retrieval. A hot topic in image retrieval is similarity-preserving binary hashing, where a representation of an image is first learned, translated to a binary hash code, and then searched against a large database of image representations. We thus reviewed existing methods in image retrieval [8, 9, 17] and identified the common idea of the articles: an embedding vector can be extracted from an artificial neural network to encode an image, and a binary hash code can be derived to efficiently retrieve images. Inspired by the image retrieval methods, we present a binary hashing approach for fast similar patient retrieval from large scale heathcare data.

3 Methods

In this section, we would first introduce the background of the case study, recapitulate the previous report on patient similarity, and introduce its converging point with our current study. We would then describe the proposed framework in the context of the case study. Major components and the workflow would be described. We then describe the metrics used to evaluate the efficiency of the framework in the case study.

3.1 The Artificial Neural Network

3.1.1 Case Study Settings

In a previous report [16], an outcome-driven clustering approach using multi-task neural network with attention was introduced for the clustering of acute coronary syndrome (ACS) patients. Our study was based on the same dataset and setting as in the previous report, which we would recapitulate here. The dataset include a cohort of 26, 986 adult hospitalized patients with a final diagnosis of ACS. Patient characteristics routinely collected for ACS patients were included in the study as features, including disease types (ACS type and Killip class), demographics, personal disease history, comorbidities, habits, laboratory test results, and procedures, which added up to 41 features as input. Feature engineering, including outlier detection, missing value imputation, feature encoding, feature normalization, was conducted as previously reported [16]. The resulting features were used as input for the neural network. Four outcomes were identified as reflective of the patient disease states from different facets, including the onset of in-hospital major adverse cardiovascular events (MACE) (measured as yes or no), which is a direct result of the disease states, and three in-hospital treatments (antiplatelet treatment, beta-blockers treatment, and statins treatment, all measured as used or not), which are reflections of the doctor's understanding of the disease states. The four outcomes were then used as labels in the multi-task neural network.

3.1.2 Structure and Training of the Artificial Neural Network

A schematic representation of the structure of the artificial neural network was illustrated in Fig. 1. The structure was adopted from the previous report [16]. The major adaptation was an additional fully connected layer (named the 'hash layer') added after the representation layer, from which we would derive the binary hash codes. The number of neurons in representation layer was set to 32. We used 'ReLU' activation for intermediate layers except for the hash layer, where 'sigmoid' activation was used to normalize the value to a range of 0 to 1. For each classification task,

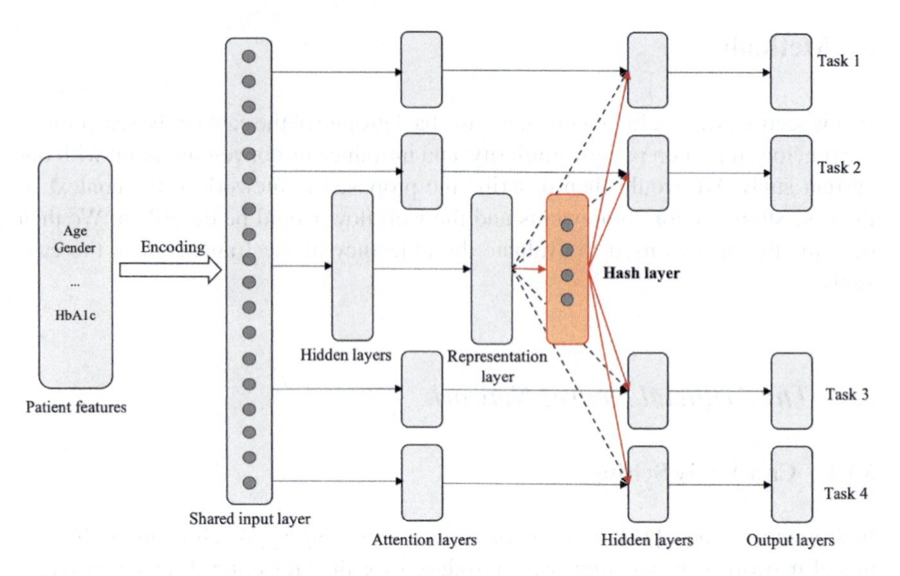

Fig. 1 A schematic representation of the structure of the artificial neural network used in this study. The neural network structure in this study was based on the multi-task neural network with attention reported previously. In the report, representation layer was directly connected with the four hidden layers for the four tasks (shown with dotted lines). In our study, we inserted a fully connected layer (called the 'hash layer') after the representation layer and before the four hidden layers (shown with red lines). Embedding vectors were extracted from the representation layer, while binary hash codes were derived from the hash layer

binary cross entropy loss was used. Weighted sum of the four losses was used as the training loss. In this study, we assign equal weights to the four tasks. Training was conducted 50 epochs with a batch size of 512 using the 'Adam' optimizer. Class weights were adjusted to deal with the biased proportion of positive and negative cases individually for all four tasks.

3.1.3 Getting the Representations

We applied the trained model to get the representations for the patients. Activation values from the representation layer were directly used as the embedding vector, which is a 32-dimension vector for each patient. Activation values from the hash layer were processed to generate the binary hash codes. For each node $i = 1...h$ (where h is the number of nodes in the hash layer), we translated the activation value $(0 \leq H^i \leq 1)$ to a binary code (B^i, called a bit) using the following function:

$$B^i = \begin{cases} 1 & 0.5 \leq H^i \leq 1, \\ 0 & 0 \leq H^i < 0.5. \end{cases} \tag{1}$$

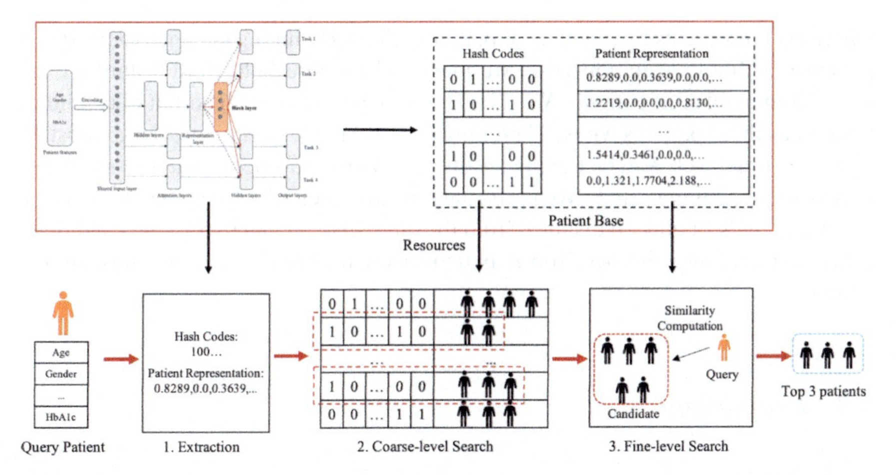

Fig. 2 Workflow of similar patient retrieval. Two major resources are used in the workflow, including a trained neural network model, and a database of patients. Three steps are included in the retrieval process: (1) Extraction of the representations of the query patient using the trained model; (2) Coarse-level search against the binary hash codes of the patients, which retrieves a small cohort of candidate similar patients; and (3) Fine-level search against the embedding vectors of the small cohort of candidates patients, which generates a ranked list of the most similar patients

3.2 *Similar Patient Retrieval*

The proposed workflow of similar patient retrieval is shown in Fig. 2. Representations, including embedding vectors and binary hash codes, of all existing patients were achieved and stored into the patient database. For a query patient, the similar patient retrieval process is conducted as follows. First, the representations of the query patient are extracted using the trained model. Second, coarse-level search is performed against the binary hash codes of existing patients, generating a small cohort of candidate patients with similar binary hash codes to the query patient. Third, fine-level search is performed against the embedding vectors of the small cohort of candidate patients, generating a ranked list of the most similar patients.

In the coarse-level search, Hamming distance is used to determine the similarity between binary hash codes. A larger Hamming distance between two binary hash codes indicates a smaller similarity. We built index for the binary hash codes using KD-Tree [2] to speed up the search. The Hamming distance threshold to form the small cohort of candidate patients is determined dynamically for each query patient. In the case we need to find the top K most similar patients for a query patient p, the threshold D is calculated as:

$$D = \min d,\ s.t. \sum_{i=0}^{d} c(i) >= K \tag{2}$$

where $c(i)$ represents the number of patients whose Hamming distance to the query patient is i. Implementation-wise, we increased the threshold from 0 until we get the desired number of patients. We therefore generated the small cohort of candidate patients as those patients whose Hamming distance to the query patient is at most D.

In the fine-level search, Euclidean distance is used to determine the similarity between embedding vectors. We performed linear search which calculated each candidate patient's Euclidean distance from the query patient and ranked the candidates based on increasing distance. Top K patients were then retrieved as the most similar patients.

3.3 *Evaluation*

3.3.1 Efficiency

The efficiency of our proposed search strategy was evaluated by calculating the average retrieval time for 1,000 randomly selected query patients. We compared the retrieval time in different patient database size settings: (1) 2,000; (2) 10,000; (3) 50,000; (4) 250,000; (5) 500,000 and (6) 1,000,000. Due to the limited size of the database, we performed bootstrap analysis, which re-samples the database with replacement to generate the desired patient database size. For each patient database size setting, 100 bootstrap analysis experiments were repeated, from which an average retrieval time was calculated. We compared the efficiency of our approach with two other approaches: (1) A conventional similarity search approach, KD-Tree, by indexing the patients in the Euclidean space before search to speed up the retrieval process; and (2) locally sensitive hashing (LSH) using FALCONN [1]. Both methods used the embedding vectors as input. In all experiments, we set the number of similar patients to retrieve (K) to 100. For our approach, we set the length of binary hash codes to 32. The experiments were carried out on the machine with Intel(R) Xeon(R) CPU E5-2690 v4 @ 2.60 GHz, 64 GB RAM.

3.3.2 Accuracy

We evaluated the similar patient retrieval accuracy based on KNN classification. For a query patient, we retrieved his K nearest neighbors. For each classification task, we calculated the proportion of positive cases in the K nearest neighbors, and used the proportion as the predicted probability. For each patient in the dataset, by taking him as the query patient, we were able to calculate the predicted probability for each classification task. Retrieval accuracy of our proposed coarse-to-fine strategy was compared with two other search strategies which directly search for patients with smallest Euclidean distance using KD-Tree and: (1) the embedding vector; and (2) the activation value from the hash layer. For all the three search strategies, we assessed a range of K settings (10, 30, 50, 70, 90, 110, 130, 150). The reported classification accuracy reported in the original article [16] was used as a benchmark. Here we set the length of binary hash codes to 32.

4 Experiments

4.1 Similar Patient Retrieval Efficiency

The comparison of retrieval time of our proposed search strategy, the KD-Tree search strategy, and locally sensitive hashing strategy using FALCONN was shown in Fig. 3. The results show that using the coarse-to-fine strategy, we achieved great reduction of time: 50.6% compared with KD-Tree search strategy and 86.8% compared with locally sensitive hashing strategy for size of 200,000. The reduction in time becomes greater when the patient database size increases. This can be explained by the fact that the coarse-level search excludes a large proportion of patients very different from the query patient, thus saving the time for KD-Tree search. Algorithmic-wise, this is justified by the fact that KD-Tree search has a logarithmic time complexity for data size, while our proposed strategy has logarithmic time complexity for average size of patients for each hashing code.

4.2 Similar Patient Retrieval Accuracy

The comparison of retrieval accuracy between our proposed search strategy and two KD-Tree search strategies (using activation values from two different hidden layers), and against the benchmark is shown in Fig. 4. The benchmark can be taken as the

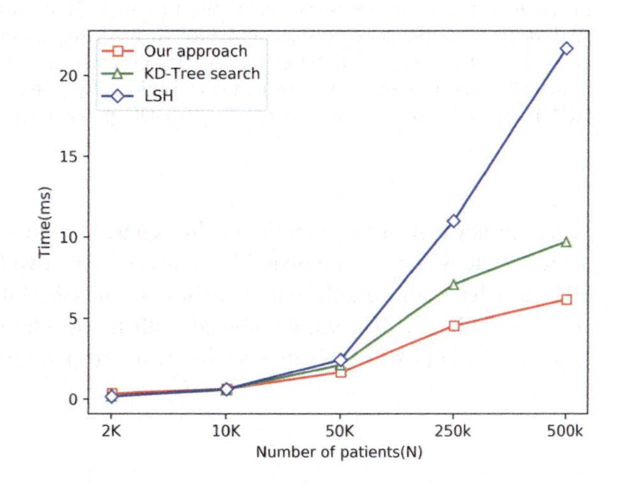

Fig. 3 Retrieval time comparison. Our coarse-to-fine search strategy was denoted as 'Our approach'. The KD-Tree search strategy was denoted as 'KD-Tree search'. Locally sensitive hashing strategy was denoted as 'LSH'

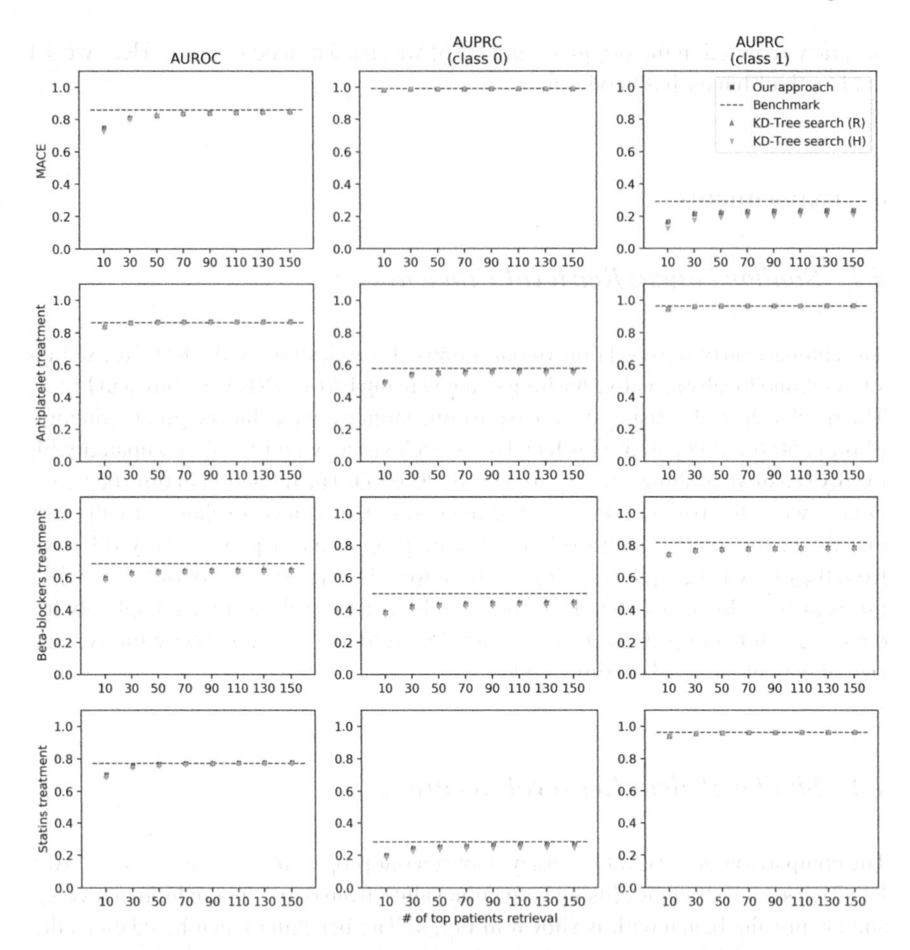

Fig. 4 Retrieval accuracy comparison. Four classification tasks are compared: MACE, antiplatelet treatment, beta-blockers treatment, and statins treatment. Three metrics were compared: AUROC, AUPRC (in predicting the class 0), and AUPRC (in predicting the class 1). Our search strategy was denoted as 'Our approach', KD-Tree search using the embedding vector was denoted as 'KD-Tree search (R)', and KD-Tree search using the activation values from the hash layer was denoted as 'KD-Tree search (H)'. Classification accuracy from the previous report was denoted as 'Benchmark'

best accuracy using representations achieved by the neural network. From Fig. 4, our search strategy receives comparable accuracy to the two KD-Tree search strategies, all of which are comparable to the benchmark when the number of top similar patients retrieved is larger than 90. Combined with the results in the efficiency, we have demonstrated our strategy's time saving without sacrificing the retrieval accuracy.

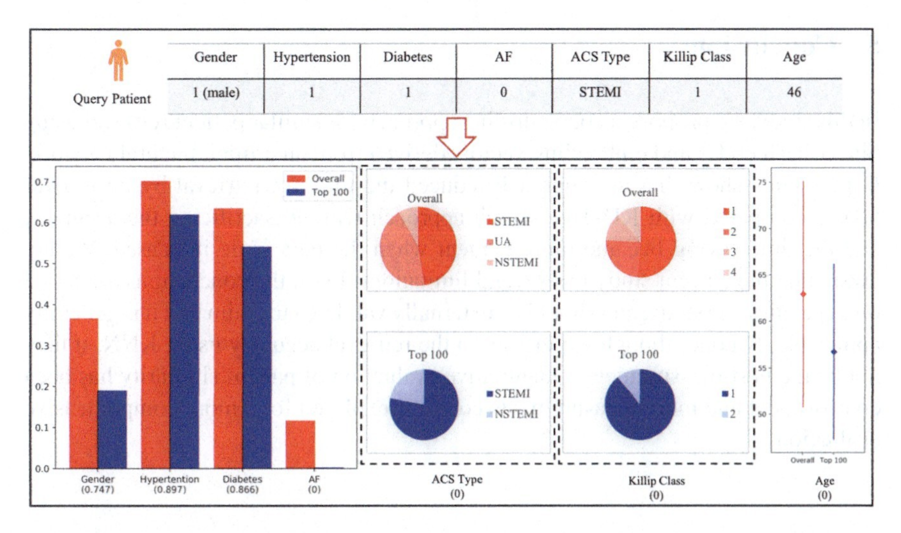

Fig. 5 Profiling of patient characteristics in similar patients compared with the general cohort. Above: a query patient with his patient characteristics. Below: comparison of patient characteristics. The similar patient cohort (labeled 'Top 100') is colored blue, and the general cohort (labeled 'Overall') is colored red. Three groups of patient characteristics are presented: binary variables, categorical variables, and a continuous variable. Gender is shown as the proportion of female patients. Hypertension, diabetes, AF are shown as the proportion of patients with the condition. ACS type and Killip class are shown as the distribution. Age is presented with the point indicating the average value and the bar indicating the standard deviation. Below each figure, the number in bracket shows the p-value between the similar patient cohort and the general cohort (using chi-square test for all variables except for age, where t-test was used)

4.3 Interpretation of the Results

To visually inspect our similar patient retrieval results, we picked a query patient from the cohort, retrieved his top 100 most similar patients, profiled the characteristics of these similar patients compared with the general cohort. The results are shown in Fig. 5. This figure includes demographics information of age and gender, important ACS disease indicators including ACS type and Killip class, as well as common comorbidities including hypertension, diabetes, and atrial fibrillation (AF). From the results, we see that the similar patient cohort has significant difference from the general cohort in terms of the proportion of AF patients, the distribution of ACS type and Killip class, and age. The two cohorts have similar proportion of hypertension and diabetes patients. As a result, for the query patient, his most similar 100 patients are characterized by young age, high proportion of STEMI patients, high proportion of patients with Killip class 1, and low proportion of AF patients.

5 Conclusions

In this study, we propose a coarse-to-fine approach for similar patient retrieval using binary hash codes and embedding vectors derived from an artificial neural network. Experiments show that the approach reduced the time for retrieval by up to over 50.6% compared with KD-Tree search approach without sacrificing the accuracy, and the time saving became more evident when the data scale increased. We are aware that the current study has several limitations. First, the conclusions are based on a specific case study, and should be externally validated to claim that it is generally applicable. Second, though we measured the retrieval accuracy using KNN, it may not be the best measurement. Quantitative evaluation of patient similarity has been controversial and more measurements could be explored for a more comprehensive evaluation.

References

1. Andoni, A., Indyk, P., Laarhoven, T., Razenshteyn, I., Schmidt, L.: Practical and optimal LSH for angular distance. In: Advances in Neural Information Processing Systems, pp. 1225–1233 (2015)
2. Bentley, J.L.: Multidimensional binary search trees used for associative searching. Commun. ACM **18**(9), 509–517 (1975)
3. Campillo-Gimenez, B., Jouini, W., Bayat, S., Cuggia, M.: Improving case-based reasoning systems by combining k-nearest neighbour algorithm with logistic regression in the prediction of patients' registration on the renal transplant waiting list. PLoS ONE **8**(9), e71991 (2013)
4. Collins, F.S., Varmus, H.: A new initiative on precision medicine. N. Engl. J. Med. **372**(9), 793–795 (2015)
5. Dasgupta, A., Kumar, R., Sarlós, T.: Fast locality-sensitive hashing. In: Proceedings of the 17th ACM SIGKDD International Conference on Knowledge discovery and Data Mining, pp. 1073–1081. ACM (2011)
6. Jee, K., Kim, G.H.: Potentiality of big data in the medical sector: focus on how to reshape the healthcare system. Healthc. Inform. Res. **19**(2), 79–85 (2013)
7. Lee, J., Maslove, D.M., Dubin, J.A.: Personalized mortality prediction driven by electronic medical data and a patient similarity metric. PLoS ONE **10**(5), e0127428 (2015)
8. Lin, K., Yang, H.F., Hsiao, J.H., Chen, C.S.: Deep learning of binary hash codes for fast image retrieval. In: Proceedings of the IEEE Conference on Computer Vision and Pattern Recognition Workshops, pp. 27–35 (2015)
9. Liu, H., Wang, R., Shan, S., Chen, X.: Deep supervised hashing for fast image retrieval. In: Proceedings of the IEEE Conference on Computer Vision and Pattern Recognition, pp. 2064–2072 (2016)
10. Ng, K., Sun, J., Hu, J., Wang, F.: Personalized predictive modeling and risk factor identification using patient similarity. In: AMIA Summits on Translational Science Proceedings 2015, p. 132 (2015)
11. Parimbelli, E., Marini, S., Sacchi, L., Bellazzi, R.: Patient similarity for precision medicine: a systematic review. J. Biomed. Inform. **83**, 87–96 (2018)
12. Sharafoddini, A., Dubin, J.A., Lee, J.: Patient similarity in prediction models based on health data: a scoping review. JMIR Med. Inform. **5**(1), e7 (2017)
13. Srinivasan, U., Arunasalam, B.: Leveraging big data analytics to reduce healthcare costs. IT Prof. **15**(6), 21–28 (2013)

14. Sun, J., Sow, D., Hu, J., Ebadollahi, S.: Localized supervised metric learning on temporal physiological data. In: 2010 20th International Conference on Pattern Recognition, pp. 4149–4152. IEEE (2010)
15. Wang, F.: Adaptive semi-supervised recursive tree partitioning: the art towards large scale patient indexing in personalized healthcare. J. Biomed. Inform. **55**, 41–54 (2015)
16. Xia, E., Du, X., Mei, J., Sun, W., Tong, S., Kang, Z., Sheng, J., Li, J., Ma, C., Dong, J., et al.: Outcome-driven clustering of acute coronary syndrome patients using multi-task neural network with attention. arXiv preprint arXiv:1903.00197 (2019)
17. Zhao, F., Huang, Y., Wang, L., Tan, T.: Deep semantic ranking based hashing for multi-label image retrieval. In: Proceedings of the IEEE Conference on Computer Vision and Pattern Recognition, pp. 1556–1564 (2015)

A Kernel to Exploit Informative Missingness in Multivariate Time Series from EHRs

Karl Øyvind Mikalsen, Cristina Soguero-Ruiz, and Robert Jenssen

Abstract A large fraction of the electronic health records (EHRs) consists of clinical measurements collected over time, such as lab tests and vital signs, which provide important information about a patient's health status. These sequences of clinical measurements are naturally represented as time series, characterized by multiple variables and large amounts of missing data, which complicate the analysis. In this work, we propose a novel kernel which is capable of exploiting both the information from the observed values as well the information hidden in the missing patterns in multivariate time series (MTS) originating e.g. from EHRs. The kernel, called TCK_{IM}, is designed using an ensemble learning strategy in which the base models are novel mixed mode Bayesian mixture models which can effectively exploit informative missingness without having to resort to imputation methods. Moreover, the ensemble approach ensures robustness to hyperparameters and therefore TCK_{IM} is particularly well suited if there is a lack of labels—a known challenge in medical applications. Experiments on three real-world clinical datasets demonstrate the effectiveness of the proposed kernel.

K. Ø. Mikalsen (✉)
University Hospital of North-Norway, Tromsø, Norway
e-mail: karl.o.mikalsen@uit.no
URL: http://machine-learning.uit.no

K. Ø. Mikalsen · C. Soguero-Ruiz · R. Jenssen
Department of Physics and Technology, UiT The Arctic University of Norway, 9037 Tromsø, Norway
e-mail: cristina.soguero@urjc.es
URL: http://machine-learning.uit.no

R. Jenssen
e-mail: robert.jenssen@uit.no
URL: http://machine-learning.uit.no

C. Soguero-Ruiz
Rey Juan Carlos University, Móstoles, Spain

© The Editor(s) (if applicable) and The Author(s), under exclusive license to Springer Nature Switzerland AG 2021
A. Shaban-Nejad et al. (eds.), *Explainable AI in Healthcare and Medicine*, Studies in Computational Intelligence 914,
https://doi.org/10.1007/978-3-030-53352-6_3

1 Introduction

The widespread growth of electronic health records (EHRs) has generated vast amounts of clinical data. Normally, an encounter-based patient EHR is longitudinal and contains clinical notes, diagnosis codes, medications, laboratory tests, etc., which can be represented as multivariate time series (MTS). As a consequence, EHRs contain valuable information about the clinical observations depicting both patients' health and care provided by physicians. However, the longitudinal and heterogeneous data sources make EHR analysis difficult from a computational perspective. In addition, the EHRs are often subject to a lack of completeness, implying that the MTS extracted from EHRs often contain massive *missing data* [24]. Missing data might occur for different reasons. It could be that physicians orders lab tests, but because of an error some results are not recorded. On the other hand, it can also happen that they decides to not order lab tests because the patient is in good shape. In the first case, the missingness is ignorable, whereas in the latter case, the missing values and patterns potentially can contain rich information about the patient's diseases and clinical outcomes. Efficient data-driven systems aiming to extract knowledge, perform predictive modeling, etc., must be capable of capturing this information.

Traditionally, missingness mechanisms have been divided into missing completely at random (MCAR), missing at random (MAR) and missing not at random (MNAR). The main difference between these mechanisms consists in whether the missingness is ignorable (MCAR and MAR) or non-ignorable (MNAR) [13]. This traditional description of missingness mechanisms is, however, not always sufficient in medical applications as the missing patterns might be correlated with additional variables, such as e.g. a disease. This means that the distribution of the missing patterns for patients with a particular disease might be different than the corresponding distribution for patients without the disease, i.e. the missingness is informative [24].

Several methods have been proposed to handle missing data in MTS [17, 22]. A simple approach is discard the MTS with missing data. Alternatively, one can do simple imputation of the missing values, e.g. using the last observation carried forward scheme (impute the last non-missing value for the following missing values) [23], zero-value imputation (replace missing values with zeros) or mean-value imputation (missing values are replaced with the mean of the observed data) [29]. A common limitation of these approaches is that they lead to additional bias, loss of precision, and they ignore uncertainty associated with the missing values [13]. This problem is to some extent solved via multiple imputation methods, i.e. by creating multiple complete datasets using single imputation independently each time. Then, by training a classifier using an ensemble learning strategy, one can improve the performance compared to simple imputation. However, this imputation procedure is an ad-hoc solution as it is performed independently of the rest of the analysis and it ignores the potential predictive value of the missing patterns [18].

Due to the limitations of imputation methods, several research efforts have been devoted to deal with missing data in MTS using alternative strategies [6, 8, 25]. Prominent examples are kernels, i.e. positive semi-definite time series similarity measures, such as the *learned pattern similarity* (LPS) [4] and the *time series cluster kernel* (TCK) [20] that can naturally deal with missing data. The former generalizes autoregressive models to local autopatterns, which capture the local dependency structure in the time series, and uses an ensemble learning (random forest) strategy in which a bag-of-words representation is created from the output of the leaf-nodes for each tree. TCK is also based on an ensemble learning approach and shares many properties with LPS. It is designed using an ensemble learning approach in which Bayesian mixture models form the base models. However, while LPS exploits the inherent missing data handling abilities of decision trees, TCK is a likelihood-based approach in which the incomplete dataset is analysed using maximum a posteriori expectation-maximization. An advantage of these methods, compared to e.g. multiple imputation that requires a careful selection of imputation model and parameters [22], is that the missing data are handled automatically and no additional tasks are left to the designer. Additionally, since the methods are based on ensemble learning, they are robust to hyperparameter choices. In particular, these properties are important in unsupervised settings, which frequently occur in medical applications where manual label annotation of large datasets often is not feasible [14].

A shortcoming of these kernel methods is, however, that they cannot exploit informative missing patterns, which frequently occur in medical MTS, and unbiased predictions are only guaranteed for ignorable missingness as MAR is an underlying assumption. Recently, several studies have focused on modeling the informative or nonignorable missigness by analyzing the observed values as well as the indicators of missingness, concluding that the missing patterns can add more insights beyond the observed values [1, 8, 24]. In this work, we present a novel time series cluster kernel, TCK_{IM}, that also represents the missing patterns using binary indicator time series. By doing so, we obtain MTS consisting of both continuous and discrete attributes. However, we do not only concatenate the binary MTS to the real-valued MTS and analyse these data in a naive way. Instead, we take a statistically principled Bayesian approach [17, 18] and model the missingness mechanism more rigorously by introducing novel mixed mode Bayesian mixture models, which can effectively exploit information provided by the missing patterns as well as the temporal dependencies in the observed MTS. The mixed mode Bayesian mixture models are then used as base models in an ensemble learning strategy to form the TCK_{IM} kernel. Experiments on three real-world datasets of patients described by longitudinal EHR data, demonstrate the effectiveness of the proposed method.

2 Time Series Cluster Kernel to Exploit Informative Missingness

Here we present the proposed TCK_{IM} kernel. The kernel is learned using an ensemble learning strategy, i.e. by training individual base models which are combined into a composite kernel in the end. As base model we introduce a novel mixed mode

Bayesian mixture model. Before we provide the details of this method, we describe the notation used throughout the paper.

Notation. We define a multivariate time series (MTS) X as a finite combination of univariate time series (UTS) x_v of length T, i.e. $X = \{x_v \in \mathbb{R}^T \mid v = 1, 2, \ldots, V\}$. The dimensionality of the MTS X is the same as the number of UTS, V, and the length of X is the same as the length T of the UTS x_v. A V–dimensional MTS, X, of length T can be represented as a matrix in $\mathbb{R}^{V \times T}$. Given a dataset of N MTS, we denote $X^{(n)}$ as the n-th MTS. In a dataset of N incompletely observed MTS, the n-th MTS is denoted by the pair $U^{(n)} = (X^{(n)}, R^{(n)})$, where $R^{(n)}$ is a binary MTS with entry $r_v^{(n)}(t) = 0$ if the realization $x_v^{(n)}(t)$ is missing and $r_v^{(n)}(t) = 1$ if it is observed.

Mixed Mode Bayesian Mixture Model. Let $U = (X, R)$ be a MTS generated from two modes, where X is a V-variate real-valued MTS ($X \in \mathbb{R}^{V \times T}$) and R is a V-variate binary MTS ($R \in \{0, 1\}^{V \times T}$). In the mixture model it is assumed that U is generated from a finite mixture density $p_u(U \mid \Phi, \Theta) = \sum_{g=1}^{G} \theta_g p_{u_g}(U \mid \phi_g)$, where G is the number of components, p_{u_g} is the density of the components parametrized by $\Phi = (\phi_1, \ldots, \phi_G)$, and $\Theta = (\theta_1, \ldots, \theta_G)$ are the mixing coefficients, $0 \leq \theta_g \leq 1$ and $\sum_{g=1}^{G} \theta_g = 1$. We formulate the mixture model in terms of a latent random variable Z, described via the one-hot vector $Z = (Z_1, \ldots, Z_G)$ with marginal distribution given by $p_z(Z \mid \Theta) = \prod_{g=1}^{G} \theta_g^{Z_g}$. The latent variable Z describes which cluster component the MTS U belongs to, i.e. $Z_g = 1$ if U belongs to cluster component g and $Z_g = 0$ otherwise. The conditional is given by $p_{u|z}(U \mid Z, \Phi) = \prod_{g=1}^{G} p_{u_g}(U \mid \phi_g)^{Z_g}$, and therefore it follows that the joint distribution is given by

$$p_{u,z}(U, Z \mid \Phi, \Theta) = p_{u|z}(U \mid Z, \Phi) p_z(Z \mid \Theta) = \prod_{g=1}^{G} \left[p_{u_g}(U \mid \phi_g) \theta_g \right]^{Z_g}. \quad (1)$$

We further assume that the parameters of each component are given by $\phi_g = (\mu_g, \Sigma_g, \beta_g)$ and

$$p_{u_g}(U \mid \phi_g) = p_{x|r}(X \mid R, \mu_g, \Sigma_g) p_r(R \mid \beta_g), \quad (2)$$

where $p_{x|r}$ is a density function given by

$$p_{x|r}(X \mid R, \mu_g, \Sigma_g) = \prod_{v=1}^{V} \prod_{t=1}^{T} \mathcal{N}(x_v(t) \mid \mu_{gv}(t), \sigma_{gv})^{r_v(t)}, \quad (3)$$

where $\mu_g = \{\mu_{gv} \in \mathbb{R}^T \mid v = 1, ..., V\}$ is a time-dependent mean, and $\Sigma_g = diag\{\sigma_{g1}^2, ..., \sigma_{gV}^2\}$ is a diagonal covariance matrix in which σ_{gv}^2 is the variance of attribute v. Hence, the covariance is assumed to be constant over time. p_r is a probability mass given by

$$p_r(R \mid \beta_g) = \prod_{v=1}^{V} \prod_{t=1}^{T} \beta_{gvt}^{r_v(t)} (1 - \beta_{gvt})^{1 - r_v(t)}, \quad (4)$$

where $\beta_{gvt} \in [0, 1]$. The idea with this formulation is to use the Bernoulli term p_r to capture information from the missing patterns and $p_{x|r}$ to capture the information from the observed data. Using Bayes' theorem we compute the conditional probability of Z given U, $P(Z_g = 1|U, \Phi, \Theta)$,

$$\pi_g = \left(\sum_{g=1}^{G} \theta_g p_{x|r}(X|R, \mu_g, \Sigma_g) p_r(R|\beta_g)\right)^{-1} \theta_g p_{x|r}(X|R, \mu_g, \Sigma_g) p_r(R|\beta_g). \tag{5}$$

To improve the capability of handling missing data, a Bayesian extension is introduced where informative priors are put over the parameters of the normal distribution as well as the Bernoulli distribution. This enforces that the cluster representatives become smooth over time even in the presence of large amounts of missing data and that the parameters of clusters with few MTS are similar to the overall mean. Towards this end, a kernel-based Gaussian prior is defined for the mean, $p_\mu(\mu_{gv}) = \mathcal{N}\left(\mu_{gv} \mid m_v, S_v\right)$, where m_v are the empirical means and $S_v = s_v \mathcal{K}$, are the prior covariance matrices. s_v are empirical standard deviations and $\mathcal{K}$ is a kernel matrix, whose elements are $\mathcal{K}_{tt'} = b_0 \exp(-a_0(t-t')^2)$, $t, t' = 1, \ldots, T$, with a_0, b_0 being user-defined hyperparameters. For the standard deviation σ_{gv}, an inverse Gamma distribution prior is introduced $p_\sigma(\sigma_{gv}) \propto \sigma_{gv}^{-N_0} e^{-N_0 s_v / 2\sigma_{gv}^2}$, where N_0 is a hyperparameter. Further, we put a Beta distribution prior on β_{gvt} $p_\beta(\beta_{gvt}) \propto \beta_{gvt}^{c_0-1}(1-\beta_{gvt})^{d_0-1}$, where c_0, d_0 are hyperparameters. We let $\Omega = \{a_0, b_0, c_0, d_0, N_0\}$ denote the set all of hyperparameters.

Given a dataset $\{U^{(n)}\}_{n=1}^{N}$, we estimate the parameters $\{\Phi, \Theta\}$ using maximum a posteriori expectation maximization (MAP-EM) [11]. The Q-function is computed as follows

$$\begin{aligned}
Q &= \mathbb{E}_{Z \mid U, \Theta, \Phi}\left[\log\left(p_{u,z}(U, Z \mid \Phi, \Theta) p(\Phi)\right)\right] \\
&= \log p(\Phi) + \sum_{n,g} \log[p_u(U^{(n)}|Z_g^{(n)} = 1, \Phi) p(Z_g^{(n)}|\Theta)]\pi_g^{(n)} \\
&= \sum_g \log\left(p_\mu(\mu_g) p_\sigma(\sigma_g) p_\beta(\beta_g)\right) + \sum_{n,g} \log\left[p_{x|r}(X^{(n)}|R^{(n)}, \mu_g, \Sigma_g) p_r(R^{(n)}|\beta_g)\theta_g\right]\pi_g^{(n)},
\end{aligned}$$

where $p_{x|r}$ is given by Eq. (3) and p_r by Eq. (4).

The E-step in MAP-EM is the same as in maximum likelihood EM and consists in updating the posterior (Eq. (5)) using the current parameter estimates, whereas the M-step consists in maximizing the Q-function wrt. the parameters $\{\Phi, \Theta\}$:

$$\{\Phi^{(m+1)}, \Theta^{(m+1)}\} = \arg\max_{\{\Phi, \Theta\}} Q(\Phi, \Theta \mid \Phi^{(m)}, \Theta^{(m)}) \tag{6}$$

Computing the derivatives of Q wrt. the parameters θ_g, μ_g, σ_g, β_g leads to Algorithm 1.

The TCK$_{IM}$ Kernel. To compute the TCK$_{IM}$ kernel, we use the mixed mode Bayesian mixture model, described above, as the base model in an ensemble approach. Key to ensure that TCK$_{IM}$ will have statistical advantages (lower variance), computational advantages (less sensitive to local optima) as well as representational advantages (increased expressiveness) compared to the individual base models, is *diversity* and *accuracy* [12]. This means that the base models should not do the same type of errors and each base model has to perform better than random guessing. Hence, to ensure diversity, we integrate multiple outcomes of the base model as it is

Algorithm 1 MAP-EM for mixed mode Bayesian mixture model

Require: Dataset $\{U^{(n)} = (X^{(n)}, R^{(n)})\}_{n=1}^{N}$, hyperparameters Ω and number of mixtures G.
1: Initialize the parameters $\Theta^{(0)} = (\theta_1^{(0)}, \ldots, \theta_G^{(0)})$ and $\Phi^{(0)} = \{\mu_g^{(0)}, \sigma_g^{(0)}, \beta_g^{(0)}\}_{g=1}^{G}$.
2: E-step. For each MTS, evaluate the posteriors $\pi_g^{(n)}$ using Eq. (5) with current parameters.
3: M-step. Update parameters using the current posteriors

$$\theta_g^{(m+1)} = N^{-1} \sum_{n=1}^{N} \pi_g^{(n)}$$
$$\sigma_{gv}^{2\ (m+1)} = \left(N_0 + \sum_{n,t=1}^{N,T} r_v^{(n)}(t)\, \pi_g^{(n)}\right)^{-1} \left(N_0 s_v^2 + \sum_{n,t=1}^{N,T} r_v^{(n)}(t)\, \pi_g^{(n)} \left(x_v^{(n)}(t) - \mu_{gv}(t)\right)^2\right)$$
$$\mu_{gv}^{(m+1)} = \left(S_v^{-1} + \sigma_{gv}^{-2} \sum_{n=1}^{N} \pi_g^{(n)} \mathrm{diag}(r_v^{(n)})\right)^{-1} \left(S_v^{-1} m_v + \sigma_{gv}^{-2} \sum_{n=1}^{N} \pi_g^{(n)} \mathrm{diag}(r_v^{(n)})\, x_v^{(n)}\right)$$
$$\beta_{gv}^{(m+1)} = \left(c_0 + d_0 - 2 + \sum_{n=1}^{N} \pi_g^{(n)}\right)^{-1} \left(c_0 - 1 + \sum_{n=1}^{N} \pi_g^{(n)} r_v^{(n)}\right)$$

4: Repeat step 2-3 until convergence.
Ensure: Posteriors $\Pi^{(n)} \equiv \left(\pi_1^{(n)}, \ldots, \pi_G^{(n)}\right)^T$ and parameter estimates Θ and Φ.

Algorithm 2 TCK$_{IM}$. Training phase.

Require: Training set of MTS $\{(X^{(n)}, R^{(n)})\}_{n=1}^{N}$, Q initializations, set of integers $\mathcal{I}_C$.
1: Initialize kernel matrix $K = 0_{N \times N}$.
2: **for** $q \in Q$ **do**
3: Compute posteriors $\Pi^{(n)}(q) \equiv (\pi_1^{(n)}, \ldots, \pi_{q_2}^{(n)})^T$, by fitting a mixed mode mixture model with q_2 clusters to the dataset and by randomly selecting (i.) hyperparameters $\Omega(q)$, (ii.) time segment $\mathcal{T}(q)$, (iii.) subset of attributes $\mathcal{V}(q)$, (iv.) subset of MTS, $\eta(q)$, and (v.) initialization of the mixture parameters $\Theta(q)$ and $\Phi(q)$.
4: Update $K_{nm} = K_{nm} + \frac{\Pi^{(n)}(q)^T \Pi^{(m)}(q)}{\|\Pi^{(n)}(q)\| \cdot \|\Pi^{(m)}(q)\|}$.
5: **end for**
Ensure: K kernel matrix, time segments $\mathcal{T}(q)$, subsets of attributes $\mathcal{V}(q)$, subsets of MTS $\eta(q)$, parameters $\Theta(q)$, $\Phi(q)$ and posteriors $\Pi^{(n)}(q)$.

trained under different, randomly chosen, settings (hyperparameters, initialization, subsampling). In more detail, the number of cluster components for the base models is sampled from a set of integers $\mathcal{I}_C = \{I, \ldots, I + C\}$. For each number of cluster components $q_2 \in \mathcal{I}_C$, we apply Q different random initial conditions and sample hyperparameters uniformly as follows: $a_0 \in [0.001, 1]$, $b_0 \in [0.005, 0.2]$, $n_0 \in [0.001, 0.2]$, $c_0, d_0 \in [0.1/N, 2/N]$. We let $Q = \{q = (q_1, q_2) \mid q_1 = 1, \ldots Q,\ q_2 \in \mathcal{I}_C\}$ be the index set keeping track of initial conditions and hyperparameters (q_1) as well as the number of components (q_2). Each base model q is trained on a random subset of MTS $\{(X^{(n)}, R^{(n)})\}_{n \in \eta(q)}$. To further increase the diversity, for each q, we select random subsets of variables $\mathcal{V}(q)$ as well as random time segments $\mathcal{T}(q)$. After having trained the individual base models using an embarrasingly parallel procedure, we compute a normalized sum of the inner products of the normalized posterior distributions from each mixture component to build the TCK$_{IM}$ kernel matrix. Details of the method are presented in Algorithm 2, whereas Algorithm 3 describes how to compute the kernel for MTS not seen during training.

Algorithm 3 TCK_{IM}. Test phase.

Require: Test set $\{X^{*(m)}\}_{m=1}^{M}$, time segments $\mathcal{T}(q)$, subsets $\mathcal{V}(q)$ and $\eta(q)$, parameters $\Theta(q)$, $\Phi(q)$ and posteriors $\Pi^{(n)}(q)$.
1: Initialize kernel matrix $K^* = 0_{N\times M}$.
2: **for** $q \in Q$ **do**
3: Compute posteriors $\Pi^{*(m)}(q)$, $m = 1, \dots, M$ using the mixture parameters $\Theta(q)$, $\Phi(q)$.
4: Update $K^*_{nm} = K^*_{nm} + \frac{\Pi^{(n)}(q)^T \Pi^{*(m)}(q)}{\|\Pi^{(n)}(q)\| \cdot \|\Pi^{*(m)}(q)\|}$.
5: **end for**
Ensure: K^* test kernel matrix.

3 Experiments

To test the performance of the proposed kernel, we considered three clinical datasets of which the characteristics are summarized in Table 1. The variables and the corresponding missing rates in the three datasets are summarized in Table 2. A more detailed description of the datasets follows below.

PhysioNet. The PhysioNet dataset is collected from the PhysioNet Challenge 2012 [26]. We extracted the first part, which consists of 4000 patient records of patients from the intensive care units (ICUs) of various hospitals. Each patient stayed in the ICU for at least 48 h and the records contain information about both vital signs and blood samples collected over time. We extracted all measurements taken within

Table 1 Description of the three real-world clinical datasets

Dataset	# of patients	# attrib (Lab,vital)	Length of MTS	Positive class	Av. missing rate
PhysioNet	4000	28 (17, 11)	48 (hours)	874 (21.9%)	76.6 %
SSI	858	11 (11, 0)	10 (days)	227 (26.5%)	80.7%
AL	402	11 (7, 4)	15 (days)	31 (7.7%)	83%

Table 2 List of variables and their corresponding missing rates

	Variables (missing rate)					
Phys	Albumin (0.99)	ALP (0.98)	AST (0.98)	Bilirubin (0.98)	BUN (0.93)	Creat. (0.93)
	Diast BP (0.11)	FiO2 (0.84)	GCS (0.68)	Glucose (0.93)	HCO3 (0.93)	HCT (0.91)
	HR (0.10)	K (0.93)	Lactate (0.96)	MAP (0.12)	Mg (0.93)	Na (0.93)
	PaCO2 (0.88)	PaO2 (0.88)	Platelet (0.93)	RespRate (0.76)	SaO2 (0.96)	SystBP (0.11)
	Temp (0.63)	Urine (0.31)	WBC (0.93)	pH (0.88)		
SSI	Albumin (0.79)	Amylase (0.95)	Creat. (0.87)	CRP (0.69)	Glucose (0.92)	Hb (0.65)
	K (0.71)	Na (0.71)	Platelet (0.92)	Urea (0.94)	WBC (0.73)	
AL	Albumin (0.80)	Creat. (0.94)	CRP (0.77)	Diast BP (0.76)	Hb (0.71)	HR (0.76)
	K (0.75)	Na (0.75)	Syst BP (0.76)	Temp (0.66)	WBC (0.80)	

48 h after admission and aligned the MTS into same-length sequences using an hourly discretization. Variables with a missing rate higher than 99% were omitted, which led to a total of 28 variables. The classification task was to predict whether the patient was recovering from surgery, which is a task that also has been considered in other work [8, 16].

Surgical Site Infection. This dataset contains data for 11 blood samples collected postoperatively for patients who underwent major abdominal surgery at the department of gastrointestinal surgery at a Norwegian university hospital in the years 2004–2012. The task considered was to detect surgical site infection (SSI), which is one of the most common types of nosocomial infections [15] and represents up to 30% of all hospital-acquired infections [19]. Patients with no recorded lab tests during the period from postoperative day 1 until day 10 were removed from the cohort, which lead to a final cohort consisting of 858 patients. The average proportion of missing data in the cohort was 80.7%. To identify the patients in the cohort who developed postoperative SSI and create ground truth labels, ICD-10 as well as NOMESCO Classification of Surgical Procedures codes related to severe postoperative complications were considered. Patients without these codes who also did not have a mention of the word "infection" in any of their postoperative text documents were considered as controls. This lead to a dataset consisting of 227 infected patients and 631 non-infected patients.

Anastomosis Leakage. Anastomosis leakage (AL) is potentially a serious complication that can occur after colon rectal cancer (CRC) surgery, of which one of the consequences is an increase in 30-day mortality [27]. It is estimated that 5–15% of the patients who undergo surgery for CRC suffer from AL [7]. Recent studies have shown that both unstructured as well structured EHR data such as measurements of blood tests and vital signs could have predictive value for AL [28]. The dataset considered in this work contains only structured data and is collected from the same hospital as the SSI dataset. It contains physiological data (blood tests and vital signs) for the period from the day after surgery to day 15 for 402 patients who underwent CRC surgery. A total of 31 of these patients got AL after surgery, but there is no information available about exactly when it happened. The classification task considered here was to detect which of the patients got AL.

Experimental Setup. We considered the following experimental setup. We performed kernel principal component analysis (KPCA) using the proposed TCK_{IM} and then trained a kNN-classifier in the low dimensional space. The dimensionality of the KPCA-representation was set to 3 to also be able to visualize the embeddings and we used 5-fold cross validation to set the number of neighbors k for the classifier. Performance was measured in terms of F1-score, sensitivity and specificity.

We compared the performance of the proposed kernel to four baseline kernels, namely the linear kernel (Lin), the global alignment kernel (GAK) [10], LPS and TCK. GAK is a positive semi-definite kernel formulation of the widely used, but non-metric, time series similarity measure called *dynamic time warping* (DTW) [5]. It has two hyperparameters, namely the kernel bandwidth and the triangular parameter, which have to be set by the user and it does not naturally deal with missing data and incomplete datasets, and therefore also requires a preprocessing step involving imputation. Therefore, we created a complete dataset using mean imputation for Lin and GAK (initial experiments showed that mean imputation worked better than last observation carried forward). In accordance with [9], for GAK we set the bandwidth σ to 0.1 times the median distance of all MTS in the training set scaled by the square root of the median length of all MTS, and the triangular parameter to 0.2 times the median length (Frobenius norm) of all MTS. In order to design baseline kernels that can exploit informative missingness, we also created baselines (referred to as Lin_{IM}, GAK_{IM} and LPS_{IM}) by concatenating the binary indicator MTS $R^{(n)}$ to $X^{(n)}$. LPS was run with default hyperparameters using the implementation provided by [3], with the exception that the minimal segment length was adjusted to account for the relatively short MTS in the datasets. For the TCK_{IM} we let $Q = 15$ and $\mathcal{I}_C = \{N/200, \ldots, N/200 + 20\}$, and, likewise, TCK was run with $Q = 15$ and $C = 20$. For all methods, except LPS which do not require standardization, we standardized each attribute to zero mean and unit standard deviation.

For PhysioNet and SSI, we did 5-fold cross validation to measure performance. The AL dataset is, however, highly imbalanced and therefore we employed an undersampling strategy by randomly sampling two patients from the negative class per positive case. 20% of this dataset was then set aside as a test set. This entire process was repeated 10 times and we reported mean and standard errors of the performance measures. The AL dataset is small, and for that reason the hyperparameters of the methods had to be adjusted accordingly. For TCK_{IM} we let $Q = 10$ and $\mathcal{I}_C = \{2, 3\}$.

4 Results and Discussion

Table 3 shows the performance of the TCK_{IM} kernel, as well as the baseline methods, on the three real-world datasets. Note that Lin and GAK, that rely on imputation, consistently perform worse than the other kernels in terms of F1-score across all datasets. We also note that these methods achieve a relatively high specificity. However, this is because they put too many patients in the negative class, which also leads to a high false negative rate and, consequently, a low sensitivity. The reasons could be that the imputation methods introduce biases and that the missingness mechanism is ignored.

TCK and LPS naturally handle missing data and perform better than the kernels that rely on imputation. TCK and LPS perform quite similarly across all 3 evaluation metrics for the SSI dataset, whereas LPS outperforms TCK on the PhysioNet and AL dataset. These methods probably perform better than the imputation methods because ignoring the missingness introduces less bias than replacing missing values with biased estimates. The performance of the baselines that account for informative missingness, Lin_{IM} and GAK_{IM}, is considerably better than Lin and GAK, respectively, for all datasets. LPS_{IM} also performs better than LPS on the PhysioNet datasets, whereas the performance of these two baselines is more or less equal on

Table 3 Performance (mean ± se) on 3 datasets

Kernel	Sensitivity	Specificity	F1-score
Phys.			
Lin	0.33 ± 0.05	0.81 ± 0.01	0.33 ± 0.05
GAK	0.31 ± 0.02	0.80 ± 0.02	0.31 ± 0.01
LPS	0.51 ± 0.07	0.95 ± 0.01	0.60 ± 0.07
TCK	0.41 ± 0.05	0.83 ± 0.02	0.41 ± 0.05
Lin_{IM}	0.56 ± 0.05	0.94 ± 0.01	0.63 ± 0.05
GAK_{IM}	0.57 ± 0.03	0.94 ± 0.01	0.64 ± 0.04
LPS_{IM}	0.61 ± 0.06	0.94 ± 0.01	0.67 ± 0.04
TCK_{IM}	**0.70 ± 0.03**	**0.98 ± 0.01**	**0.79 ± 0.03**
SSI			
Lin	0.48 ± 0.04	0.88 ± 0.04	0.53 ± 0.07
GAK	0.64 ± 0.06	0.92 ± 0.04	0.69 ± 0.03
LPS	0.69 ± 0.04	0.93 ± 0.04	0.73 ± 0.03
TCK	0.68 ± 0.07	0.92 ± 0.02	0.72 ± 0.07
Lin_{IM}	0.70 ± 0.01	**0.94 ± 0.03**	0.76 ± 0.02
GAK_{IM}	0.72 ± 0.07	0.94 ± 0.03	0.76 ± 0.04
LPS_{IM}	0.65 ± 0.06	0.93 ± 0.02	0.70 ± 0.04
TCK_{IM}	**0.77 ± 0.02**	0.93 ± 0.01	**0.79 ± 0.02**
AL			
Lin	0.41 ± 0.27	0.91 ± 0.12	0.47 ± 0.24
GAK	0.43 ± 0.22	**0.94 ± 0.08**	0.52 ± 0.21
LPS	**0.84 ± 0.11**	0.84 ± 0.08	0.78 ± 0.06
TCK	0.70 ± 0.15	0.89 ± 0.11	0.72 ± 0.08
Lin_{IM}	0.74 ± 0.22	0.86 ± 0.1	0.72 ± 0.1
GAK_{IM}	0.73 ± 0.23	0.93 ± 0.05	0.76 ± 0.16
LPS_{IM}	0.80 ± 0.12	0.86 ± 0.09	0.77 ± 0.09
TCK_{IM}	**0.84 ± 0.18**	0.86 ± 0.09	**0.79 ± 0.11**

the two other datasets. The proposed TCK_{IM} performs considerably better than all baselines, and in particular compared to TCK (the kernel which it is an improvement of) for the PhysioNet and SSI datasets in terms of F1-score, and it performs better or comparable than the other kernels on the AL dataset. This demonstrates that the missing patterns in clinical time series are often informative and the TCK_{IM} can exploit this information very efficiently.

Figure 1 shows the KPCA embeddings obtained using five kernels (Lin, GAK, LPS, TCK and TCK_{IM}). In general, the classes are more separated in the representations obtained using TCK_{IM} than in the other representations.

Limitations, Future Work and Conclusions. In this paper, we presented a MTS kernel capable of exploiting informative missingness along with the temporal depen-

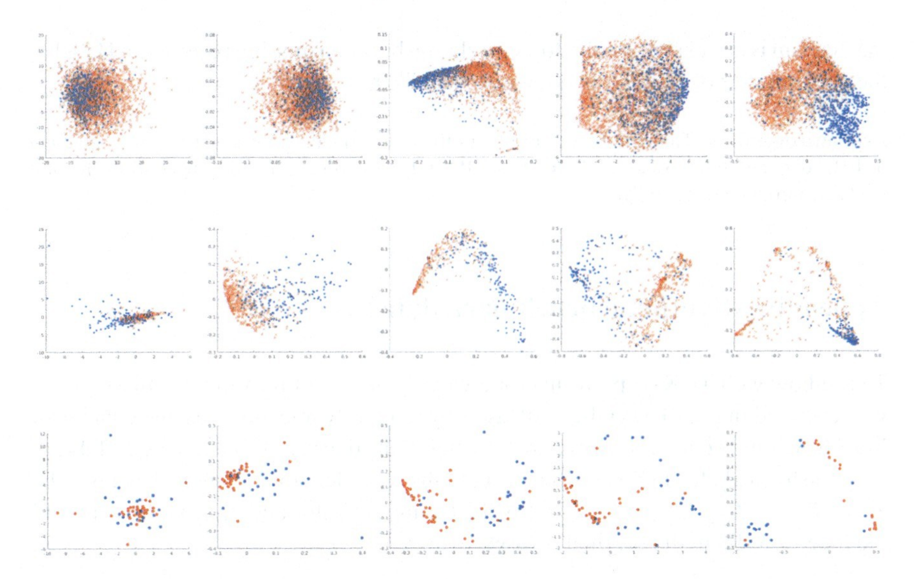

Fig. 1 KPCA plots obtained using Lin, GAK, LPS, TCK and TCK_{IM}, respectively. Row 1: Physio., row 2: SSI, row 3: AL

dencies in the observed data. We showed that TCK_{IM} can learn good representations that can be exploited both in supervised and unsupervised tasks, even when the percentage of missing data is high. In this work, the representations learned using TCK_{IM} were evaluated visually (Fig. 1) and using a supervised scheme by training a classifier (Table 3). The experimental results suggested that TCK_{IM} achieved superior performances compared to baselines on three real datasets.

The experiments presented in this work focused on binary classification tasks, both of patients at the ICU and patients who had undergone colonrectal cancer surgery. However, we believe that TCK_{IM} is also a very good choice in applications where there is a lack of labels, which often is the case in medical applications, thanks to the ensemble learning strategy that makes the kernel robust to hyperparameter choices. In fact, since it is a kernel, it can be used in many different applications, including classification as well as clustering tasks, benefiting from the vast body of work in the field of kernel methods. In future work, we would like to test TCK_{IM} in a realistic unsupervised task from the medical domain.

A limitation of TCK_{IM} is that it is only designed for MTS of the same length. In further work, we would therefore like to design a time series cluster kernel that can also deal with varying length MTS. It should also be pointed out that if the missing patterns are not informative, i.e. the missingness is not correlated the particular medical condition(s) of interest, the performance gain of TCK_{IM} compared to TCK is low. It is therefore an advantage if the user has some understanding about the underlying missingness mechanism in the data. On the other hand, our experiments on benchmark datasets (see Appendix) demonstrate that in cases when the missingness mechanism is almost ignorable (low correlation between missing patterns and labels), the performance of TCK_{IM} is not worse than TCK.

Acknowledgements The authors would like to thank K. Hindberg for assistance on extraction of EHR data and the physicians A. Revhaug, R.-O. Lindsetmo and K. M. Augestad for helpful guidance throughout the study.

Appendix—Synthetic benchmark datasets

To test how well TCK_{IM} performs for a varying degree of informative missingness, we generated in total 16 synthetic datasets by randomly injecting missing data into 4 MTS benchmark datasets. The characteristics of the datasets are described in Table 4. We transformed all MTS in each dataset to the same length, T, where T is given by $T = \lceil T_{max} / \lceil T_{max}/25 \rceil \rceil$. Here, $\lceil \; \rceil$ is the ceiling operator and T_{max} is the length of the longest MTS in the original dataset.

Datasets. The following procedure was used to create 8 synthetic datasets with missing data from the Wafer and Japanese vowels datasets. We randomly sampled a number $c_v \in \{-1, 1\}$ for each attribute $v \in \{1, \ldots, V\}$, where $c_v = 1$ indicates that the attribute and the labels are positively correlated and $c_v = -1$ negatively correlated. Thereafter, we sampled a missing rate γ_{nv} from $\mathcal{U}[0.3 + E \cdot c_v \cdot (y^{(n)} - 1), 0.7 + E \cdot c_v \cdot (y^{(n)} - 1)]$ for each MTS $X^{(n)}$ and attribute. The parameter E was tuned such that the Pearson correlation (absolute value) between the missing rates for the attributes γ_v and the labels $y^{(n)}$ took the values $\{0.2, 0.4, 0.6, 0.8\}$, respectively. By doing so, we could control the amount of informative missingness and because of the way we sampled γ_{nv}, the missing rate in each dataset was around 50% independently of the Pearson correlation. Further, the following procedure was used to create 8 synthetic datasets from the uWave and Character trajectories datasets, which both consist of only 3 attributes. We randomly sampled a number $c_v \in \{-1, 1\}$ for each attribute $v \in \{1, \ldots, V\}$. Attribute(s) with $c_v = -1$ became negatively correlated with the labels by sampling γ_{nv} from $\mathcal{U}[0.7 - E \cdot (y^{(n)} - 1), 1 - E \cdot (y^{(n)} - 1)]$, whereas the attribute(s) with $c_v = 1$ became positively correlated with the labels by sampling γ_{nv} from $\mathcal{U}[0.3 + E \cdot (y^{(n)} - 1), 0.6 + E \cdot (y^{(n)} - 1)]$. The parameter

Table 4 Characteristics of benchmark datasets. Attr is number of attributes, Train and Test number of training and test samples. N_c is the number of classes, T_{min} and T_{max} length of shortest and longest MTS, and T is the MTS length after the transformation

Datasets	Attr	Train	Test	N_c	T_{min}	T_{max}	T	Source
uWave	3	200	4278	8	315	315	25	[2]
Char.tra.	3	300	2558	20	109	205	23	[2]
Wafer	6	298	896	2	104	198	25	[21]
Jap.vow.	12	270	370	9	7	29	15	[2]

Table 5 Performance (accuracy) of TCK_{IM} and three baselines

Corr.	TCK	TCK_B	TCK_0	TCK_{IM}	TCK	TCK_B	TCK_0	TCK_{IM}
	Wafer				*Japanese vowels*			
0.2	0.95	0.95	0.95	**0.96**	0.94	**0.95**	0.95	0.94
0.4	**0.96**	0.95	0.96	**0.96**	0.93	0.94	0.94	**0.94**
0.6	0.96	0.90	0.97	**1.00**	0.92	0.95	0.92	**0.96**
0.8	0.96	0.89	0.96	**1.00**	0.92	0.92	0.94	**0.97**
	uWave				*Character trajectories*			
0.2	0.76	0.46	0.76	**0.84**	**0.85**	0.74	0.85	0.85
0.4	0.81	0.59	0.81	**0.86**	0.85	0.79	0.84	**0.87**
0.6	0.83	0.67	0.84	**0.87**	0.83	0.79	0.82	**0.87**
0.8	0.83	0.70	0.84	**0.88**	0.84	0.71	0.85	**0.90**

E was computed in the same way as above. Then, we computed the mean of each attribute μ_v over the complete dataset and let each element with $x_v^{(n)}(t) > \mu_v$ be missing with probability γ_{nv}. This means that the probability of being missing is dependent on the value of the missing element, i.e. the missingness mechanism is MNAR within each class. Hence, this type of informative missingness is not the same as the one we created for the Wafer and Japanese vowels datasets.

Baselines. Three baseline models were created. The first baseline, namely ordinary TCK, ignores the missingness mechanism. In the second one, refered to as TCK_B, we modeled the missing patterns naively by concatenating the binary missing indicator MTS R to the MTS X and creating a new MTS U with $2V$ attributes. Then, ordinary TCK was trained on the datasets consisting of $\{U^{(n)}\}$. In the third baseline, TCK_0, we investigated how well informative missingness can be captured by imputing zeros for the missing values and then training the TCK on the imputed data.

Results. Table 5 shows the performance of the proposed TCK_{IM} and the three baselines for all of the 16 synthetic datasets. We see that the proposed TCK_{IM} achieves the best accuracy for 14 out of 16 datasets and is the only method which consistently has the expected behaviour, namely that the accuracy increases as the correlation between missing values and class labels increases. It can also be seen that the performance of TCK_{IM} is similar to TCK when the amount of information in the missing patterns is low, whereas TCK is clearly outperformed when the informative missingness is high. This demonstrates that TCK_{IM} can effectively exploit informative missingness.

References

1. Agniel, D., et al.: Biases in electronic health record data due to processes within the healthcare system: retrospective observational study. BMJ **361**, k1479 (2018)

2. Bagnall, A., et al.: The UEA multivariate time series classification archive 2018. *arXiv preprint*arXiv:1811.00075 (2018)
3. Baydogan, M.: LPS Matlab implementation (2014). http://www.mustafabaydogan.com/. Accessed 06 Sept 2019
4. Baydogan, M.G., Runger, G.: Time series representation and similarity based on local autopatterns. Data Min. Knowl. Disc. **30**(2), 476–509 (2016)
5. Berndt, D.J., Clifford, J.: Using dynamic time warping to find patterns in time series. In: 3rd International Conference on Knowledge Discovery and Data Mining, pp. 359–370. AAAI Press (1994)
6. Bianchi, F.M., et al.: Learning representations of multivariate time series with missing data. Patt. Rec. **96**, 106973 (2019)
7. Branagan, G., Finnis, D.: Prognosis after anastomotic leakage in colorectal surgery. Dis. Colon Rectum **48**(5), 1021–1026 (2005)
8. Che, Z., et al.: Recurrent neural networks for multivariate time series with missing values. Sci. Rep. **8**(1), 6085 (2018)
9. Cuturi, M., Fast global alignment kernel Matlab implementation (2011). http://www.marcocuturi.net/GA.html. Accessed 02 Sept 2019
10. Cuturi, M.: Fast global alignment kernels. In: Proceedings of the 28th International Conference on Machine Learning, pp. 929–936 (2011)
11. Dempster, A.P., Laird, N.M., Rubin, D.B.: Maximum likelihood from incomplete data via the EM algorithm. J. Roy. Stat. Soc. Ser. B (Methodol.) **39**, 1–38 (1977)
12. Dietterich, T.G.: Ensemble methods in machine learning. In: International Workshop on Multiple Classifier Systems, pp. 1–15 (2000)
13. Donders, A.R., et al.: Review: a gentle introduction to imputation of missing values. J. Clin. Epidemiol. **59**(10), 1087–1091 (2006)
14. Halpern, Y., et al.: Electronic medical record phenotyping using the anchor and learn framework. J. Am. Med. Inform. Assoc. **23**(4), 731–40 (2016)
15. Lewis, S.S., et al.: Assessing the relative burden of hospital-acquired infections in a network of community hospitals. Infect. Control Hosp. Epidemiol. **34**(11), 1229–1230 (2013)
16. Li, Q., Xu, Y.: VS-GRU: a variable sensitive gated recurrent neural network for multivariate time series with massive missing values. Appl. Sci. **9**(15), 3041 (2019)
17. Little, R.J., Rubin, D.B.: Statistical Analysis with Missing Data. Wiley, Hoboken (2014)
18. Ma, Z., Chen, G.: Bayesian methods for dealing with missing data problems. J. Korean. Stat. Soc. **47**(3), 297–313 (2018)
19. Magill, S.S., et al.: Prevalence of healthcare-associated infections in acute care hospitals in Jacksonville. Florida. Infect. Control **33**(03), 283–291 (2012)
20. Mikalsen, K.Ø., et al.: Time series cluster kernel for learning similarities between multivariate time series with missing data. Pattern Recogn. **76**, 569–581 (2018)
21. Olszewski, R.T.: Generalized feature extraction for structural pattern recognition in time-series data. Ph.D. thesis, Carnegie Mellon University, Pittsburgh, PA, USA (2001)
22. Schafer, J.L., Graham, J.W.: Missing data: our view of the state of the art. Psychol. Methods **7**(2), 147 (2002)
23. Shao, J., Zhong, B.: Last observation carry-forward and last observation analysis. Stat. Med. **22**(15), 2429–2441 (2003)
24. Sharafoddini, A., et al.: A new insight into missing data in intensive care unit patient profiles: observational study. JMIR Med Inform. **7**(1), e11605 (2019)
25. Shukla, S.N., Marlin, B.: Interpolation-prediction networks for irregularly sampled time series. In: ICLR (2019)
26. Silva, I., et al.: Predicting in-hospital mortality of ICU patients: the physionet/computing in cardiology challenge 2012. In: 2012 Computing in Cardiology, pp. 245–248. IEEE (2012)
27. Snijders, H., et al.: Anastomotic leakage as an outcome measure for quality of colorectal cancer surgery. BMJ Qual. Saf. **22**(9), 759–767 (2013)
28. Soguero-Ruiz, C., et al.: Predicting colorectal surgical complications using heterogeneous clinical data and kernel methods. J. Biomed. Inform. **61**, 87–96 (2016)
29. Zhang, Z.: Missing data imputation: focusing on single imputation. Ann. Transl. Med. **4**(1), 9 (2016)

Machine Learning Discrimination of Parkinson's Disease Stages from Walker-Mounted Sensors Data

Nabeel Seedat and Vered Aharonson

Abstract Clinical methods that assess Parkinson's Disease (PD) gait are mostly qualitative. Quantitative methods necessitate costly instrumentation or cumbersome wearables, limiting usability. This study applies machine learning to discriminate six stages of PD. The data was acquired by low cost walker-mounted sensors at a movement disorders clinic. A large set of features were extracted and three feature selection methods were compared using a Random Forest classifier. The feature subset selected by the ANOVA method provided performance similar to the full feature set: 93% accuracy, with a significantly shorter computation time. Compared to PCA, it enabled clinical interpretability of the features, an essential attribute in healthcare. All selected-feature sets were dominated by information theoretic and statistical features and offer insights into the characteristics of PD gait deterioration. The results indicate a feasibility of machine learning to accurately classify PD severity stages from kinematic signals acquired by low-cost, walker-mounted sensors and can aid medical practitioners in quantitative assessment of PD progression. The study presents a solution to the small and noisy data problem, typical to sensor-based healthcare assessments.

Research supported by NRF Block Grant (UID: 111755).

N. Seedat · V. Aharonson (✉)
University of the Witwatersrand, Johannesburg, South Africa
e-mail: vered.aharonson@wits.ac.za

N. Seedat
e-mail: seedatnabeel@gmail.com

N. Seedat
Shutterstock, New York, USA

© The Editor(s) (if applicable) and The Author(s), under exclusive license to Springer Nature Switzerland AG 2021

A. Shaban-Nejad et al. (eds.), *Explainable AI in Healthcare and Medicine*, Studies in Computational Intelligence 914,
https://doi.org/10.1007/978-3-030-53352-6_4

1 Introduction

Automated disease diagnosis has the potential to reduce labor and cost in healthcare, as well as offer an augmented accuracy which may improve treatment efficacy. Machine learning methods that could provide automated disease assessment have been extensively studied on medical imaging and physiological signals datasets [11, 17]. However, when deep learning methods are employed, very large datasets are needed in order to provide accurate detection or classification of diseases. Such datasets are not available in many healthcare contexts.

Datasets of patients suffering from Parkinson's disease (PD); a debilitating neurodegenerative disease with increasing prevalence; are typically scarce and small. Many of the physical symptoms of PD pertain to gait; the manner in which a person walks [14]. Gait impairments reduces mobility, significantly limits a person's functionality and independence and can cause instability and fall hazards. Thus, gait assessment is important to discriminate the severity of PD [7, 8]. The Hoehn and Yahr scale is commonly employed by clinicians to reflect PD severity, with stages between 1 and 5, with a "mid-stage" of 2.5 added for granularity [9]. This and other PD scales are predominantly based on observational symptoms which are challenging to reliably acquire in an automated manner. Healthcare professionals and patients could greatly benefit from automated disease assessment tools since observation- or reporting-based data are qualitative and subjective, and hence are prone to bias and inaccuracy, which may lead to incorrect treatment.

Attempts to automate PD gait assessment typically use strap-on or wearable sensors such as accelerometers and gyroscopes that capture patients' motor data [10, 12]. Although sensors have the potential to efficiently record large quantities of quantitative data, practical considerations often limit their usage in clinical settings: Many sensors and devices are cumbersome and uncomfortable, and/or are not suitable for severe stages of PD, when patients cannot walk without support [1, 6]. The small data size may also be the reason that most previous works discriminate between patients and controls, without attempting the multi-class discrimination of disease stages [3, 13, 16]. An accurate and timely discrimination of disease stages is very important for efficient treatment and patient care.

This paper makes the following contributions:

- A novel sensor-based system that offers a solution to the aforementioned limitations. It collects adequate data on PD symptoms and uses machine learning for quantitative multi-class discrimination of PD stages.
- A machine learning method that provides in a translucent way the most relevant features that discriminate between PD stages, to enable clincians to interpret the system's output and employ it patient-care planning.
- Solutions for the application of machine learning on small-scale datasets of noisy sensors' signals through an integration of signal processing techniques.

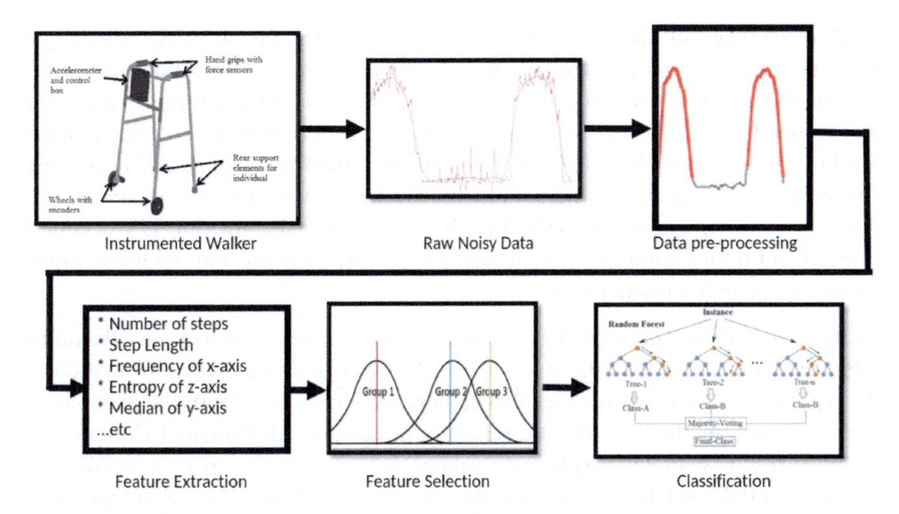

Fig. 1 System Pipeline outlining the six stages of processing from raw data acquired from the instrumented walker till the classification of PD severity stage

2 Our Approach: Machine Learning Enabled PD Stage Discrimination from Low-Cost Walker Mounted Sensor

We present a novel solution where kinematic signals acquired by walker-mounted sensors coupled with machine learning is used for PD stage discrimination. The walker method provides a patient-centric approach that combines reliability and usability, even for severe stages of the disease. The analysis pipeline outlined in Fig. 1 is implemented as follows:

Dataset: The data was acquired in a movement disorders clinic by an aluminum walking frame retrofitted with low-cost and off-the-shelf kinematic sensors, monitoring gait unobtrusively, whilst simultaneously supporting walking. The sensors include a tri-axial accelerometer, wheel distance encoders, and force sensors on the handles that measure grip strength. Sixty-seven PD patients (spread across severity groups) and nineteen age-matched healthy controls recruited from the clinic were subjected to a simple and clinically adopted walking test (Timed-Up-and-Go), where they walked 3 m and turned around and walked back while using the walker. PD severity scores were assigned by a neurologist using the modified Hoehn and Yahr severity scores of 1, 2, 2.5, 3 and 4.

Data Pre-processing: Pre-processing is necessary for low-cost, noisy sensors and in the case of walker-mounted sensors, where the sensors do not directly move with the subject's body. The pre-processing of the walker's signals include noise and artifact removal, walking phase segmentation and a novel footfall detection algorithm [15] using adaptive Empirical Mode Decomposition (EMD).

Feature Extraction: The extracted feature set provides a quantitative representation of a subject's gait and are an amalgamation of novel features pertinent to the system and features used in previous studies. A set of 211 features are extracted and grouped into 4 categories: spatio-temporal features (quantify the characteristics used by clinicians, such as step length, walk time, number of steps and turn time), statistical features (e.g. skewness, kurtosis, mean, median), frequency domain features (e.g. mean frequency, 3 dB bandwidth, 99% occupied bandwidth, 99% occupied power,) and information theoretic features (e.g. cross-entropy, mutual information, correlation, and harmonic ratio).

Feature Selection: The small data size requires feature selection to reduce the initial high, 211-dimensional feature space. The selection can also convey information on the types or categories of features that capture the characteristics of the phenomena studied. Three methods of feature selection are examined: Principal Components Analysis (PCA) with 95% explained variance, feature selection using ANOVA with a p-value of 0.05 and the embedded feature selection performed by a Random Forest (based on permutation feature importance).

Classification: A Random Forest (RF) classifier [5] is implemented to discriminate between the six stages of PD (i.e. controls and Hoehn and Yahr severity scores of 1, 2, 2.5, 3 and 4). The RF is both interpretable and as an ensemble is suited to the small data problem. An RF classifier of 100 trees (providing the best trade-off of accuracy and execution time) is used on all feature sets. Given the small data size problem, separate training, validation and test sets are not possible for evaluation, thus 5-fold cross-validation is performed.

3 Experimental Results

The effectiveness of our method for automated discrimination of PD stages is evaluated based on 1. Discrimination accuracy and execution time, 2. The information gained on the features relevant to this discrimination, which is important for clinical interpretability. The performance of the Random Forest classifier in discriminating the six classes (Healthy controls and five PD Hoehn and Yahr stages) is presented in Table 1.

Figure 2 displays the confusion matrix for the Random Forest classifier using the ANOVA selected features as it has high accuracy, low execution time and retains

Table 1 Accuracy for the different feature sets in the discrimination of PD stages

Test set	All features	ANOVA selected	RF selected	PCA
Accuracy (%)	94 ± 1.02	$\mathbf{93 \pm 1.3}$	90 ± 1.7	95 ± 1.1
Mean execution time (s)	41.4	**15.2**	6.36	10.6

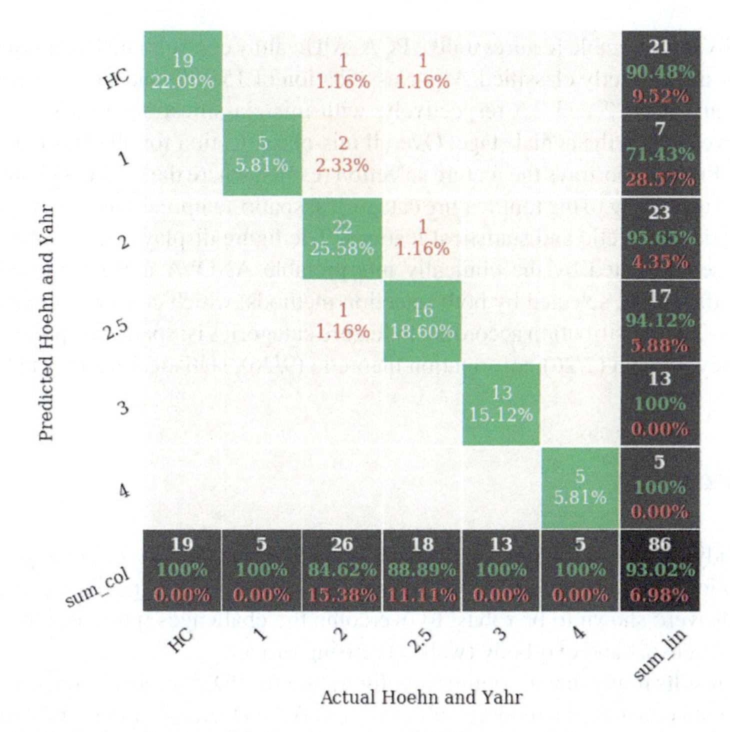

Fig. 2 Confusion matrix of the RF classifier using ANOVA selected features, where the x-axis: actual labels and y-axis: predicted labels

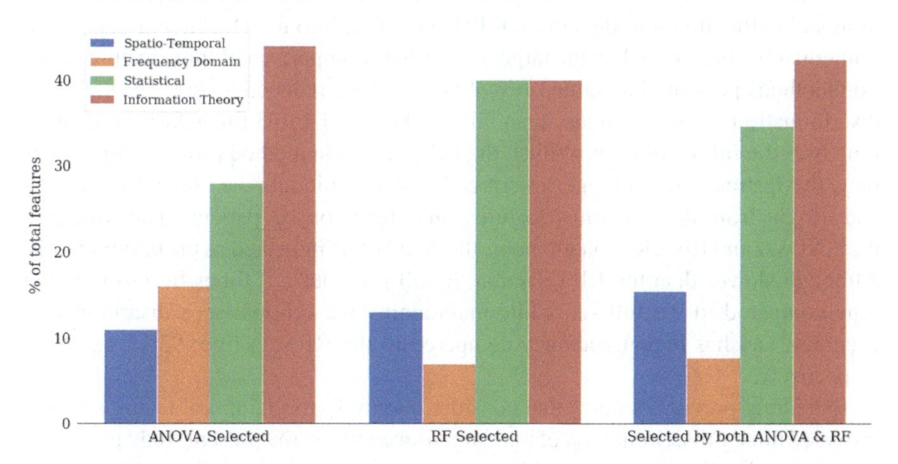

Fig. 3 Selected features of the ANOVA method, the RF method and ANOVA and RF, grouped by feature category

clinically interpretable features unlike PCA. All healthy controls and Hoehn and Yahr 1, and 4 are correctly classified. Mis-classification of 15.38% and 11.1% is seen for Hoehn and Yahr 2 and 2.5 respectively, with mis-classifications to a stage lower (less severe) than the actual stage. Overall mis-classification for all class labels was 6.98%. Figure 3 portrays the feature selection results, where the selected features are grouped according to the four feature categories: spatio-temporal, frequency domain, information theoretic and statistical features. The figure displays the distribution of the subsets selected by the clinically interpretable ANOVA and RF methods, as well as the subset selected by both selection methods, which consists of twenty six features. Their distribution according to feature categories is: spatio-temporal (4/26), frequency domain (2/26), information theoretic (9/26), statistical features (11/26).

4 Discussion

This study presents an automated machine learning discrimination of PD stages using kinematic signals obtained from walker-mounted sensors. The signal pre-processing methods were shown to be robust to overcome the challenges introduced by noisy, low-cost sensors and exo-body (walker) sensing of gait.

The results imply that the optimal configuration for PD stage discrimination based on this data consists of a reduced subset of ANOVA-selected features, and a random forest classifier of 100 trees. The mean accuracy achieved by this configuration is 93%. The confusion matrix in Fig. 2 indicates that classification errors typically result in mis-classification into the adjacent PD stage (i.e. into the Hoehn and Yahr stage immediately above or below the target stage). The comparison of three feature selection methods presented in Table 1 reveal that all three reduced feature sets yielded a discrimination accuracy greater than 90% and offered 3 to 5 times faster execution time over the full feature set. Whilst, the PCA selection method provides high accuracy, the features are no longer interpretable which hinders the clinical usefulness and insight from the individual features' manifestations in patient's gait. Between the ANOVA and RF selection methods, the ANOVA is indicated as the best trade-off. Although slower than the RF selection, it still provides a 3-times faster execution time compared to the full set, while maintaining the original set's discrimination accuracy, and has higher accuracy compared to the RF selection: 93%, and 90%, respectively.

Extending previous studies that consider mainly spatio-temporal features and/or provided binary discrimination of patients from controls, the present study presents a multi-class discrimination of healthy control subjects and five different PD severity stages. Previous studies focused on spatio-temporal features and specifically step time, step length and step velocity [1, 2, 4, 12, 16]. Whilst, these spatio-temporal features like step length and step velocity are among the significant features in our analysis, corroborating earlier studies, a wider variety of features to characterize PD is also extracted and the features are evaluated by their prevalence in the feature selection subsets. The percentage of information theory and statistical features in all selected feature sets is much larger than the spatio-temporal ones. This finding implies that gait analysis of PD should not be limited to the observable, time-based features, but need also include these more abstract and mathematical features. Including all four feature categories may yield broader insight into PD gait deterioration with disease severity. The analysis of feature importance also indicates that although the Hoehn and Yahr scale is based on other motor assessments in addition to gait, it could be accurately classified using only gait features, implying a significant impact of gait impairment characteristics in the Hoehn and Yahr scale.

The combined usability design, signal processing and machine learning discrimination proposed in this study has the potential to assist healthcare professionals in PD severity evaluation and facilitate a patient-centred care at a low-cost. The method is tested in a clinic, but has the potential to be used in the future by patients at home as part of an eHealth monitoring scheme enabling quantitative PD severity assessment.

References

1. Aharonson, V., Schlesinger, I., McDonald, A., Dubowsky, S., Korczyn, A.: A practical measurement of Parkinson's patients gait using simple walker-based motion sensing and data analysis. J. Med. Devices **12**, 011012 (2018)
2. Akbari, A., Dewey, R., Jafari, R.: Validation of a new model-free signal processing method for gait feature extraction using inertial measurement units to diagnose and quantify the severity of Parkinson's disease. In: International Conference on Computer Communication and Networks (ICCCN), pp. 1–5, July 2017
3. Arora, S., Venkataraman, V., Donohue, S., Biglan, K., Dorsey, E., Little, M.A.: Detecting and monitoring the symptoms of Parkinson's disease using smartphones: A pilot study. Parkinsonism Rel. Disord. **21**, 03 (2015)
4. Ballesteros, J., Urdiales, C., Martinez, A.B., Tirado, M.: Automatic assessment of a rollator-users condition during rehabilitation using the i-walker platform. IEEE Trans. Neural Syst. Rehabil. Eng. **25**(11), 2009–2017 (2017)
5. Breiman, L.: Random forests. Mach. Learn. **45**(1), 5–32 (2001)
6. Bryant, M., Rintala, D., Graham, J., Hou, J., Protas, E.: Determinants of use of a walking device in persons with Parkinson's disease. Arch. Phys. Med. Rehabil. **95**(10), 1940–5 (2014)
7. Chen, P.H., Wang, R.L., Liou, D., Shaw, J.S.: Gait disorders in Parkinson's disease: assessment and management. Int. J. Gerontol. **7**(4), 189–193 (2013)
8. Dingwell, J., Cusumano, J.: Nonlinear time series analysis of normal and pathological human walking. Chaos: Interdiscipl. J. Nonlinear Sci. **10**(4), 848–863 (2000)
9. Goetz, C., Poewe, W., Rascol, O., Cristina Sampaio, C., Stebbins, G., Counsell, C., Giladi, N., Holloway, R., Moore, C., Wenning, G., Yahr, M., Seidl, L.: Movement disorder society task force report on the Hoehn and yahr staging scale: status and recommendations. Movement Dis. **19**(9), 1020–1028 (2004)
10. Goschenhofer, J., Pfister, F.M.J., Yuksel, K.A., Bischl, B., Fietzek, U., Thomas, J.: Wearable-based Parkinson's disease severity monitoring using deep learning. ArXiv, arXiv:abs/1904.10829 (2019)
11. Havaei, M., et al.: Brain tumor segmentation with deep neural networks. Med. Image Anal. **35**, 18–31 (2017)
12. Jarchi, D., Pope, J., Lee, T.K.M., Tamjidi, L., Mirzaei, A., Sanei, S.: A review on accelerometry-based gait analysis and emerging clinical applications. IEEE Rev. Biomed. Eng. **11**, 177–194 (2018)

13. Keijsers, N.L., Horstink, M.W., Gielen, S.C.: Automatic assessment of levodopa-induced dyskinesias in daily life by neural networks. Mov. Disord.: Official J. Mov. Disord. Soc. **18**(1), 70–80 (2003)
14. Perumal, S., Sankar, R.: Gait and tremor assessment for patients with Parkinson's disease using wearable sensors. ICT Express **2**(4), 168–174 (2016). Special Issue on Emerging Technologies for Medical Diagnostics
15. Seedat, N., Beder, D., Aharonson, V., Dubowsky, S.: A comparison of footfall detection algorithms from walker mounted sensors data. In: 2018 IEEE EBBT, pp. 1–4, April 2018
16. Tahir, N., Manap, H.: Parkinson disease gait classification based on machine learning approach. J. Appl. Sci. **12**, 180–185 (2012)
17. Tomašev, N., Glorot, X., Rae, J., Zielinski, M., Askham, H., Saraiva, A., Mottram, A., Meyer, C., Ravuri, S., Protsyuk, I.: A clinically applicable approach to continuous prediction of future acute kidney injury. Nature **572**(7767), 116 (2019)

Personalized Dual-Hormone Control for Type 1 Diabetes Using Deep Reinforcement Learning

Taiyu Zhu, Kezhi Li, and Pantelis Georgiou

Abstract We introduce a dual-hormone control algorithm for people with Type 1 Diabetes (T1D) which uses deep reinforcement learning (RL). Specifically, double dilated recurrent neural networks are used to learn the control strategy, trained by a variant of Q-learning. The inputs to the model include the real-time sensed glucose and meal carbohydrate content, and the outputs are the actions necessary to deliver dual-hormone (basal insulin and glucagon) control. Without prior knowledge of the glucose-insulin metabolism, we develop a data-driven model using the UVA/Padova Simulator. We first pre-train a generalized model using long-term exploration in an environment with average T1D subject parameters provided by the simulator, then adopt importance sampling to train personalized models for each individual. *In-silico*, the proposed algorithm largely reduces adverse glycemic events, and achieves time in range, i.e., the percentage of normoglycemia, 93% for the adults and 83% for the adolescents, which outperforms previous approaches significantly. These results indicate that deep RL has great potential to improve the treatment of chronic diseases such as diabetes.

T. Zhu (✉) · P. Georgiou
Department of Electrical and Electronic Engineering, Imperial College London, London SW7 2AZ, UK
e-mail: taiyu.zhu17@imperial.ac.uk

P. Georgiou
e-mail: pantelis@imperial.ac.uk

K. Li
Institute of Health Informatics, University College London, London NW1 2DA, UK
e-mail: ken.li@ucl.ac.uk

© The Editor(s) (if applicable) and The Author(s), under exclusive license to Springer Nature Switzerland AG 2021

A. Shaban-Nejad et al. (eds.), *Explainable AI in Healthcare and Medicine*, Studies in Computational Intelligence 914,
https://doi.org/10.1007/978-3-030-53352-6_5

1 Introduction

Diabetes is a lifelong condition that affects an estimated 451 million people worldwide [1]. Delivering an optimal insulin dose to T1D subjects has been one of the long-standing challenges since the 1970s [2].

In the past, the quality of life of people with Type 1 Diabetes (T1D) has relied heavily on the accuracy of human-defined models & features of the delivery strategy. Recently, however, deep learning has provided new ideas and solutions to many healthcare problems [3]. This has been empowered by the increasing availability of medical data and rapid progress of analytic tools, e.g. deep reinforcement learning (RL). However, several reasons hinder the building of efficient RL models to solve problems in chronic diseases. In most cases, RL requires medical data collected from a dynamic interaction between the human and environment, which is limited and expensive [4]. In addition, it is different to playing Atari in a virtual environment [5]; RL costs heavily to 'explore' the possibilities on humans in terms of price and safety. Finally, the variability of physiological responses to the same treatment can be very large for different people with T1D [6]. These are the reasons why there has been little progress in using RL in treatemt of chronic diseases.

To overcome these obstacles, we propose a two-step framework to apply deep RL in chronic diseases, and use T1D as a case study. T1D is chosen because it is a typical disease that requires dynfamic treatment consistently. A generalized deep RL model for a hormone delivery strategy in diabetes is pre-trained using a variant of Q-learning as the first step. Secondly, by prioritizing the transitions in experience memory, importance sampling is implemented to train personalized models with individual data [7]. As there are many emerging deep learning applications to diabetes management [8, 9], it has been shown that dilated recurrent neural networks (DRNN) perform extremely well in processing long-term dependencies and future glucose prediction [10, 11]. Thus, we employ DRNNs to build double deep Q-networks (DQN) for multi-dimensional medical time series, in which basal dual-hormone delivery (each at five minutes intervals) is considered as the action determined by DQN's policy. Glucose levels and corresponding time in range (TIR) are considered as the reward. TIR presents the time percentage of glucose values within a target range considered to be normoglycemia. It is a key derived metric in for glyceamic control [12].

We use the UVA/Padova T1D Simulator, a credible glucose-insulin dynamics simulator which has been accepted by the Food and Drug Administration [13], as the environment. It can generate glucose data for T1D subjects with high variability depending on meal intake, body conditions and other factors. During the training, the agent interacts with the T1D environment to obtain the optimal policy for the closed-loop dual-hormone delivery, as shown in Fig. 1. Then 10 adults and 10 adolescents are tested within a 6-month period of time, respectively. The results show that TIRs achieve 93% for adults and 83% for adolescents *in-silico*, which is a significant improvement over current state-of-the-art methods.

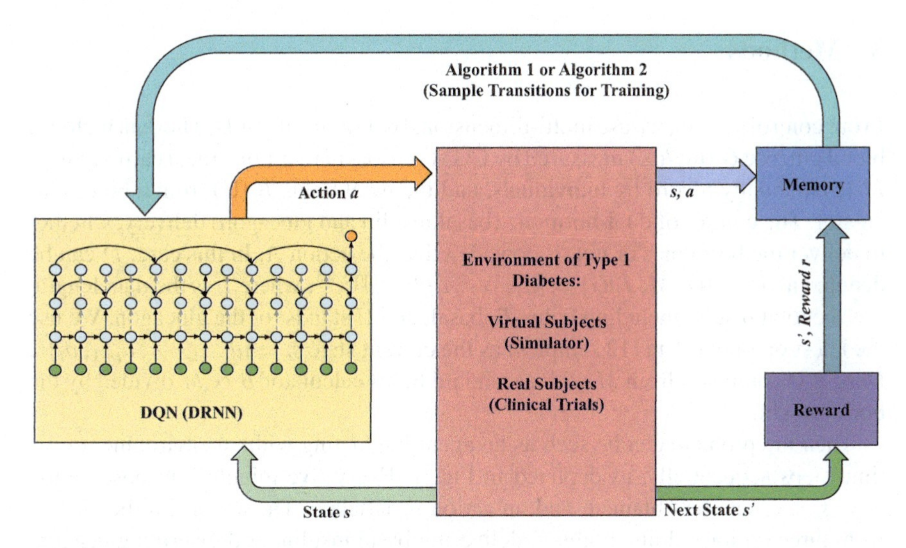

Fig. 1 The system architecture to apply deep RL on T1D

2 Related Work and Preliminaries

The rapid growth of continuous glucose monitoring (CGM) and insulin pump therapy have motivated use of a closed-loop system, known as the artificial pancreas (AP) [14, 15]. Many algorithms are developed and verified as closed-loop single/dual hormone delivery controllers [16], but most of them are based on conventional control algorithms [17, 18].

In recent work, researchers have investigated RL routines to update several parameters of glucose controller in gradient [19, 20]. Moreover, we find an RL environment was built in a simulator of the 2008 version [21]. Based on this simulator, a more recent paper has introduced deep RL to improve average risk index [22]; the method uses DQNs to control single insulin delivery. However, we use an updated simulator, the 2013 version (S2013), with new features such as glucagon kinetics that allows dual-hormone actions [13].

We see the problem an infinite-state Markov decision process (MDP) with noise. An MDP can be defined by a tuple $\langle S, A, R, \mathcal{T}, \gamma \rangle$ with state S, action A, reward function $R : S \times A \mapsto [R_{\min}, R_{\max}]$, transition function $\mathcal{T}$, and discount factor $\gamma \in [0, 1]$. At each time step, the agent takes an action $a \in A$, causes the environment from some state $s \in S$ to transit to state s' with probability $T(s', s, a) = P(s'|s, a)$. A policy π specifies the strategy of selecting an action. RL's goal is to find the policy π mapping states to actions that maximizes the expected long-term reward. Thus, we define Q-function $Q^{\pi}(s, a)$ for state-action values and the optimal $Q^*(s, a) = \max_{\pi} Q^{\pi}(s, a) = \mathbb{E}_{s'}\left[R(s, a) + \gamma \max_{a'} Q^*(s', a')\right]$.

3 Methods

In our control system, we use multi-dimensional data as the input D. The data includes blood glucose G (mg/dL) measured by CGM sensors in real-time, meal carbohydrate M (g) manually record by individuals, and the meal bolus B (U) from a bolus calculator. The control of dual-hormone (basal insulin and glucagon) delivery, whether to deliver the hormones or not, is considered as the action A. In this case, D can be denoted as $D = \{G, M, I, C\} = [d_1, \cdots, d_L]^{\mathbb{T}} \in \mathbb{R}^{\mathbb{L}\times 4}$, where L is the data length, I is the total insulin including bolus B, basal, and C stands for the glucagon. We use the latest one-hour data (12 samples) as the current state $s_t = [d_{t-11} \cdots, d_{t-1}, d_t]^{\mathbb{T}}$. Here B is computed from M with a standard bolus calculator $B \propto M$ divided by the body weight.

Then the problem can be seen as an agent interacting with an environment over time steps sequentially, as depicted in Fig. 1. Every five minutes, an observation $o_t = s_t + e_t$ can be obtained, and an action a_t is taken. The action can be chosen from three options: doing nothing, delivering basal insulin, or delivering glucagon. The amount of basal insulin and glucagon is a small constant that determined by the subject profile in advance. To maintain the BG in a target range, we define a reward carefully that the agent receives in each time step, as shown in Eq. 1 and Fig. 2.

$$r_t = \begin{cases} -0.6 + (G_{t+1} - 70)/100 & 30 \leq G_{t+1} < 70 \\ 0 & 70 \leq G_{t+1} < 90 \\ 1 & 90 \leq G_{t+1} < 140 \\ 0 & 140 \leq G_{t+1} < 180 \\ -0.4 - (G_{t+1} - 180)/200 & 180 \leq G_{t+1} \leq 300 \\ -1 & \text{else} \end{cases} \tag{1}$$

The goal of the agent is to learn a personalized basal insulin and glucagon delivery strategy within a short period of time (one month) and with limited data for each individual. Therefore, we propose a two-step learning approach: A generalized model is pre-trained using long-term exploration in an in an environment with average T1D subject parameters provided by the simulator, and then we obtain personalized models for each subject by fine-tuning the generalized model. From the intrinsic perspective of T1D in a real-life scenario, we cannot use a trial and error process with poor initial performance, so the first step needs to be done in the simulator. For the second step, we could adopt the personalized training in real clinical trials, with a generalized model as the agent's initial policy. Meanwhile, proper safety constrains, such as insulin/glucagon suspension, are still required during the training process.

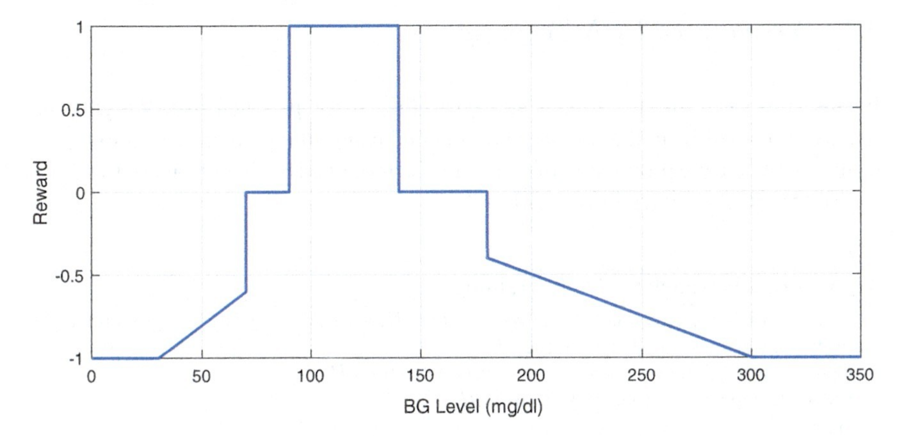

Fig. 2 Reward function

3.1 Generalized DQN Training

With the interactive environment in the simulator, the DRNN is employed as the main module in the DQNs, because it has larger receptive field that is crucial for glucose time series processing [11]. Double DQN weights θ_1, θ_2 in the simulator are trained because it has been proved as a robust approach to solve overestimations [23]. With the loss $J_{DQ}(Q) = \left(R(o, a) + \gamma Q(o', \arg\max_a Q(o', a; \theta_1); \theta_2) - Q_i(o, a; \theta_1)\right)^2$, action network θ_1 is trained at each time step. The pseudo-code is sketched in Algorithm 1.

Algorithm 1 Generalized DQN Training

1: Inputs: Initializing environment E, update frequency T, two DRNNs of random weights θ_1, θ_2, respectively.
2: **repeat**
3: Select action from $a \sim \pi(Q_{\theta_1}, \varepsilon)$, observe (o', r) in $E(Is)$
4: Store (o, a, r, o') into replay buffer $\mathcal{B}$
5: Sample a mini-bath uniformly from $\mathcal{B}$ and calculate loss $J_{DQ}(Q)$
6: Perform a gradient descent to update θ_1
7: **if** $t \bmod T = 0$ **then** $\theta_2 \leftarrow \theta_1$ **end if**
8: **until** converge or reach the number of iterations

The agent explores random hormone delivery actions under policy π that is ε-greedy with respect to Q_{θ_1} in the simulator. Some human intervention/demonstration at the beginning of the RL process can reduce the training time slightly, but *in-silico* it is not necessary.

3.2 Personalized DQN Training

In this step we refine the model and customize it for personal use. Weights and features obtained from the last step are updated using limited data with importance sampling, where the probability of sampling Pr is defined by the priority of transition as [7]. Details are shown in Algorithm 2.

Algorithm 2 Personalized DQN Training

1: Inputs: Initializing environment E, historical data H, generalized Q-function Q with weights θ_1, target weights θ_2, update frequency T
2: Generate replay buffer $\mathcal{D}$ with experience collected from H
3: Calculate importance probability Pr from H
4: **repeat**
5: Select action from policy $a \sim \pi(Q_{\theta_1}, \varepsilon)$, observe (o', r)
6: Store (o, a, r, o') in $\mathcal{D}$, overwriting the samples from H
7: Sample a mini-batch from $\mathcal{D}$ with Pr
8: Update Pr based on prioritization
9: Calculate loss $J(Q) = J_{DQ}(Q)$
10: Perform a gradient descent step to update θ_1
11: **if** t mod $T = 0$ **then** $\theta_2 \leftarrow \theta_1$ **end if**
12: **until** converge or reach the number of iterations

Table 1 Performance on 10 adult subjects

Method	Normo (TIR)	Hypo	Hyper
CB	$81.91 \pm 8.66^{\ddagger}$	$5.29 \pm 3.93^{\ddagger}$	$12.80 \pm 8.67^{\ddagger}$
ISCR	$87.62 \pm 7.57^{\ddagger}$	$2.36 \pm 1.44^{\ddagger}$	$10.01 \pm 7.35^{\ddagger}$
G-DQN	89.16 ± 5.04	1.92 ± 1.36	8.92 ± 5.38
P-DQN	$\mathbf{93.12 \pm 4.48}$	$\mathbf{1.25 \pm 1.32}$	$\mathbf{5.63 \pm 3.29}$

$^{*}p \leq 0.05$ $^{\dagger}p \leq 0.01$ $^{\ddagger}p \leq 0.005$

Table 2 Performance on 10 adolescent subjects

Method	Normo (TIR)	Hypo	Hyper
CB	$61.68 \pm 10.95^{\ddagger}$	$9.04 \pm 7.22^{\ddagger}$	$29.28 \pm 11.16^{\ddagger}$
ISCR	$74.55 \pm 9.61^{\dagger}$	$2.38 \pm 1.82^{\dagger}$	$23.07 \pm 7.26^{\ddagger}$
G-DQN	74.89 ± 8.58	2.36 ± 2.19	22.75 ± 8.63
P-DQN	$\mathbf{83.39 \pm 8.03}$	$\mathbf{2.10 \pm 1.56}$	$\mathbf{14.51 \pm 9.98}$

$^{*}p \leq 0.05$ $^{\dagger}p \leq 0.01$ $^{\ddagger}p \leq 0.005$

4 Experiment Results

We compare the results with the following experimental setup: 1. constant basal insulin (CB); 2. insulin suspension and carbohydrate recommendation (ISCR) [24]; 3. generalized DQN (G-DQN, Algorithm 1); 4. personalized DQN (P-DQN, Algorithm 1, 2). CB is the baseline method in the simulator, as the standard hormone control of T1D, and ISCR is based on conventional control algorithms.

In experiments, we used the TIR ([70, 180] mg/dL), the percentage of hypoglycemia (<70 mg/dL) and hyperglycemia (>180 mg/dL) as the metrics to measure the performance. In general, either higher TIR or lower Hypo/Hyper indicates better glycemia control. Table 1 presents the overall glycemia performance on the adult subjects. It is noted that the personalized DQN (P-DQN) achieves the best performance and increases the mean TIR by 11.21% ($p \leq 0.005$), compared to the CB setup. For the adolescent case in Table 2, the best TIR of 83.39% is also obtrained by P-DQN. In the two cases, the personalized model outperforms the baseline methods on both TIR and Hypo/Hyper results with considerable improvements. Moreover, the percentage of hypoglycemia in time out of the target range significantly decreases, and most adverse events are hyperglycemia. Compared to hypoglycemia, slight hyperglycemia is preferable. It is because that there are severe risks of hypoglycemia, including short-term complications (e.g., paresis, convulsions and encephalopathy), and long-term intellectual impairment [25]. Nevertheless, hyperglycemia is inevitable for T1D subjects in some situations, such as the intake of a large amount of carbohydrate.

In Fig. 3, we visualize the TIR performance through 30-day personalized training, and specific BG values of two subjects, as average cases, over a 6-month testing period. We have explored conventional neural networks (NN) as the DQNs, using five fully-connected layers. Although there are increasing trends of TIR for both the DRNN and the NN during the training, the TIR performance of the NN is not as stable as that of the DRNN. The BG trajectories of the adult are basically in accordance with statistical results in Table 1, and the DQN model avoids many hypoglycemia events during the night. For the adolescent, it is observed that the DQN model also helps reduce the average BG level and improve TIR significantly.

5 Conclusion

We propose a new dual-hormone delivery algorithm and employ deep RL for glucose management. DRNNs are used in the architecture of the double DQN. The reward function is defined by the glucose level at the successive time step. With the 2-step learning framework, a generalized model is trained by long-term exploration. Then the personalized models for each T1D subject are developed using importance sampling in a period of 30 days. We then evaluated the models on 10 adults and 10 adolescents over 180 days, comparing the performance to CB and ISCR baseline methods. The proposed algorithm increases the TIR from 82% to 93% for the

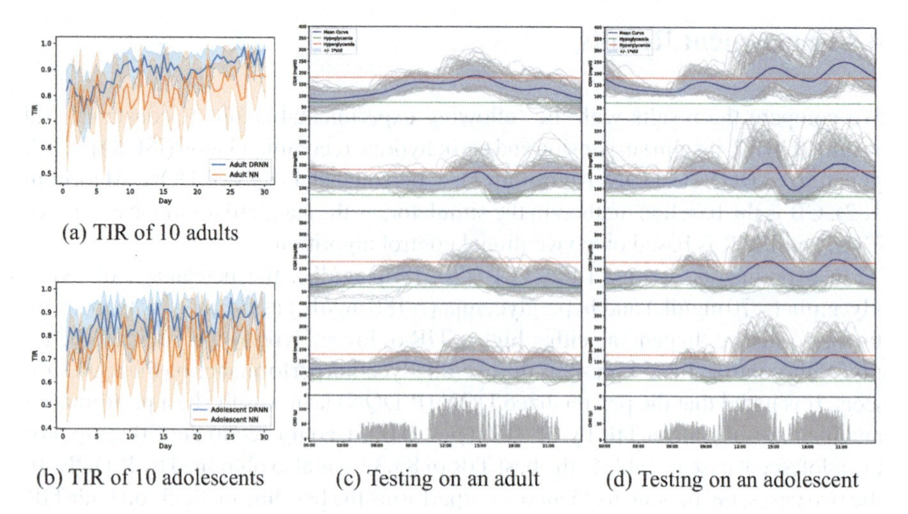

(a) TIR of 10 adults

(b) TIR of 10 adolescents

(c) Testing on an adult

(d) Testing on an adolescent

Fig. 3 **Left:** The performance of personalized training with confidence intervals on 10 subjects over a one-month period, using DRNN or NN. **Middle and Right:** The testing performance of four setup over a 6-month period: (Top-to-bottom) CB, ISCR, generalized DQN, personalized DQN, distribution of meal carbohydrate. The average BG levels for 180 days are shown in solid blue lines, and the hypo/hyperglycemia regions are shown in dotted green/red lines. Each gray line stands for a glucose trajectory over one day (totally 180 ensembles), and the blue shaded regions indicate the standard deviation

adult cohort, and from 62% to 83% for the adolescents. These promising results show the significance of using deep RL in control of diabetes, achieving a notable improvement in glycemic control and outperforming existing works. This has the potential to improve the management of diabetes and can be generalised to also provide therapeutic benefits to other chronic diseases.

Acknowledgements The work is supported by EPSRC EP/P00993X/1 and the President's PhD Scholarship at Imperial College.

References

1. Cho, N., Shaw, J., Karuranga, S., Huang, Y., da Rocha Fernandes, J., Ohlrogge, A., Malanda, B.: IDF diabetes atlas: global estimates of diabetes prevalence for 2017 and projections for 2045. Diabet. Res. Clin. Pract. **138**, 271–281 (2018)
2. Reddy, M., Pesl, P., Xenou, M., Toumazou, C., Johnston, D., Georgiou, P., Herrero, P., Oliver, N.: Clinical safety and feasibility of the advanced bolus calculator for type 1 diabetes based on case-based reasoning: a 6-week nonrandomized single-arm pilot study. Diabet. Technol. Therapeut. **18**(8), 487–493 (2016)
3. Jiang, F., Jiang, Y., Zhi, H., Dong, Y., Li, H., Ma, S., Wang, Y., Dong, Q., Shen, H., Wang, Y.: Artificial intelligence in healthcare: past, present and future. Stroke Vascular Neurol. **2**, 230–243 (2017)

4. Artman, W.J., Nahum-Shani, I., Wu, T., Mckay, J.R., Ertefaie, A.: Power analysis in a SMART design: sample size estimation for determining the best embedded dynamic treatment regime. Biostatistics **21**(3), 432–448 (2018)
5. Mnih, V., Kavukcuoglu, K., Silver, D., Rusu, A.A., Veness, J., Bellemare, M.G., Graves, A., Riedmiller, M., Fidjeland, A.K., Ostrovski, G., et al.: Human-level control through deep reinforcement learning. Nature **518**(7540), 529 (2015)
6. Vettoretti, M., Facchinetti, A., Sparacino, G., Cobelli, C.: Type 1 diabetes patient decision simulator for in silico testing safety and effectiveness of insulin treatments. IEEE Trans. Biomed. Eng. **65**(6), 1281–1290 (2018)
7. Schaul, T., Quan, J., Antonoglou, I., Silver, D.: Prioritized experience replay. International Conference on Learning Representations, vol. abs/1511.05952 (2015)
8. Zhu, T., Li, K., Herrero, P., Chen, J., Georgiou, P.: A deep learning algorithm for personalized blood glucose prediction. In: KHD@ IJCAI, pp. 74–78 (2018)
9. Li, K., Liu, C., Zhu, T., Herrero, P., Chen, J., Georgiou, P.: GluNet: a deep learning framework for accurate glucose forecasting. IEEE J. Biomed. Health Inform. **24**(2), 414–423 (2019)
10. Chang, S., Zhang, Y., Han, W., Yu, M., Guo, M., Tan, W., Cui, X., Witbrock, M., Hasegawa-Johnson, M.A., Huang, T.S.: Dilated recurrent neural networks. In: Advances in Neural Information Processing Systems, pp. 77–87 (2017)
11. Chen, J.. Li, K., Herrero, P., Zhu, T., Georgiou, P.: Dilated recurrent neural network for short-time prediction of glucose concentration. In: KHD@ IJCAI, pp. 69–73 (2018)
12. Vigersky, R.A., McMahon, C.: The relationship of hemoglobin a1c to time-in-range in patients with diabetes. Diabet. Technol. Therapeut. **21**(2), 81–85 (2018)
13. Man, C.D., Micheletto, F., Lv, D., Breton, M., Kovatchev, B., Cobelli, C.: The UVA/PADOVA type 1 diabetes simulator. J. Diab. Sci Technol. **8**, 26–34 (2014)
14. Cobelli, C., Renard, E., Kovatchev, B.: Artificial pancreas: past, present, future. Diabetes **60**(11), 2672–2682 (2011)
15. Hovorka, R.: Closed-loop insulin delivery: from bench to clinical practice. Nat. Rev. Endocrinol. **7**, 385 (2011)
16. Bergenstal, R.M., Garg, S., Weinzimer, S.A., Buckingham, B.A., Bode, B.W., Tamborlane, W.V., Kaufman, F.R.: Safety of a hybrid closed-loop insulin delivery system in patients with type 1 diabetes. JAMA **316**(13), 1407–1408 (2016)
17. Facchinetti, A.: Continuous glucose monitoring sensors: past, present and future algorithmic challenges. Sensors **16**(12), 2093 (2016)
18. Haidar, A.: The artificial pancreas: how closed-loop control is revolutionizing diabetes. IEEE Control Syst. Mag. **36**, 28–47 (2016)
19. Holubová, A., Phuong, D.N., Wei, S., Godtliebsen, F.: Control of blood glucose for type-1 diabetes by using reinforcement learning with feedforward algorithm. Comput. Math. Meth. Med. **2018**, 1–8 (2018)
20. Herrero, P., Bondia, J., Oliver, N., Georgiou, P.: A coordinated control strategy for insulin and glucagon delivery in type 1 diabetes. Comput. Methods Biomech. Biomed. Eng. **20**(13), 1474–1482 (2017)
21. Xie, J.: Simglucose v0.2.1 (2018). https://github.com/jxx123/simglucose (2018)
22. Fox, I., Wiens, J.: Reinforcement learning for blood glucose control: challenges and opportunities. In: RL4RealLife Workshop, ICML 2019 (2019)
23. Van Hasselt, H., Guez, A., Silver, D.: Deep reinforcement learning with double Q-learning. In: Thirtieth AAAI Conference on Artificial Intelligence (2016)
24. Liu, C., Avari, P., Oliver, N., Georgiou, P., Vinas, P.H.: Coordinating low-glucose insulin suspension and carbohydrate recommendation for hypoglycaemia minimization. Diab. Technol. Therapeut. **21**, A85–A85 (2019)
25. Yale, J.F., Paty, B., Senior, P.A.: Hypoglycemia. Can. J. Diab. **42**, S104–S108 (2018)

A Generalizable Method for Automated Quality Control of Functional Neuroimaging Datasets

Matthew Kollada, Qingzhu Gao, Monika S. Mellem, Tathagata Banerjee, and William J. Martin

Abstract Over the last twenty five years, advances in the collection and analysis of functional magnetic resonance imaging (fMRI) data have enabled new insights into the brain basis of human health and disease. Individual behavioral variation can now be visualized at a neural level as patterns of connectivity among brain regions. As such, functional brain imaging is enhancing our understanding of clinical psychiatric disorders by revealing ties between regional and network abnormalities and psychiatric symptoms. Initial success in this arena has recently motivated collection of larger datasets which are needed to leverage fMRI to generate brain-based biomarkers to support the development of precision medicines. Despite numerous methodological advances and enhanced computational power, evaluating the quality of fMRI scans remains a critical step in the analytical framework. Before analysis can be performed, expert reviewers visually inspect individual raw scans and preprocessed derivatives to determine viability of the data. This Quality Control (QC) process is labor intensive, and the inability to adequately automate at large scale has proven to be a limiting factor in clinical neuroscience fMRI research. In this paper, we present a novel method for automating the QC of fMRI scans. We train machine learning classifiers using features derived from brain MR images to predict the "quality" of those images, which is based on the ground truth of an expert's opinion. Specifically, we emphasize the importance of these classifiers' ability to generalize their predictions across data collected in different studies. To address this, we propose a novel approach entitled "FMRI preprocessing Log mining for Automated, Generalizable Quality Control" (FLAG-QC), in which features derived from mining runtime logs are used to train the classifier. We show that classifiers trained on FLAG-QC features perform much better (AUC = 0.79) than previously proposed feature sets (AUC = 0.56) when testing their ability to generalize across studies.

M. Kollada (✉) · Q. Gao · M. S. Mellem · T. Banerjee · W. J. Martin
BlackThorn Therapeutics, San Francisco, CA 94103, USA
e-mail: matt.kollada@blackthornrx.com

© The Editor(s) (if applicable) and The Author(s), under exclusive license to Springer Nature Switzerland AG 2021

A. Shaban-Nejad et al. (eds.), *Explainable AI in Healthcare and Medicine*, Studies in Computational Intelligence 914,
https://doi.org/10.1007/978-3-030-53352-6_6

Keywords Neuroimaging · Precision medicine · Automated quality control

1 Introduction

Brain imaging offers a window into the structure and function of the brain. Functional magnetic resonance imaging (fMRI) is a non-invasive method for measuring endogenous contrast in a blood oxygenation dependent (BOLD) signal as a proxy for brain activation. Changes in the patterns of the BOLD signal can be used to track brain activity, during tasks or at rest, holding the promise of bringing objective measures of symptom severity to the field of clinical psychiatry which is a basis for precision medicine. For example, results from recent fMRI studies predicted individual symptom severity [15], revealed therapeutic normalization of brain activity [12], and further subtyped patients within diagnostic categories [6]. Recent empirical and simulation work has highlighted the importance of increasing the number of samples to maintain enough statistical power when analyzing fMRI data [4]. Researchers have heeded these findings, as numerous large consortia have recently collected fMRI data at continuously growing orders of magnitude [2, 5, 7]. The UK Biobank exemplifies this new scale of brain imaging, having already reported collecting fMRI data, among other modalities, from 10,000 study participants with the stated intention of pushing that number to above 100,000 [1]. However, with this new scale of data comes new challenges.

Raw fMRI images must undergo a complex set of computational transformations, often termed preprocessing, before being used in any statistical analysis. These raw and preprocessed images are commonly manually assessed for quality by expert reviewers in a process referred to as Quality Control (QC). These reviewers, often in multiple steps, visualize the preprocessed images, and inspect them for apparent errors that may erroneously bias future analysis [14, 20]. Many evaluation schemes for QC have been proposed [1, 8], however, one simple, clear strategy is determining whether the scan "passes" and is therefore usable, or "fails" and is discarded from further analysis. We use this scheme in our study.

The labor-intensive and time-consuming nature of QC serves as a clear bottleneck to the analysis of fMRI images at scale. QC of an fMRI dataset with hundreds of scans can take weeks to months of manual assessment from a single expert reviewer before analysis can begin. As discussed above, many recent fMRI studies have collected data at or even above that scale [1, 2, 5, 7], providing compelling motivation for the field to develop a scalable QC framework to reduce the burden on individual researchers and standardize quality control of fMRI data.

One tangible, potential solution is to train machine learning classifiers to predict image quality based on expert reviewer ground-truth QC labels. This could massively decrease manual reviewer QC time and overcome the current bottleneck. Researchers have made certain progress in creating automatic QC classifiers for structural Magnetic Resonance Imaging (sMRI), which examines the anatomical structure of the brain and must also undergo a QC process. In these paradigms, feature sets derived from the raw and preprocessed sMRI images have been used to train machine learning classifiers to predict manual QC labels. Here, Esteban et al. [8] made important progress, testing a predictive framework for automatic sMRI QC, and importantly, evaluating the model's ability to generalize across datasets.

Recent literature, however, demonstrates that fMRI can either successfully augment or individually outperform sMRI in many crucial neuropsychiatric applications [10, 13, 15]. Nonetheless, little effort has been made to apply the QC framework from sMRI to fMRI, providing clear incentive to develop automatic QC classification on fMRI data.

In this study we present our work in building automatic QC classifiers for fMRI data. Using two large, open-source fMRI datasets we build these classifiers and evaluate a range of feature sets, including one novel to this work entitled, "FMRI preprocessing Log mining for Automated, Generalizable Quality Control," or FLAG-QC. Specifically, we evaluate the ability of these classifiers to generalize across fMRI data collected within different studies.

Our results demonstrate that we are able to achieve this generalization using only the novel FLAG-QC feature set proposed within this work (Fig. 1).

2 Background

Automation of QC for brain imaging scans has received attention in structural MRI. Some of the first approaches attempted to quantify measures of image quality by creating indices, often termed "Image Quality Metrics" (IQMs), through which the quality of a raw sMRI scan can be judged in comparison to other scans. Woodard and Carley-Spencer [22] explored the adequacy of IQMs derived from Natural Scene Statistics and those defined by the JPEG Consortium. Their work identified the abil-

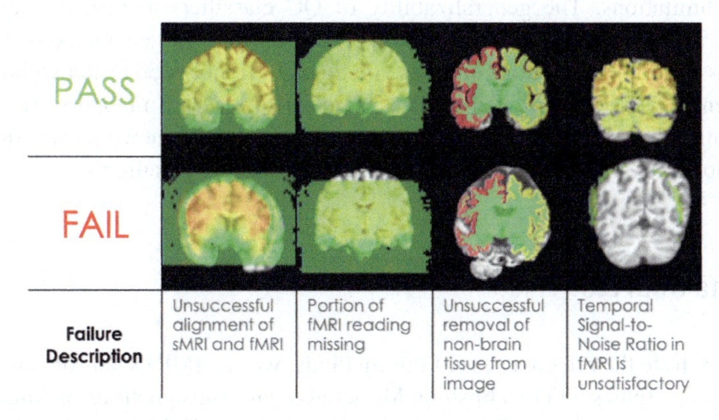

Fig. 1 Examples of 2D Slices of preprocessed fMRI Images labelled as both passing and failing by expert QC reviewers

ity of IQMs to discriminate between raw structural MR images and those that had been artificially distorted. Similarly, Mortamet et al. [16] proposed a pair of sMRI quality indices based on the analysis of the background air (non-brain) volumetric pixels (voxels) of a brain sMRI scan, which proved to be useful in identifying poor quality scans in Alzheimer patients. The above-mentioned studies facilitated the understanding of certain attributes or features that are able to contribute to the discrimination between high and low quality scans, but stopped short of testing of a predictive framework evaluated on data held-out during training.

Pizzaro et al. [17] built on this approach by introducing their own set of features, engineered to target specific artifacts found in brain sMRI scans. These features were used to train a Support Vector Machine (SVM) classifier to predict manual expert QC labels on 1457 3D-MRI volumes within cross validation experiments. Similarly, Alfaro-Almagro et al. [1], in working with the first set of imaging data to come out of the UK Biobank study, created automatic QC classifiers with 5800 structural scans. They developed a set of 190 features to train an ensemble machine learning classifier to predict manual QC labels. Together, these studies expanded the space of features explored in automatic MRI QC and tested predictive modeling frameworks. However, neither evaluated the generalizability of their classifiers across datasets.

Esteban et al. [8] illuminated a limitation in across-dataset generalizability, testing whether classifiers could predict QC labels when trained on a dataset from one study and evaluated on data from an entirely different one. Their aim was to successfully train an sMRI QC classifier with scans from the Autism Brain Imaging Data Exchange [5] (ABIDE) (N = 1001) dataset to be evaluated on scans from the UCLA Consortium for Neuropsychiatric Phenomics LA5c [18] (CNP) (N = 265) dataset. The study revealed that their classifier performed quite well in cross-validation prediction within ABIDE. However, the model's predictive power dropped notably when trained on ABIDE and tested across datasets on CNP. This work brought to light the challenge of building robust automatic QC prediction of any MRI data.

As a whole, the study of automatic MRI QC has made progress, but still faces notable limitations. The generalizability of QC classifiers to new datasets must improve if practical applications are to be built on top of these methods. Further, as discussed in the **Introduction**, these methods have not been tested within fMRI data even though it has proved itself as a powerful research modality. To address these limitations we evaluate the existing automatic QC frameworks within fMRI, in addition to exploring new feature spaces and modeling paradigms.

3 Data Sources

To demonstrate the effectiveness of our methods, we use fMRI scans obtained from two separate studies: (1) Establishing Moderators and Biosignatures of Antidepressant Response for Clinical Care for Depression [21] (EMBARC), (2) UCLA Consortium for Neuropsychiatric Phenomics LA5c [18] (CNP). The CNP study was also used in previously mentioned work of Esteban et al. [8].

3.1 EMBARC

The EMBARC dataset was collected to examine a range of biomarkers in patients with depression to understand how they might be able to inform clinical treatment decisions. The study enrolled 336 patients aged 18–65, collecting demographic, behavioral, imaging, and wet biomarker measures for multiple visits over a period of 14 weeks. Data were acquired from the National Data Archive (NDA) repository on June 19, 2018 with a license obtained by Blackthorn Therapeutics.

Our study only analyzes data from sMRI and fMRI scans collected during patients' first and second visit to the study site. Specifically, we use T1-weighted structural MRI scans and T2*-weighted blood-oxygenation-level-dependent (BOLD) resting-state functional MRI scans that were labelled as "run 1". In total we analyze 324 structural-functional MRI scan pairs from the first site visit and 288 pairs from the second, producing a total of 612 scan pairs.

3.2 CNP

The CNP dataset was collected to facilitate discovery of the genetic and environmental bases of variation in psychological and neural system phenotypes, to elucidate the mechanisms that link the human genome to complex psychological syndromes, and to foster breakthroughs in the development of novel treatments for neuropsychiatric disorders. The study enrolled a total of 272 participants aged 21–50. Within the participant group, there were 138 healthy individuals, 58 diagnosed with schizophrenia, 49 diagnosed with bipolar disorder, and 45 diagnosed with ADHD. All data were collected in a single visit per participant and included demographic, behavioral, and imaging measures. We used version 1.0.5 of the study data downloaded from openneuro.org on January 4, 2018 with a license obtained by Blackthorn Therapeutics.

Similar to EMBARC, we use data from participants that have both T1-weighted sMRI and T2*-weighted BOLD resting-state fMRI scans that were labelled "run 1". This amounts to 251 structural-functional MRI scan pairs.

4 Methods

We define the problem addressed in this study within a classification framework. Using features derived from brain fMRI scans, we train machine learning classifiers that predict QC outcome labels provided by expert reviewers. In this section, we briefly discuss each of the steps we take in our study.

4.1 MRI Preprocessing Pipeline

In the study of brain MRI, preprocessing refers to a set of transformations that a raw image must undergo to prepare it for data analysis. All MRI data used in our analysis was preprocessed by a pipeline with two distinct modules, one for preprocessing T1-weighted sMRI scans and one for T2*-weighted resting-state fMRI scans. The sMRI module uses the FreeSurfer Software Suite [9, 19], specifically the *recon-all* command, to segment the structural image into various structures of the brain (parcellated gray and white matter regions of the cortex and subcortical regions). The fMRI module then uses the Analysis of Functional Neuroimages (AFNI) Software Suite [3] to preprocess the raw resting-state fMRI image and the sMRI image further. This pipeline includes steps to remove noise in collected images, standardize scans to a template space for across scan comparison, segment the brain image into different tissue types and regions of interest, and extract numerical features from the preprocessed brain image for further analysis (see [15] for the full details of our sMRI and fMRI preprocessing steps).

4.2 Feature Sets

The features that we use to train QC classifiers come from two distinct pipelines: (1) FLAG-QC Features, a feature set novel to this study, and (2) MRIQC Features, those generated by the *MRIQC* software suite [8]. A high-level block diagram showing the process for creating each set of features is shown in Fig. 2.

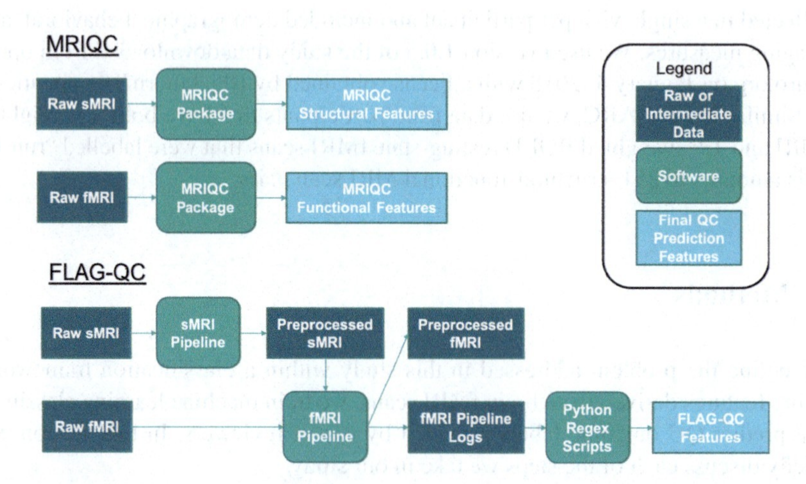

Fig. 2 Flow charts detailing the processes for creating the individual feature sets used in automatic QC prediction

4.2.1 FLAG-QC Features

The FLAG-QC features are derived from AFNI software commands run within our fMRI preprocessing pipeline. These commands are responsible for transforming the fMRI image into the final outputs that undergo manual QC, making information about their outcomes useful for automated QC decisions. While an AFNI command is executing, it emits runtime logs. Our preprocessing pipeline copies those logs and saves them to text files. These logs contain a large assortment of information, some of it pertaining to results of final or intermediate steps of a given command. We use the Python regular expression (re) library to parse the text files and extract potentially informative features. An example of a portion of an fMRI pipeline log file is shown in Fig. 3.

The collected FLAG-QC Features can be divided into four subgroups; Step Runtimes, Voxel Counts, Brain Coordinates, and Other Metrics. (1) Step Runtime features quantify how long a given step, or set of steps, in the pipeline took to run. (2) Voxel Count features measure the size of the output of a given step in the pipeline in terms of "voxels", or volumetric 3D pixels. (3) Brain Coordinate features simply refer to the X, Y, and Z coordinates of the bounding box of the brain image. (4) Other Metrics are miscellaneous values that quantify the outcome of a certain step of the preprocessing pipeline. An example of one of these Other Metrics is the cost function value associated with the step of the pipeline that aligns the structural and functional scans. In total, there are 38 FLAG-QC features.

The full explanation of each individual feature falls outside the scope of this paper, but the authors welcome contact to discuss the features further.

4.2.2 MRIQC Features

MRIQC is a software developed by the Poldrack Lab at Stanford University [8]. One of its features is the ability to generate measures of image quality from raw MRI images. These Image Quality Metrics (IQMs) are used in Esteban et al. [8] to predict manual QC labels on sMRI scans. The metrics are designated as "no-reference", or

```
+ -- Parameters = 8.9273 0.0422 -2.6384 0.2276 -0.9048 3.8872 1.014
+ - cost(#7)=-0.194017
+ -- Parameters = -8.3974 1.3055 9.8740 -2.4815 1.2345 1.5863 0.994
+ - case #2 is now the best
+ - Initial  cost = -0.229625
+ - Initial fine Parameters = -8.7082 1.2242 9.9995 -1.1902 0.9805
+ - Finalish cost = -0.229902 ; 139 funcs
+ - Final    cost = -0.229969 ; 103 funcs
+ Final fine fit Parameters:
      x-shift=-8.7153   y-shift= 1.2831    z-shift=10.0000
      z-angle=-1.2502   x-angle= 0.9799    y-angle= 4.5120
      x-scale= 0.9922   y-scale= 1.0066    z-scale= 0.9742
    y/x-shear=-0.0053 z/x-shear= 0.0101 z/y-shear=-0.0227
```

Fig. 3 A portion of an individual scan's fMRI preprocessing log text file. The highlighted region displays a FLAG-QC feature mined from this log file for each scan having no ground-truth correct value. Instead, the metrics generated from one image can be judged in relation to a distribution of these measures over other sets of images. *MRIQC* generates IQMs from both structural and functional raw images.

The structural IQMs are divided into four categories: measures based on noise level, measures based on information theory, measures targeting specific artifacts, and measures not covered specifically by the other three.

The functional IQMs are broken down into three categories: measures for spatial structure, measures for temporal structure, and measures for artifacts and others.

In total there are 112 features generated by *MRIQC*, 68 structural features and 44 functional features. A full list of the features generated by *MRIQC* can be found at mriqc.readthedocs.io. The software can be run as either a Python library or Docker container. We used the Docker version to generate IQMs on EMBARC and CNP.

4.3 Feature Selection

In this work we use a two-phase feature-selection approach:

1. In the first phase, we apply a model-independent approach. Specifically we use Hilbert-Schmidt Independence Criterion Lasso (HSIC Lasso) based feature selection [23]. HSIC Lasso utilizes a feature-wise kernelized Lasso for capturing non-linear input-output dependency. A globally optimal solution can be efficiently calculated making this approach computationally inexpensive.
2. In the second phase, we apply a model-dependent Forward Selection approach [11].

We choose this two-phase approach because it offers a good balance of classifier performance, fast computation and generalization. The actual number of features selected depends on cross-validation performance.

4.4 Expert Manual QC Labels

Our classifiers attempt to predict QC outcome labels that come from a trained reviewer who has years of experience working with MRI data. It is common practice in the field that expert QC reviewers examine raw MRI scans and preprocessed images to determine if the quality is sufficient for further analysis [14, 20]. For volumetric data, the 3D preprocessed MRI images are spatially sampled as 2D images for easier assessment by the reviewer. Examples of 2D images that both "pass" and "fail" QC are shown in Fig. 1 with common failure points, such as mis-alignment of structural and functional MRI scans or unsuccessful automatic removal of non-brain tissue, highlighted. In our study, following assessment of image quality of raw data and across the multiple preprocessing steps, the reviewer makes a binary "pass" or "fail" decision for each subject's fMRI scan. Thus, an fMRI scan is tagged as useable (pass) or not (fail), and these labels serve as the ground-truth decisions on which our classifier is trained.

For the data presented here, the ground-truth QC outcome labels were generated by a single expert reviewer who performed QC on 612 scans from EMBARC and 251 scans in CNP. Our expert reviewer labelled 73.9% (452/612) of the scans from EMBARC as passing QC and 58.6% (147/251) as passing for CNP.

Inter-rater reliability has been noted as a source of label noise in other automatic MRI QC efforts [8]. To assess this possible noise in our data, a second expert reviewer independently performed QC on the 251 scans from CNP. The reviewers' pass/fail QC decisions agreed on 87.6% (220/251) of the scans. That there is some disagreement is not surprising, given that the current standard of manual fMRI QC has certain subjective aspects, but the overall high congruence of the two raters predictions gives us confidence that reasonable predictions can be made on these labels.

Going forward, we will use only the labels of the first reviewer mentioned, who performed QC on both EMBARC and CNP, to train and test our classifiers.

5 Results

In this section, we demonstrate that our classifiers can accurately predict manual QC labels on fMRI scans within one data source using any of the feature sets mentioned above, but that only the novel FLAG-QC feature set proposed in this study successfully generalizes to data of another independent study. To address these two scenarios, we structure our results in two folds. Data collected from the same study will be referred to as "within dataset" samples, while data collected from a study upon which a given model has not been trained will be referred to as "unseen study" data.

5.1 *Predictive Models*

To predict fMRI QC labels, four different predictive models are evaluated using the *sci-kit learn* Python library; Logistic Regression, Support Vector Machines (SVM), Random Forest, and Gradient Boosting classifiers. We tune the hyperparameters for the SVM, Random Forest, and Gradient Boosting models using 5-Fold Grid Search Cross Validation.

5.2 *Within Dataset Cross-Validation*

We first attempt to predict manual QC labels for held out sets of scans within datasets collected in a single study. We train and test Logistic Regression, SVM, Random

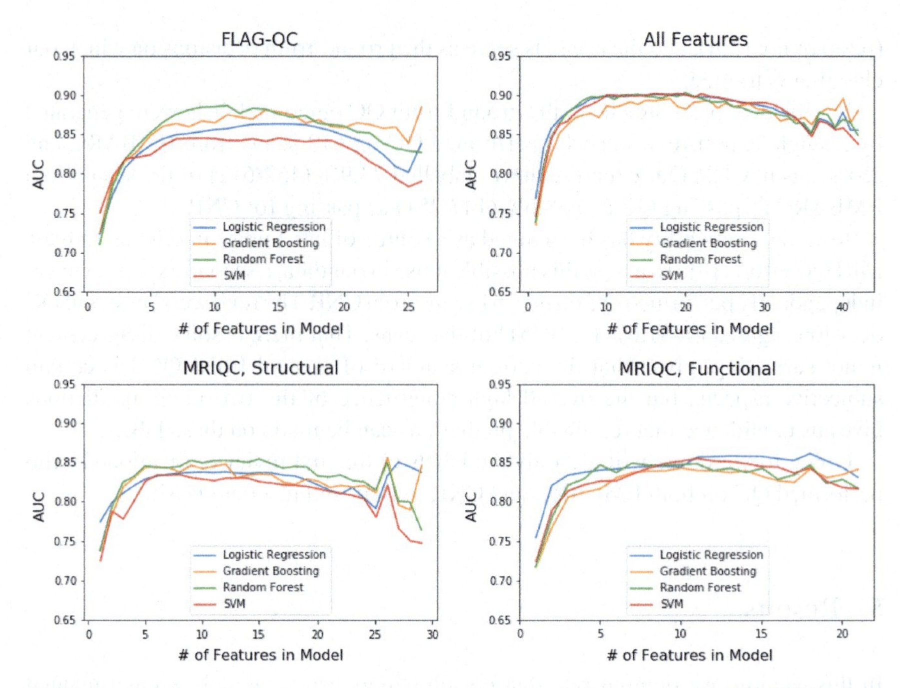

Fig. 4 Forward feature selection classification results on set of features chosen by HSIC Lasso—EMBARC

Forest, and Gradient Boosting classifiers separately on each of the three feature sets mentioned before, labelled "FLAG-QC", "MRIQC, Functional", and "MRIQC, Structural", as well as the ensemble of all, labelled "All Features". To do so, we evaluate each feature-model pair within a 5-fold Cross Validation scheme, first using HSIC Lasso to reduce the dimensionality of the feature space. We then run Forward Feature Selection and report the mean AUC across folds for each number of selected features. The results using these methods for the EMBARC dataset are shown in Fig. 4, and summary results for both EMBARC and CNP are reported in Table 1.

In the EMBARC dataset, after Forward Feature Selection, we find that our FLAG-QC feature set achieves an AUC of 0.89. The other individual feature sets perform slightly worse with the AUC being 0.86 for the MRIQC, Functional features and 0.86 for the MRIQC, Structural features. We see that using all of the features together creates the classifier with the best performance, achieving an AUC of 0.90. Although, there is variability in which model performed best on each feature set, all models performed reasonably well across all feature sets, with the lowest feature-model AUC being 0.83 (MRIQC, Structural—SVM).

We replicate the same procedure on the CNP dataset, resulting in an AUC of 0.93 for the FLAG-QC (SVM); 0.79 for the MRIQC, Functional Features (SVM); 0.85 for MRIQC, Structural (Random Forest); and 0.97 for the ensemble feature sets (SVM). We see a similar pattern that the FLAG-QC features outperform the

Table 1 Summary of within dataset forward feature selection classification results

	EMBARC			CNP		
Feature set	Classifier w/ Max AUC	# of Features @ Max AUC	Max AUC	Classifier w/ Max AUC	# of Features @ Max AUC	Max AUC
FLAG-QC	Random Forest	11	0.89	SVM	7	0.93
MRIQC, Functional	Logistic Regression	18	0.86	SVM	13	0.79
MRIQC, Structural	Gradient Boosting	26	0.86	Random Forest	10	0.85
All Features	Random Forest	20	0.90	SVM	9	0.97

MRIQC feature sets (though by a larger magnitude this time), and the combination of all feature sets outperforms any individual set. Also again all feature-model pairs perform reasonably accurately (min AUC 0.77 with MRIQC, Functional features using Gradient Boosting).

5.3 *Unseen Study Dataset as Test Set*

In our next experiment, we apply the same modeling framework to predict QC labels on one dataset, while the classifier is trained on data collected from a completely separate study. All 612 labelled scans from the EMBARC dataset are used as our training set. We use our results from EMBARC within dataset cross validation prediction with Forward Feature Selection to select the model that will be evaluated on the test set, CNP. For each feature set, we select the classifier with the highest AUC. The classifiers selected for each feature set are shown in Table 1.

Within each feature set, we again start by running HSIC Lasso on the training dataset for an initial model independent feature selection, and then perform Forward Feature Selection to choose the final set of features to be tested on CNP data. Finally, we perform one last 5-fold CV parameter grid search to tune and train the model specifically for the final selected set of features. Using this framework, we predict manual QC labels on scans from the CNP dataset to evaluate our model's performance.

The FLAG-QC features perform much better when predicting on the unseen study data from the CNP dataset than any other set of features, attaining an AUC of 0.79. The ROC curves from these predictions shown in Fig. 5 clearly display a difference in performance between our novel feature set and those previously proposed. The individual MRIQC feature sets perform much worse on the unseen study, each only reaching an AUC of 0.56. Additionally, the second best performing set of features is "All Features" with an AUC of 0.64. This set of course contains the FLAG-

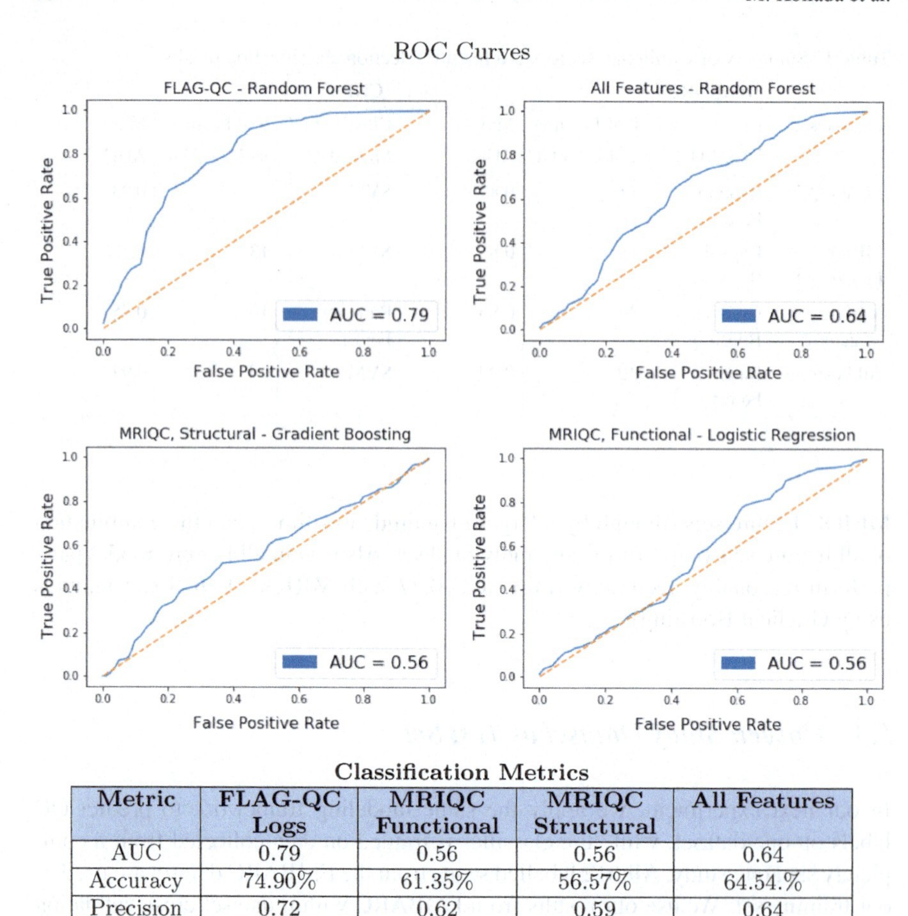

Classification Metrics

Metric	FLAG-QC Logs	MRIQC Functional	MRIQC Structural	All Features
AUC	0.79	0.56	0.56	0.64
Accuracy	74.90%	61.35%	56.57%	64.54.%
Precision	0.72	0.62	0.59	0.64
Recall	0.95	0.85	0.83	0.93

Fig. 5 Prediction on unseen study data: Training Set—EMBARC, Test Set—CNP

QC features, further highlighting the importance of the FLAG-QC features in the classifiers' ability to generalize across datasets. The pronounced drop in performance in unseen study prediction associated with all models that include MRIQC features implies that these features may lead to greater overfitting on the training set as compared to that of FLAG-QC. The results achieved using the FLAG-QC features demonstrates our classifier's generalizability in predicting fMRI QC labels in unseen studies, a main goal of this work.

6 Conclusion

In our study we produced a number of novel findings. We demonstrate the first successful classifiers that predict manual QC labels on fMRI scans. Additionally, we develop a novel set of features that proves useful in making these QC predictions. And finally, our classifiers generalize to an unseen dataset with high accuracy and do so only when trained on our novel feature set.

We'd like to note that the across-dataset generalizability of this method was only able to be tested on one unseen dataset. These methods should be further tested on additional datasets to confirm the robustness of this generalization.

In order to realize the full potential of automatic fMRI QC, we see a need to evaluate its impact. Before automatic QC is actively used in fMRI research we must assess how applying these techniques instead of manual QC affects the outcomes of later fMRI analysis. A clear question for further work arises: What is the performance bar for automatic QC at which subsequent analysis will yield similar results to data that has undergone manual QC? Answering questions like this will be vital to the practical implementation of automatic QC for fMRI analysis.

Automating the QC of fMRI data has the potential to meaningfully reduce the time spent by expert reviewers labelling the quality of these scans, unlocking new magnitudes of analysis. The impact of these methods are continually magnified by the growing scale of collected fMRI datasets. The weeks to months spent labelling datasets of this scale can instead be spent analyzing this powerful data modality that is steadily proving its ability to help researchers understand the brain basis of human health and disease. Our work demonstrates the feasibility of this paradigm, opening the door to exciting new insights generated from fMRI data toward the advancement of cognitive neuroscience, clinical psychiatry, and precision medicine.

Acknowledgements We want to thank Humberto Gonzalez, Parvez Ahammad, and Yuelu Liu for their important discussions in the conceptualization of the automation of fMRI QC. Additionally, Yuelu Liu was instrumental in the design of the manual fMRI QC process discussed in this work.

References

1. Alfaro-Almagro, F., et al.: Image processing and quality control for the first 10,000 brain imaging datasets from UK biobank. NeuroImage **166**, 400–424 (2018). https://doi.org/10.1016/j.neuroimage.2017.10.034. http://www.sciencedirect.com/science/article/pii/S1053811917308613
2. Casey, B., et al.: The adolescent brain cognitive development (ABCD) study: imaging acquisition across 21 sites. Dev. Cognit. Neurosci. **32**, 43–54 (2018). https://doi.org/10.1016/j.dcn.2018.03.001. http://www.sciencedirect.com/science/article/pii/S1878929317301214. The Adolescent Brain Cognitive Development (ABCD) Consortium: Rationale, Aims, and Assessment Strategy
3. Cox, R.W.: AFNI: software for analysis and visualization of functional magnetic resonance neuroimages. Comput. Biomed. Res. **29**(3), 162–173 (1996). https://doi.org/10.1006/cbmr.1996.0014. http://www.sciencedirect.com/science/article/pii/S0010480996900142

4. Cremers, H.R., Wager, T.D., Yarkoni, T.: The relation between statistical power and inference in fMRI. PLoS ONE **12**(11), 1–20 (2017). https://doi.org/10.1371/journal.pone.0184923
5. Di Martino, A., et al.: The autism brain imaging data exchange: towards a large-scale evaluation of the intrinsic brain architecture in autism. Mol. Psychiatry **19**(6), 659–667 (2014). https://doi.org/10.1038/mp.2013.78
6. Drysdale, A.T., et al.: Resting-state connectivity biomarkers define neurophysiological subtypes of depression. Nat. Med. **23**, 28 EP (2016). https://doi.org/10.1038/nm.4246. Article
7. Essen, D.V., et al.: The human connectome project: a data acquisition perspective. NeuroImage **62**(4), 2222–2231 (2012). https://doi.org/10.1016/j.neuroimage.2012.02.018. http://www.sciencedirect.com/science/article/pii/S1053811912001954. Connectivity
8. Esteban, O., et al.: MRIQC: advancing the automatic prediction of image quality in MRI from unseen sites. PLoS ONE **12**(9), 1–21 (2017). https://doi.org/10.1371/journal.pone.0184661
9. Fischl, B., et al.: Whole brain segmentation: automated labeling of neuroanatomical structures in the human brain. Neuron **33**(3), 341–355 (2002). https://doi.org/10.1016/S0896-6273(02)00569-X. http://www.sciencedirect.com/science/article/pii/S089662730200569X
10. Gao, S., Calhoun, V.D., Sui, J.: Machine learning in major depression: from classification to treatment outcome prediction. CNS Neurosci. Ther. **24**(11), 1037–1052 (2018). https://doi.org/10.1111/cns.13048. https://onlinelibrary.wiley.com/doi/abs/10.1111/cns.13048
11. Hastie, T., Tibshirani, R., Friedman, J.: The Elements of Statistical Learning: Data Mining, Inference, and Prediction. Springer, New York (2009)
12. Liu, Y., et al.: Machine learning identifies large-scale reward-related activity modulated by dopaminergic enhancement in major depression. Biol. Psychiatry Cognit. Neurosci. Neuroimaging (2019). https://doi.org/10.1016/j.bpsc.2019.10.002
13. Liu, Y., et al.: Highly predictive transdiagnostic features shared across schizophrenia, bipolar disorder, and adhd identified using a machine learning based approach. bioRxiv (2018). https://doi.org/10.1101/453951. https://www.biorxiv.org/content/early/2018/12/18/453951
14. Lu, W., Dong, K., Cui, D., Jiao, Q., Qiu, J.: Quality assurance of human functional magnetic resonance imaging: a literature review. Quant. Imaging Med. Surg. **9**(6) (2019). http://qims.amegroups.com/article/view/25794
15. Mellem, M.S., et al.: Machine learning models identify multimodal measurements highly predictive of transdiagnostic symptom severity for mood, anhedonia, and anxiety. bioRxiv (2018). https://doi.org/10.1101/414037. https://www.biorxiv.org/content/early/2018/09/12/414037
16. Mortamet, B., et al.: Automatic quality assessment in structural brain magnetic resonance imaging. Magn. Reson. Med. **62**(2), 365–372 (2009). https://doi.org/10.1002/mrm.21992. https://onlinelibrary.wiley.com/doi/abs/10.1002/mrm.21992
17. Pizarro, R.A., et al.: Automated quality assessment of structural magnetic resonance brain images based on a supervised machine learning algorithm. Front. Neuroinform. **10**, 52 (2016). https://doi.org/10.3389/fninf.2016.00052. https://www.frontiersin.org/article/10.3389/fninf.2016.00052
18. Poldrack, R.A., et al.: A phenome-wide examination of neural and cognitive function. Sci. Data **3**(1), 160110 (2016). https://doi.org/10.1038/sdata.2016.110
19. Reuter, M.: Freesurfer (2013). http://surfer.nmr.mgh.harvard.edu/
20. Soares, J.M., et al.: A hitchhiker's guide to functional magnetic resonance imaging. Front. Neurosci. **10**, 515 (2016). https://doi.org/10.3389/fnins.2016.00515. https://www.frontiersin.org/article/10.3389/fnins.2016.00515
21. Trivedi, M.H., et al.: Establishing moderators and biosignatures of antidepressant response in clinical care (embarc): Rationale and design. J. Psychiatr. Res. **78**, 11–23 (2016). https://doi.org/10.1016/j.jpsychires.2016.03.001. http://www.sciencedirect.com/science/article/pii/S0022395616300395
22. Woodard, J.P., Carley-Spencer, M.P.: No-reference image quality metrics for structural MRI. Neuroinformatics **4**(3), 243–262 (2006). https://doi.org/10.1385/NI:4:3:243
23. Yamada, M., Jitkrittum, W., Sigal, L., Xing, E.P., Sugiyama, M.: High-dimensional feature selection by feature-wise kernelized lasso. Neural Comput. **26**(1), 185–207 (2014). https://doi.org/10.1162/neco_a_00537

Uncertainty Characterization for Predictive Analytics with Clinical Time Series Data

Yang Guo, Zhengyuan Liu, Savitha Ramasamy, and Pavitra Krishnaswamy

Abstract Recurrent neural networks (RNNs) offer state-of-the-art (SOTA) performance on a variety of important prediction tasks with clinical time series data. However, meaningful translation to actionable decisions requires capability to quantify confidence in the predictions, and to address the inherent ambiguities in the data and the associated modeling process. We propose a Bayesian LSTM framework using Bayes by Backprop to characterize both modelling (epistemic) and data related (aleatoric) uncertainties in prediction tasks for clinical time series data. We evaluate our approach on mortality prediction tasks with two public Intensive Care Unit (ICU) data sets, namely, the MIMIC-III and the PhysioNet 2012 collections. We demonstrate the potential for improved performance over SOTA methods, and characterize aleatoric uncertainty in the setting of noisy features. Importantly, we demonstrate how our uncertainty estimates could be used in realistic prediction scenarios to better interpret the reliability of the data and the model predictions, and improve relevance for decision support.

Keywords Bayesian LSTM · Uncertainty quantification · Time series

S. Ramasamy and P. Krishnaswamy - Equal Contribution

Y. Guo · Z. Liu · S. Ramasamy · P. Krishnaswamy (✉)
Institute for Infocomm Research Agency for Science, Technology and Research (A*STAR), Singapore, Singapore
e-mail: pavitrak@i2r.a-star.edu.sg

© The Editor(s) (if applicable) and The Author(s), under exclusive license to Springer Nature Switzerland AG 2021

A. Shaban-Nejad et al. (eds.), *Explainable AI in Healthcare and Medicine*, Studies in Computational Intelligence 914,
https://doi.org/10.1007/978-3-030-53352-6_7

1 Introduction

The increasing adoption of electronic health records has resulted in the accumulation of large collections of longitudinal records of patient state. This presents rich opportunities for predictive modeling and decision support. Example applications include risk assessment for triage, proactive and personalized management of adverse events, or decision support for effective resource allocation. Traditional approaches for such problems relied on statistical learning and dynamical systems methodologies [1, 4, 20]. Increasingly, deep learning approaches using recurrent neural networks (RNNs) [13] are shown to provide state-of-the-art performance [2, 6, 10, 16] for a variety of clinical needs such as prediction of mortality, decompensation, and condition monitoring. However, meaningful translation of RNN predictions to actionable clinical decisions remains challenging.

One key difficulty is that clinical decision support applications have a critical need to provide quantitative measures of uncertainty alongside the prediction [5]. For instance, clinical practitioners making life-and-death decisions often need to weigh risk-benefit ratio of possible interventions. Another key difficulty is that observational clinical data collected during routine care delivery may exhibit quality limitations, high prevalence of missing or noisy feature values, and sometimes non-apparent labels [11]. Hence, it is important to model the inherent ambiguity in the data in the estimation process, and adapt to limitations in data quality. Despite this need, few investigations have studied the feasibility and benefit of transparently accounting for, adapting to and reasoning under uncertainty with RNN models for clinical time series prediction tasks.

Bayesian methods offer a means to represent, understand and adapt to uncertainty within deep learning models [17, 18]. Broadly, Bayesian approaches model two types of uncertainties: (a) *epistemic uncertainty* to account for modeling error or lack of perfect knowledge about parameters of the model; (b) *aleatoric uncertainty* to represent noise inherent in the observation process that generates the data. Accordingly, Bayesian approaches have been integrated with deep learning to address uncertainty in computer vision [15] and natural language processing applications [8, 9, 22]. Here, we focus on investigating the benefits of quantifying data and model uncertainties for prediction tasks involving clinical time series data.

The main contributions of our work are as follows: We first present an integrated Bayesian LSTM framework to characterize both epistemic and aleatoric uncertainty for prediction tasks involving clinical time series data. Next, we evaluate the framework on mortality prediction with two publicly available real-world datasets, namely: the MIMIC-III Intensive Care Unit dataset [14] and the PhysioNet 2012 dataset [19], and show the potential to improve performance over SOTA baselines. Then, we show that the uncertainty terms can effectively characterize data quality or ambiguity, and quantify the modeling confidence. Finally, we demonstrate the value of the modeled uncertainties to improve practical translation of predictions in scenarios with feature noise and missingness.

2 Methods

Problem Formulation: We consider a dataset $\mathcal{D}$ of N samples. For the nth sample, the input $\mathbf{X}^{(n)} \in \Re^{F \times T}$ is a multivariate time series comprising F features and T time steps, with associated output $\mathbf{y}^{(n)}$. The objective is to use $\mathcal{D}$ to train a model that can (a) predict $\mathbf{y}^*$ given $\mathbf{X}^*$ for a new input denoted by $*$, and (b) quantify the associated epistemic and aleatoric uncertainties.

Bayesian LSTM: An LSTM model $f^{\mathbf{W}}(.)$ of a chosen network architecture $f(.)$ learns the point estimate of weights $\mathbf{W}$. However, in practice, there is a degree of uncertainty associated with the weights. Such uncertainty can be addressed by using a Bayesian LSTM that represents the weights as a distribution [12]. During training, the Bayesian LSTM seeks to estimate the posterior distribution of $\mathbf{W}$ given the dataset D and a prior distribution of $\mathbf{W}$. The posterior is denoted as $p(\mathbf{W}|D)$ and prior as $p(\mathbf{W})$. Once the posterior is estimated, the prediction for a new (test) input sample is given by:

$$p(\mathbf{y}^*|\mathbf{X}^*, \mathcal{D}) = \int_{\mathbf{W}} p(\mathbf{y}^*|f^{\mathbf{W}}(\mathbf{X}^*))p(\mathbf{W}|\mathcal{D})d\mathbf{W} \tag{1}$$

As the exact solution is intractable, it is common to estimate the true posterior $p(\mathbf{W}|D)$ with an approximation $q(\mathbf{W}|\theta)$ parameterized by θ. This approximation is obtained by minimizing the Kullback-Leibler (KL) divergence between the two distributions:

$$\theta^* = \arg\min_{\theta} [\text{KL}\,[q\,(\mathbf{W}|\theta)\,||p\,(\mathbf{W}|\mathcal{D})]] \tag{2}$$

$$\theta^* = \arg\min_{\theta}[\text{KL}[q(\mathbf{W}|\theta)||p(\mathbf{W})] - E_{(q(\mathbf{W}|\theta))}[log(p(\mathcal{D})|\mathbf{W})]] \tag{3}$$

For large-scale problems, the above approximation can be performed with Bayes by Backprop [3]. Once approximated, the weights can be drawn from the approximate posterior $\widehat{\mathbf{W}} \sim q(\mathbf{W}|\theta)$.

Uncertainty Quantification: Recently, Bayes by Backprop (BBB) has been adapted to provide epistemic uncertainty estimates [8]. We, therefore, employ the BBB scheme from [8] to represent epistemic uncertainty, and introduce regularization terms into the loss function to quantify aleatoric uncertainty and model the inherent data noise. To cater to the quality variations between different input samples in our prediction task, we consider input data dependent or *heteroscedastic* aleatoric uncertainty. This treats the aleatoric uncertainty as distinct for every input sample $\mathbf{X}^{(\mathbf{n})}$. For the nth sample $\mathbf{X}^{(n)}$, the Bayesian LSTM outputs a logits vector (denoted as S_n) and an indicator of aleatoric uncertainty (denoted as U_n). Then,

$$[S_n, U_n] = f^{\widehat{\mathbf{W}}}(\mathbf{X}^{(n)}) \tag{4}$$

$$P_n = \text{SoftMax}(S_n) \tag{5}$$

$$\sigma_n^2 = \text{SoftPlus}(U_n) \tag{6}$$

where P_n is the probability vector for the output $\mathbf{y}$ and σ_n^2 denotes the estimated aleatoric uncertainty. With the estimate of σ_n, we use Monte Carlo integration to distort the logit prediction S_n by $\epsilon_m \sim N(0, \sigma_n^2)$ for the mth of M total Monte Carlo iterations, and correspondingly distort the loss function ([7, 15]):

$$\widetilde{S}_n = S_n + \epsilon_m \tag{7}$$

$$L_{\text{d}} = \sum_n \frac{1}{M} \sum_m \exp\left(\widetilde{S}_{n,c,m} - \log \sum_{c'} \exp(\widetilde{S}_{n,c',m})\right) \tag{8}$$

Here, c denotes the observed class and c' is the c'^{th} element in S_n.

Experimental Procedures. For our experiments, we define the prior distribution as a mixture of two Gaussian distributions ($\sigma_1 = e^{-1}, \sigma_2 = e^{-7}$) with mixing probability $\pi = 0.25$). We initialize the weights $\mathbf{W}$ randomly [8] and update the weights using back-propagation. However, instead of initializing each batch with the hidden states of prior batches [8], we initialize each batch independently. We add a regularization term $e^{\sigma_n^2} - 1$ to the loss function to prevent the σ_n^2 from railing to infinity. We use Monte Carlo approximation to sample the weights during training, and use the mean of the weight distribution for inference during testing.

3 Experiments and Results

We evaluate our method on mortality prediction tasks on two benchmark data sets, and perform a series of characterizations to demonstrate usefulness of quantifying the aleatoric and epistemic terms. We first describe the data set and the performance metrics used in the study, and then present our observations.

Data: We conduct experiments on mortality prediction tasks from two benchmark data sets: namely PhysioNet 2012 Challenge dataset [19] and MIMIC-III collection [14]. Both these data sets involve predicting the inpatient mortality of patients using multi-variate time series clinical parameters recorded, in the Intensive Care Unit, over 48 hourly time steps. PhysioNet 2012 Challenge [19] contains 4,000 admissions with 35 numerical features while MIMIC-III collection [14] contains 17,917 admissions with 12 numerical features and 5 categorical features. The imbalance factor, which is the ratio between number of samples in the minority class and total number of samples, is around 13% for both data sets.

Performance Metrics: For each data set, we train all models for the in-hospital mortality prediction task and evaluate the models on a held out test set by computing the area under the receiver operating characteristic curve (AUROC) and area under the precision recall curve (AUPRC). We perform training and evaluation in a manner that is consistent with the performance evaluations of prior works. For the MIMIC-III dataset, we bootstrap on 3,236 samples in the test set. For the Physionet dataset, we perform 5-fold cross validation with the 4,000 samples. We compare against state-of-the-art (SOTA) baselines: Logistic Regression (LR) and standard LSTM.

Table 1 Performance comparison for predictions

Methods	MIMIC-III (Bootstrap)		PhysioNet (Cross Validation)	
	AUROC	AUPRC	AUROC	AUPRC
LR	0.848 (0.828, 0.868)	0.474 (0.419, 0.529)	/	/
LSTM	0.855 (0.835, 0.873)	0.485 (0.431, 0.537)	0.817 (0.813, 0.821)	0.468 (0.451, 0.485)
BLSTM	0.855 (0.835, 0.874)	0.503 (0.450, 0.555)	0.824 (0.821, 0.827)	0.475 (0.466, 0.484)
BLSTM-E	0.853 (0.833, 0.872)	**0.505** (**0.452, 0.556**)	**0.825** (**0.821, 0.829**)	0.475 (0.463, 0.487)
BLSTM-A	0.855 (0.835, 0.874)	0.498 (0.434, 0.551)	0.820 (0.815, 0.825)	0.472 (0.460, 0.484)
BLSTM-EA	**0.856** (**0.836, 0.875**)	0.502 (0.449, 0.551)	0.823 (0.819, 0.827)	**0.476** (**0.463, 0.489**)

3.1 Performance Results

Table 1 presents the performance results of our proposed method and those of other SOTA methods in each case. Specifically, we compare the performances of the SOTA baselines against 4 variants of our Bayesian LSTM formulation: Bayesian LSTM (BLSTM), Bayesian LSTM with Epistemic term (BLSTM-E), Bayesian LSTM with the aleatoric term (BLSTM-A) and the Bayesian LSTM with both uncertainty terms (BLSTM-EA). We report results of the BLSTM and BLSTM-A with the mean of the weights, and base the results of BLSTM-E and BLSTM-EA on test set inference by sampling the posterior distribution of weights over 500 independent runs.

We observe that although the AUROC of the BLSTM is comparable to that of the LSTM classifier, the BLSTM methods outperform the LSTM classifier in terms of AUPRC. We infer that the Bayesian framework helps to classify the minority class better. These results show that our Bayesian LSTM framework performs at least as good as, if not better than the SOTA baselines. At the same time, the Bayesian approaches provide quantification of the epistemic and aleatoric uncertainties, and therefore have the potential to enable further insights about the data set, model and predictions. We therefore perform some characterization studies to uncover these insights. For these studies, we focus only on the mortality prediction task with the MIMIC-III data set.

3.2 Epistemic Uncertainty

First, we characterize the ability of BLSTM-E to characterize the epistemic uncertainty. We train the BLSTM-E classifier for the inpatient mortality prediction task

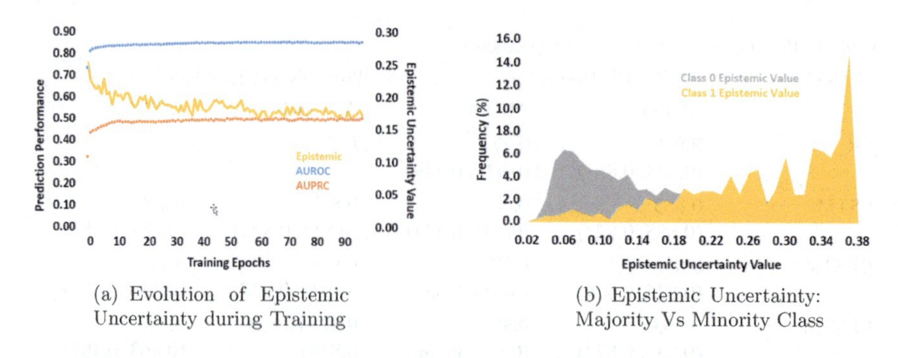

(a) Evolution of Epistemic Uncertainty during Training

(b) Epistemic Uncertainty: Majority Vs Minority Class

Fig. 1 Characterization of epistemic uncertainty

using the MIMIC-III data set, over 100 epochs. At the end of training in each epoch, we compute the AUROC, the AUPRC and the epistemic uncertainty of the BLSTM-E on the test data set. We highlight that the epistemic uncertainty in each epoch is obtained through the model inference, on the test set, for 500 independent runs of sampling the posterior distribution of weights.

Epistemic Uncertainty as a Function of Training: The AUROC, the AUPRC and the epistemic uncertainty are plotted in Fig. 1(a). We observe that the performance of the BLSTM classifier (AUROC, AUPRC) increases as the training epochs proceed. On the other hand, the epistemic uncertainty is very high in the initial epochs, and it decreases with training. Thus, as training progresses, the model gains higher confidence in its predictions and the predictions become more accurate. This suggests the epistemic uncertainty could be used to additionally inform model training beyond the standard validation approaches.

Epistemic Uncertainty as a Function of Class: Next, we compare the epistemic uncertainty for samples from the two classes. We bin the epistemic uncertainty values into intervals of 0.01, and obtain the percent of samples for each class in each bin. Results are in Fig. 1(b). We see that in general, while the majority class samples are characterized by lower epistemic uncertainties, the samples in the minority class have higher epistemic uncertainty. This is consistent with the fact that epistemic uncertainty is higher when we have lesser data. Accordingly, the model is more certain about its prediction on samples in the majority class, while being relatively less certain about its prediction on samples in the minority class.

3.3 Aleatoric Uncertainty

Second, we characterize how the performance of the BLSTM-A model varies with data noise. Specifically, we characterize the aleatoric uncertainty as a function of increasing number of noisy observations, and then as a function of increasing noise levels. To guide these experiments, we perform a feature importance study on the

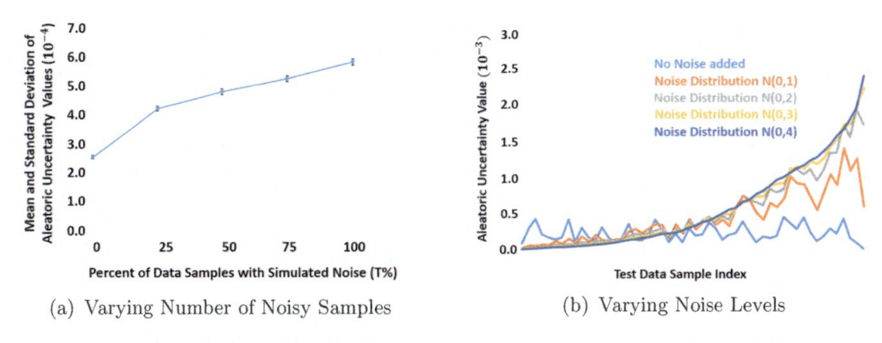

(a) Varying Number of Noisy Samples

(b) Varying Noise Levels

Fig. 2 Characterization of aleatoric uncertainty

mortality prediction task using the MIMIC-III data set with logistic regression, and observe that the four most important features for this task are temperature, Glascow coma scale verbal response, Glascow coma scale motor response, Glascow coma scale eye opening. We note that except for the temperature, the remaining features are categorical features. We randomly add Gaussian noise ($\mathcal{N}(0, 1)$) to these 4 features, and present the characterization results in Fig. 2.

Aleatoric Uncertainty vs Number of Noisy Samples. We vary the number of corrupted samples by introducing noise to 25%, 50%, 75% and 100% of samples in the testing set. We quantify the mean and standard deviation of the aleatoric term as a function of varying numbers of corrupted test samples. The results are in Fig. 2(a). We observe that the aleatoric uncertainty increases in proportion to the number of noisy observations. This suggests that the aleatoric uncertainty can serve as an indicator of the number of test set samples with noisy feature values.

Aleatoric Uncertainty vs Noise Levels. Next, we randomly choose 10% of the testing samples and add noise at varying levels to the most important numerical feature, namely, temperature. In doing so, we increase the noise levels by sampling noise from normal distributions with increasing variance: $\mathcal{N}(0, 1)$, $\mathcal{N}(0, 2)$, $\mathcal{N}(0,3)$, and $\mathcal{N}(0, 4)$. We then plot the aleatoric uncertainties of individual samples in Fig. 2(b). We observe that the aleatoric uncertainty increases with noise levels in the data, and the rate of increase drops with increasing noise levels. This suggests that the aleatoric uncertainty can serve as an indicator of the noise in features of the test data.

In addition, we study the effect of adding noise at various time points in the series (earlier vs. later parts of the time series). As the MIMIC-III data set comprises a time series of 48 hourly readings for inpatient mortality prediction task, we add noise from a standard normal distribution $\mathcal{N}(0, 1)$ in a randomly chosen subset of 10% of the samples for the first 12 h and for the last 12 h. We observe that the average aleatoric uncertainty of the test data set for noise injection in the first 12 h is $2.6 \times 10^{(-4)}$, and that for noise injection in the last 12 h is $3.9 \times 10^{(-4)}$. Thus, having a noisy observation at earlier times in the series has lower influence on the uncertainty of the overall sample for the purposes of prediction.

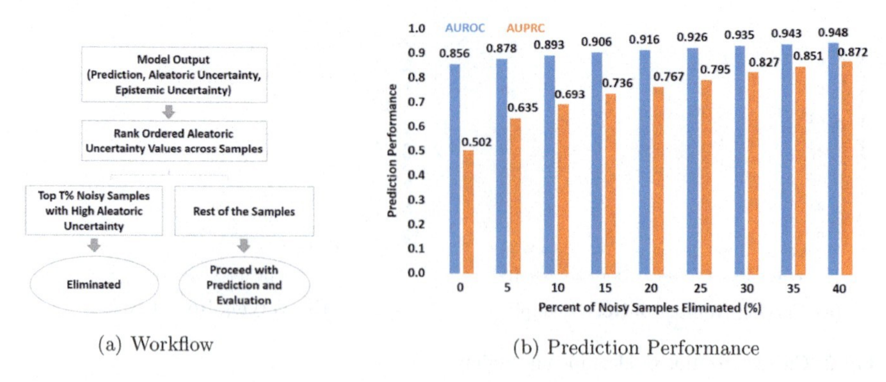

(a) Workflow (b) Prediction Performance

Fig. 3 Performance changes with elimination of uncertain test samples

3.4 Implications for Prediction Performance

We now integrate the above observations with a study on how the uncertainty quantification relates to prediction performance. Specifically, we rank test data samples by their aleatoric uncertainty levels; eliminate 5%, 10%, 15%, 20%, 25%, 30%, 35% and 40% of the samples with highest aleatoric uncertainty values; and in each case characterize performance of the BLSTM-EA model on the remaining samples in the test data (Fig. 3). We observe that the performance of the model increases as we eliminate more uncertain samples from the test data. In particular, there is a stark increase of over 5% in AUROC and 30% in the AUPRC with elimination of uncertain samples. This suggests that the aleatoric uncertainty estimates could be used to rank the test data samples by their inherent noise and quality. In addition, the uncertainty estimates would enable an end user to transparently have a sense of the ambiguity in any given data sample and therefore inform the use of model predictions for decision making.

Finally, we study the effect of missing values by simulating Missingness at Random (MAR) in the test data. Specifically, we gradually increase the MAR in the test data and compare the performance of the LSTM and the BLSTM-EA. We observe that the BLSTM-EA model outperforms the LSTM by upto 1% in AUPRC even when upto 50% of the samples are missing. This suggests that the uncertainty quantification can enhance robustness of the model predictions in light of missing feature values, which are very common in clinical time series data sets.

4 Conclusions

We have developed and demonstrated a novel integrated Bayesian LSTM framework that models both aleatoric and epistemic uncertainty for prediction tasks involving clinical time series data. The framework incorporates regularization for aleatoric uncertainty atop a Bayes By Backprop in LSTM. While one prior study considered epistemic uncertainty in the context of clinical data [21], it did not explore data uncertainty and offer prescriptive insights on use of the uncertainty measures in practice. In contrast, we show that both epistemic and aleatoric uncertainty terms effectively characterize model and data related uncertainties, in the setting of noisy features on two datasets for in-hospital mortality prediction tasks. We also demonstrated the ability to use these uncertainty estimates to better interpret the reliability of the data and associated model predictions. As quality of observational time series data recorded during routine clinical care is highly variable, incorporation of such uncertainty estimates in real-world clinical predictive analytics tasks would have important implications for decision support.

Acknowledgments Research efforts were supported by funding and infrastructure from A*STAR, Singapore (Grant Nos. SSF A1818g0044 and IAF H19/01/a0/023). The authors would also like to acknowledge inputs from Wu Jiewen and Ivan Ho Mien on the clinical datasets. We also thank Vijay Chandrasekhar for his support, and are grateful for the constructive feedback from the anonymous reviewers.

References

1. Adams, J.R., Marlin, B.M.: Learning time series detection models from temporally imprecise labels. In: Proceedings of Machine Learning Research, vol. 54 (2017)
2. Baytas, I.M., Xiao, C., Zhang, X., Wang, F., Jain, A.K., Zhou, J.: Patient subtyping via time-aware LSTM networks. In: Proceedings of the 23rd ACM SIGKDD International Conference on Knowledge Discovery and Data Mining, pp. 65–74. ACM (2017)
3. C. Blundell, J. Cornebise, K.K., Wierstra, D.: Weight uncertainty in neural networks. arXiv preprint arXiv:1505.05424 (2015)
4. Caballero Barajas, K.L., Akella, R.: Dynamically modeling patient's health state from electronic medical records: a time series approach. In: Proceedings of the 21st ACM SIGKDD International Conference on Knowledge Discovery and Data Mining, vol. 31, pp. 69–78 (2015)
5. Challen, R., et al.: Artificial intelligence, bias and clinical safety. BMJ Qual. Saf. **28**(3), 231–237 (2019)
6. Choi, E., et al: Doctor AI: predicting clinical events via recurrent neural networks. In: Machine Learning for Healthcare Conference, pp. 301–318 (2016)
7. Depeweg, S., Miguel, J., Hernández-Lobato, Doshi-Velez, F., Udluft, S.: Decomposition of uncertainty in Bayesian deep learning for efficient and risk-sensitive learning. arXiv preprint arXiv:1710.07283 (2018)
8. Fortunato, M., Blundell, C., Vinyals, O.: Bayesian recurrent neural networks. arXiv:1704.02798v4 (2019)
9. Gal, Y., Ghahramani, Z.: A theoretically grounded application of dropout in recurrent neural networks. In: Neural Information Processing Systems (2016)
10. Ghassemi, M., et al: A multivariate timeseries modeling approach to severity of illness assessment and forecasting in ICU with sparse, heterogeneous clinical data. In: Association for the Advancement of Artificial Intelligence, pp. 446–453 (2015)
11. Ghassemi, M., et al: Opportunities in machine learning for healthcare. arXiv:1806.00388v3 (2019)
12. Graves, A.: Practical variational inference for neural networks, pp. 2348–2356 (2011)
13. Hochreiter, S., Schmidhuber, J.: Long short-term memory. Neural Comput. **9**(8), 1735–1780 (1997)

14. Johnson, A.E., Pollard, T.J., Shen, L., Lehman, L.W., Feng, M.M., Ghassemi, M., Moody, B., Szolovits, P., Celi, L.A., Mark, R.G.: MIMIC-III, a freely accessible critical care database. Sci. Data **3**, 160035 (2016)
15. Kendall, A., Gal, Y.: What uncertainties do we need in Bayesian deep learning for computer vision. In: Neural Information Processing Systems, vol. 30 (2017)
16. Lipton, Z.C., Wetzel, R., et al: Learning to diagnose with LSTM recurrent neural networks. In: International Conference on Learning Representations (2016)
17. MacKay, D.J.C.: A practical Bayesian framework for backpropagation networks. Neural Comput. **4**(3), 448–472 (1992)
18. Neal, R.: Bayesian Learning for Neural Networks, vol. 118. Springer, New York (2012)
19. PhysioBank, P.: PhysioNet: components of a new research resource for complex physiologic signals. Circulation **101**(23), e215–e220 (2000)
20. Quinn, J.A., et al.: Factorial switching linear dynamical systems applied to physiological condition monitoring. IEEE Trans. Pattern Anal. Mach. Intell. **31**, 1537–1551 (2009)
21. van der Wasthuizen, J., Lasenby, J.: Bayesian LSTMs in medicine. arXiv:1706.01242 (2017)
22. Xiao, Y., Wang, W.Y.: Quantifying uncertainties in natural language processing tasks. In: Association for the Advancement of Artificial Intelligence, pp. 1019–1027 (2019)

A Dynamic Deep Neural Network for Multimodal Clinical Data Analysis

Maria Hügle, Gabriel Kalweit, Thomas Hügle, and Joschka Boedecker

Abstract Clinical data from electronic medical records, registries or trials provide a large source of information to apply machine learning methods in order to foster precision medicine, e.g. by finding new disease phenotypes or performing individual disease prediction. However, to take full advantage of deep learning methods on clinical data, architectures are necessary that (1) are robust with respect to missing and wrong values, and (2) can deal with highly variable-sized lists and long-term dependencies of individual diagnosis, procedures, measurements and medication prescriptions. In this work, we elaborate limitations of fully-connected neural networks and classical machine learning methods in this context and propose AdaptiveNet, a novel recurrent neural network architecture, which can deal with multiple lists of different events, alleviating the aforementioned limitations. We employ the architecture to the problem of disease progression prediction in rheumatoid arthritis using the Swiss Clinical Quality Management registry, which contains over 10.000 patients and more than 65.000 patient visits. Our proposed approach leads to more compact representations and outperforms the classical baselines.

1 Introduction

Driven by increased computational power and larger datasets, deep learning (DL) techniques have successfully been applied to process and understand complex data [11]. In recent years, the adoption of electronic medical records (EMRs), reg-

M. Hügle (✉) · G. Kalweit · J. Boedecker
Neurorobotics Lab, Department of Computer Science, University of Freiburg, Freiburg, Germany
e-mail: hueglem@cs.uni-freiburg.de

T. Hügle
Department of Rheumatology, University Hospital Lausanne, CHUV, Lausanne, Switzerland
e-mail: Thomas.Hugle@chuv.ch
© The Editor(s) (if applicable) and The Author(s), under exclusive license to Springer Nature Switzerland AG 2021

A. Shaban-Nejad et al. (eds.), *Explainable AI in Healthcare and Medicine*,
Studies in Computational Intelligence 914,
https://doi.org/10.1007/978-3-030-53352-6_8

istries and trial datasets has heavily increased the amount of captured patient data from thousands up to millions of individuals patients. Those datasets can be used to predict individual disease progression and outcomes in medicine to assist patients and doctors. Especially deep learning methods are more and more applied to process clinical data [5, 20, 24, 27] and learn from former experiences on a large scale as a potential tool to guide treatment and surveillance [9]. To establish high-quality decision making systems for clinical datasets, machine learning (ML) methods have to be able to deal with the varying structure of the data, containing variable-sized lists of diagnosis, procedures, measurements and medication prescriptions. Capturing long-term dependencies and handling irregular event sequences with variable time spans in between is crucial to model complex disease progressions of patients in personalized medicine.

Classical ML models, such as Random Forests (RFs) [2], and fully-connected networks (FCNs) or recurrent neural networks (RNNs) were applied successfully to predict disease progression [30] and disease outcomes [10]. However, classical FCN architectures[1] are limited to a fixed number of input features. As consequence, these models can only consider a fixed number of events, such as visits or medication prescriptions and adjustments. As a simple workaround, only the last N visits and M medication prescriptions can be considered and older entries ignored. In case patients have less than N entries for visits or M medication adjustments, dummy values have to be used to guarantee a fixed number of input features. However, depending on the choice of N and M, the full patient history is not considered.

Classical recurrent neural networks, such as long short-term memories (LSTMs) [6] can deal with variable-sized inputs and can handle irregularly timed events. In previous work, LSTMs were used for outcome prediction in intensive care units [15], heart failure prediction [19], Alzheimer's disease progression prediction [12] and other applications [1, 23]. However, the proposed architectures can only deal with one variable-sized list of one feature representation, which is limiting when working with clinical data.

In this work, we aim at exploiting and unifying all available patient data. The resulting neural network architecture has to be able to deal with multiple variable-sized lists of different events like visits, medication adjustments or imaging, which all have different feature representations. In [16, 17], this problem was approached by using large one-hot encodings, which contain features for all different event types. However, this approach leads to a high amount of dummy values and has limited flexibility. We propose AdaptiveNet, a recurrent neural network architecture that can be trained in an end-to-end fashion. The scheme of the architecture is shown in Fig. 1. AdaptiveNet can deal with multiple variable-sized lists of different event types by using multiple fully-connected encoding network modules to project all event types to the same latent space. Then, the sorted list of latent event representations is fed to a recurrent unit in order to compute a latent representation describing the full patients

[1]As architecture we understand the full specification of interactions between different network modules, in- and outputs. In the following, we denote classical fully-connected neural networks without any extensions as FCN.

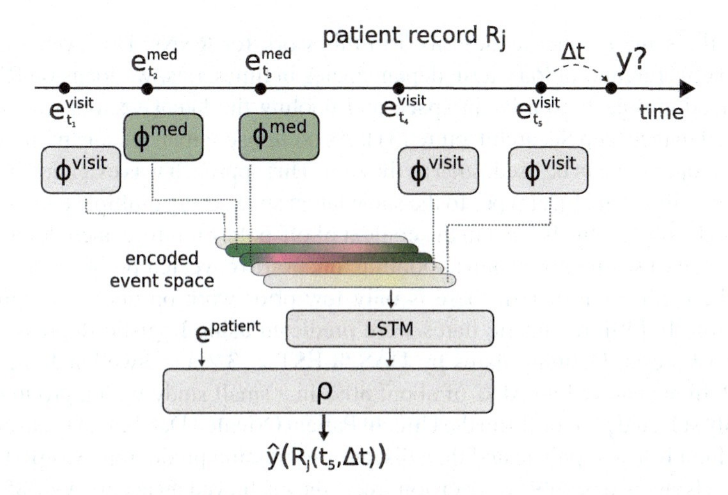

Fig. 1 Scheme of AdaptiveNet, which projects visits and medication adjustments to the same latent space using encoder networks ϕ^{visit} and ϕ^{med}, where the output vectors $\phi^{(\cdot)}(\cdot)$ have the same length. The sorted list of encoded events are pooled by an LSTM to compute a fixed-length encoded patient history. The final output $\hat{y}$ is computed by the network module ρ

history. We employ AdaptiveNet for the problem of disease progression prediction in rheumatoid arthritis (RA) and evaluate the performance on the Swiss Clinical Quality Management (SCQM) dataset.

2 Related Work

Data in health care and biology often contains a variety of missing and wrong values. Previous work already addressed missing values in time series [26], but did not provide good performance when the missing rate was too high and inadequate samples were kept. While classical methods can omit missing data by performing analysis only on the observed data, for deep neural network architectures this is not straightforward. A possible solution is to fill in the missing values with substituted values, e.g. by smoothing or interpolation [21]. In [25], it was shown that missing values can provide useful information about target labels in supervised learning tasks. In [3] this fact was exploited by using recurrent neural networks that are aware of missing values.

Other network architectures, such as convolutional neural networks (CNNs) [22], Transformer architectures [13] or graph neural networks (GNNs) [32] were also used to process clinical data [4]. In contrast to CNNs, which are limited to their initial grid size, Transformers and GNNs are able to deal with variable-sized input lists. In general, these architectures could be extended to deal with multiple variable-sized lists in the same manner as described in this work for RNNs. However, since we want to cover temporal long-term dependencies in timeseries, we focus on RNNs.

Projecting objects to a latent space and pooling the latent vectors was already proposed in the Deep Set architecture [31]. As pooling component, any permutation-invariant operator can be used, such as the *sum*. This approach was extended in [7] by projecting different object types to the same latent space using multiple encoders and pooling the objects by the *sum* in the context of off-policy reinforcement learning. In this work, we use a recurrent unit as pooling function to cover temporal dependencies.

In the application of RA, there is only few prior work on disease progression prediction. In [30], rheumatic flares were predicted using Logistic Regression and Random Forests. Defining flares by DAS28-EST ≥ 3.2 and Swollen Joint Count $28 \geq 2$, they achieved an AUC of about 80% in a small study with a group of 314 carefully selected patients from the Utrecht Patient Oriented Database. Disease detection, which is less sophisticated than disease progression prediction, was performed in [28]. Using an ensemble of decision trees, they achieved an accuracy of 85% and a sensitivity and specificity of 44% and 74% on a dataset of around 2500 patients referred to a rheumatology clinic in Iran. [14] performed disease detection based on classical ML methods trained on >2500 clinical notes and lab values. With an SVM, they achieved the best performance with an AUC score of 0.831.

3 Methods

In this section, we describe the disease progression prediction problem and how input samples can be generated from clinical datasets. Finally, the architecture of AdaptiveNet is explained in detail.

3.1 Disease Progression Prediction

Disease progression prediction based on a clinical dataset can be modeled as time-series prediction, where for a patient at time point t the future disease activity at time point $t + \Delta t$ is predicted. The dataset consists of records for a set of patients $\mathcal{R} = \bigcup_j \mathcal{R}_j$. Records can contain general patient information (e.g. age, gender, antibody-status) and multiple list of events. The subset $R_j(t) \subseteq \mathcal{R}_j$, denotes all records of a patient j collected until time point t with

$$R_j(t) = \{e^{\text{patient}}(t)\} \cup E^{\text{visit}}(t) \cup E^{\text{med}}(t),$$

where $e^{\text{patient}}(t)$ contains general patient information collected until time point t. $E^k(t)$ is a list of events collected until time point t, where

$$E^k(t) = \{e^k(t_e) \mid e^k(t_e) \in \mathcal{R}_j \text{ and } t_e \leq t\},$$

Algorithm 1: Sample Generation from Records $\mathcal{R}$

```
1  Initialize feature array X and labels y.
2  for patient record R_j ∈ R do
3  |  for visit time point t ∈ T_j^visit do
4  |  |  // For all follow up visits before the next medication adjustment:
5  |  |  T_j^follow-up ← {t' ∈ T_j^visit | t' > t and
6  |  |                  ∄t_m ∈ T_j^med s.t. t ≤ t_m ≤ t'}
7  |  |  for follow up t' ∈ T_j^follow-up do
8  |  |  |  Δt ← t' − t
9  |  |  |  add R_j(t, Δt) to X
10 |  |  |  add score_j(t + Δt) − score_j(t) to y
```

for all event types $k \in \{\text{visit}, \text{med}\}$. Visit events contain information like joint swelling or patient reported outcomes (e.g. joint pain, morning stiffness and HAQ) and lab values (e.g. CRP, BSR). Medication events contain adjustment information, such as drug type and dose. In the same manner, further lists of other event types could be added, such as imaging data (e.g. MRI, radiograph). Considering records $R_j(t)$ as input, we aim to learn a function $f : (R_j(t), \Delta t) \to \mathbb{R}$ that maps the records at time point t to the expected change of the disease level score_j until time $t + \Delta t$ with

$$f(R_j(t), \Delta t) = \text{score}_j(t + \Delta t) - \text{score}_j(t).$$

To account for variable time spans, for all events, the time distance Δt to the prediction time point is added as additional input feature. Records with included time feature are denoted as $R_j(t, \Delta t)$ and event lists with $E^{(\cdot)}(t, \Delta t)$. If not denoted explicitly in the following, we assume that Δt is included in all records and lists.

3.2 *Sample Generation from Clinical Data*

To train ML models on a clinical dataset in a supervised fashion, input samples and the corresponding labels are constructed over all patient records $\mathcal{R}_j$ by iterating over the list of visit time points $\mathcal{T}_j^{\text{visit}}$ and over all follow-up visits until the next medication adjustment. The list of time points for medication treatments is denoted as $\mathcal{T}_j^{\text{med}}$. The sample generation procedure is shown in Algorithm 1.

3.3 *AdaptiveNet*

In order to deal with the above defined patient records $R_j(t)$ and the corresponding variable-sized event lists $E^{(\cdot)}(t)$ for a time point t, we propose the neural network architecture AdaptiveNet, which is able to deal with K input sets $E^1, ..., E^K$, where every set can have variable length and different feature representations. With neural network modules $\phi^1, ..., \phi^K$, every element of the K lists can be projected to a latent space and a sorted list of latent events can be computed as

$$\Psi(R_j(t)) = \Psi(E^1, ..., E^K) = sort\left(\bigcup_k \bigcup_{e \in E^k(t)} \phi^k(e)\right),$$

with $1 \leq k \leq K$, sorted according to the time points of the events. The output vectors $\phi^k(\cdot) \in \mathbb{R}^F$ of the encoder networks ϕ^k have the same length F. These network modules can have an arbitrary architecture. In this work, we use fully-connected network modules to deal with numerical and categorical input features. We additionally propose to share the parameters of the last layer over all encoder networks. Then, $\phi^k(\cdot)$ can be seen as a projection of all input objects to the same encoded *event space* (effects of which we investigate further in the results). The prediction $\hat{y}$ of the network is then computed as

$$\hat{y}(R_j(t)) = \rho\left(\text{LSTM}\left[\Psi(R_j(t))\right] \,||\, e^{\text{patient}}\right),$$

where $||$ denotes concatenation of the vectors and ρ is a fully-connected network module. In this work, we use an LSTM [6] to pool the events by a recurrent unit. The scheme of the architecture is shown in Fig. 1. To tackle the disease progression prediction problem, in this work, we use two encoder modules ϕ^1, ϕ^2 for the set of event types {visit, med}. It is straightforward to add other event types, for example imaging data. In this case, the encoder module could consist of a convolutional neural network, which can optionally be pre-trained and have fixed weights.

4 Experimental Setup

In this section, we first explain the dataset and describe how to perform disease progression prediction in RA. After that, we explain baselines and training details, such as hyperparameter optimization and architectural choices.

4.1 Data Set

In this work, we use the Swiss Clinical Quality Management (SCQM) database [29] for rheumatic diseases, which includes data of over 10.000 patients with RA, assessed during consultations and via the mySCQM online application. The database consists of general patient information, clinical data, disease characteristics, ultrasound, radiographs, lab values, medication treatments and patient reported outcome (HAQ, RADAI-5, SF12, EuroQol). Patients were followed-up with one to four visits yearly and clinical information was updated every time. The data collection was approved by a national review board and all individuals willing to participate, signing an informed consent form before enrolment in accordance with the Declaration of Helsinki.

4.2 Disease Progression Prediction in RA

To represent the disease level, we use the hybrid score DAS28-BSR as prediction target, which contains DAS28 (Disease Activity Score 28) and the inflammation bloodmarker BSR (blood sedimentation rate). DAS28 defines the disease activity based on 28 joints (number of swollen joints, number of painfull joints and questionnaires). For training and evaluation, we consider only visits with available DAS28-BSR score. We focus on 13 visit features, selected by a medical expert. All visit features are shown in Table 1. Additionally, we consider eight medications. The corresponding features are listed in Table 2 and general patient features in Table 3.

4.3 Baselines

As baselines, we consider a **Naive Baseline** where we set $f(R_j(t), \Delta t) \approx 0$, which means no change in the disease level. Further, we compare to a **Random Forest** (ensemble of decision trees) and a classical **Fully-Connected Network**. In order to deal with the variable patient history lengths of the clinical dataset, for both RF and the FCN architecture the inputs have to be padded with dummy values (e.g. -1). Since we want to consider the full patient history without any loss of information, we pad the input until the maximum number of visit features is reached for the patient with the longest history, and analogously for medication features. Considering a maximum history length over many years can lead to a huge input size and large amount of dummy values, complicating reliable estimation of relevant correlations considerably. For a history length of 5 years, the input sizes of the RF and FCN are 1178, due to a maximum of 35 visits for one patient. In contrast, AdaptiveNet has 8 input features for general patient information, 21 input features for visits and 18 features for medications. For a history length of 5 years, patients have in mean 6.3 ($\pm$5.3) visits and 2.5 ($\pm$2.7) medication adjustments.

Table 1 Visit features. The values for Δt are shown for a prediction horizon of 1 year and a maximum history length of 5 years. (*) This score is a rheumatoid arthritis disease activity index (RADAI)

Numerical	Missing [%]	Mean (± Std.)
Minimal disease activity	1.6	1.3 (±1.1)
Number swollen joints*	6.6	3.3 (±4.6)
Number painfull joints*	6.9	3.5 (±5.3)
BSR	14.6	18.5 (±17.1)
DAS28BSR score	16.4	3.2 (±1.4)
Pain level*	22.4	3.3 (±2.7)
Disease activity index*	22.7	3.4 (±2.7)
Haq score	27.8	0.8 (±0.7)
Weight [kg]	36.5	70.7 (±15.6)
Height [cm]	40.8	165.3 (±12.2)
CRP	45.4	7.32 (±12.7)
Δt (5y history)	0.0	2.2 (±1.4)
Categorical		**Values [%]**
Morning	22.7	all day (1.9%)
stiffness*		<0.5 h (15.4%)
		0.5–1 h (12.0%)
		>4 h (1.6%)
		12 h (6.1%)
		24 h (3.5%)
		no (36.8%)
smoker	60.2	current (9.3%)
		former (12.3%)
		never (18.2%)

4.4 Training

To train our models, we considered only patients with more than two visits and a minimum prediction horizon of 3 months up to a maximum of one year. In total, we trained on 28601 samples. For preprocessing, we scaled all features in the range (0, 1). The final architecture of AdaptiveNet is shown in Table 4. As activation function, rectified linear units (ReLU) were used for all hidden layers. Additionally, we used $l1$-regularization for the weights of the network. For the FCN, we used dropout of 0.1, $l1$-regularization and three hidden layers of hidden dimension 32. For the RF, 100 tree estimators were used with a maximum depth of 12. To train the neural networks, weights w are updated in supervised fashion via gradient descent on a minibatch of samples X and labels y of size B using the update rule:

$$w \leftarrow w - \alpha \nabla_w L(w),$$

Table 2 Medication features

Categorical	Value	Pct. [%]
Drug	Dmard mtx	24.1
	Prednison	16.8
	Adalimumab	7.9
	Etanercept	7.3
	Tocilizumab	4.0
	Abatacept	4.0
	Rituximab	3.5
	Golimumab	2.4
	Other	30.1
Type	Prednison	16.8
	Dmard	24.1
	Biologic	29.0
	Other	30.1
Dose	No	41.3
	<10 [mg]	9.6
	10–15 [mg]	12.6
	>15 [mg]	36.5
Numerical	Missing [%]	Mean (± Std.)
Δt (5y history)	0.0	2.2 (±1.3)

Table 3 General patient features

Numerical	Missing [%]	Mean (± Std.)
Age	0.0	58.8 (±13.0)
Disease duration	2.7	12.2 (±9.5)
Categorical		Values [%]
Gender	0.0	Male (26.0%)
		Female (74.0%)
R-factor	9.1	Yes (62.9%)
		No (28.0%)
Anti-ccp	31.6	Yes (42.4%)
		No (26.0%)

with learning rate α and loss L. The loss is computed as:

$$L(w) = \frac{1}{B}\sum_{i=1}^{B}(y_i - \hat{y}(X_i))^2.$$

Table 4 Architecture of AdaptiveNet, where FC denotes fully-connected layers, $seq^{(\cdot)}$ variable-sized lists of events and B the batch size. (*) A second FC(100) layer is used in experiments with parameter-sharing for the encoders

AdaptiveNet
Input($B \times seq^{visit} \times 21$) and Input($B \times seq^{med} \times 18$)
ϕ^{visit}: FC(100)*, ϕ^{med}: FC(100)*
LSTM(·)
concat(·, Input($B \times 8$))
FC(100), FC(100)
Linear(1)

Table 5 Configuration spaces of the different approaches. The best performing architectural choices are shown in bold. (*) For experiments with parameter-sharing in the last layer, one additional layer of the same hidden dimension is added

Model	Parameter	Config. space
RF	Max depth	[8, 10, **12**, 15]
FCN	Num hidden layers	[2, **3**, 4]
	Hidden dim	[**32**, 64, 100]
	Dropout rate	[0.0, **0.1**, 0.25]
AdaptiveNet	$\phi^{(\cdot)}$: hidden dim	[32, 64, **100**]
	$\phi^{(\cdot)}$: num hidden layers	[**1**, 2]*
	ρ: hidden dim	[64, **100**, 200]
	ρ: num hidden layers	[1, **2**, 3]
	Dropout rate	[**0.0**, 0.1, 0.25]

All architectures were trained with a batch size of $B = 256$ for 7000 steps (batches of samples). For optimization, we used the Adam optimizer [8] with a learning rate of $1e^{-4}$. To evaluate fairly, all models were optimized using grid search, including the baselines. The configuration spaces for all methods can be found in Table 5.

5 Results

To evaluate the performance of all models, we use 5-fold cross validation, where samples and corresponding labels were generated as shown in Algorithm 1. The results are shown in Fig. 2 for a prediction horizon of one year for all methods and different maximum history lengths of 6 months to 5 years.

All ML methods outperform the Naive Baseline significantly, which has a MSE of 1.369. The performance of the RF decreases with increasing history size from a MSE 0.983 for 6 months to a MSE of 1.058, probably due to the huge amount of input features. In contrast, both neural network architectures are able to profit from longer

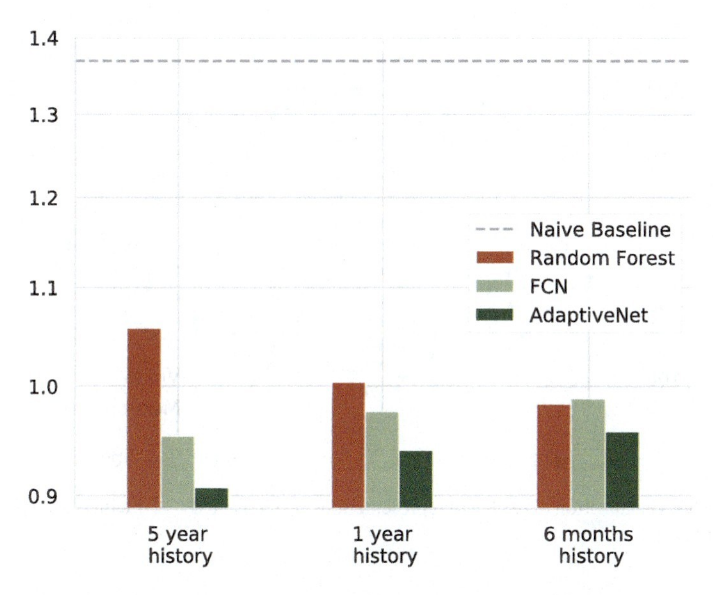

Fig. 2 Mean squared error of the disease progression prediction for different maximum history lengths in a range from 5 years to 6 months. The prediction horizon is 1 year

histories. With a MSE of 0.988 for history of 6 months, the FCN shows slightly worse performance than the RF. However, for longer histories, the FCN outperforms the RF, showing MSEs of 0.953 for 1 year and 1.058 for 5 years. The best performance over all history lengths is achieved by AdaptiveNet with a MSE of 0.957 for 6 months up to 0.907 for a history length of 5 years, which corresponds to an error of 7.94% in the range of the target value (change of DAS28-BSR).

For further evaluation, we use the trained regression model to perform classification. Defining two classes as *active disease* (DAS28-BSR ≤ 2.6) and *in remission* (DAS28-BSR < 2.6), we can classify between future disease levels by estimating the absolute disease level with $f^{\text{abs}}(R_j(t), \Delta t) = \text{score}_j(t) + f(R_j(t), \Delta t)$. AdaptiveNet achieves an accuracy of 76% and an Area Under the Curve (AUC) score of 0.735. Please note that these results can not directly be compared to the performance of the flare prediction approach shown in [30] due to different definitions of active disease levels and different datasets (10.000 patients in this study vs. 314 patients).

Using parameter sharing in the last layer in the $\phi^{(\cdot)}$ modules, the network performs slightly better for long histories. With parameter sharing, the MSE of the disease progression prediction for AdaptiveNet for a considered maximum history of 5 years decreases from 0.923 to 0.899. The fact, that the individual latent representations for the different events are separated in our architecture makes it possible to characterize the structure of the encoded event space, which would not be possible for the other methods. We visualize the latent vectors $\phi^k(\cdot)$ using the t-distributed Stochastic Neighbor Embedding (t-SNE) [18] algorithm, which is a nonlinear dimensionality reduction method to reduce high-dimensional data to lower dimensions for visual-

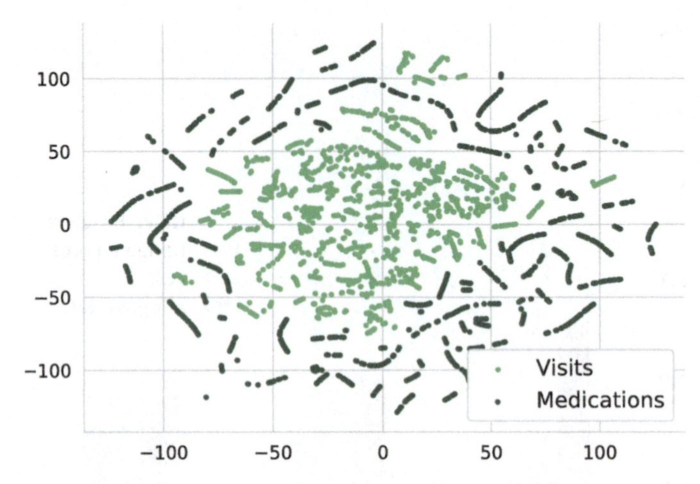

Fig. 3 t-SNE visualization of the latent representations $\phi^{\text{visit}}(\cdot)$ and $\phi^{\text{med}}(\cdot)$ with shared parameters in the last layer

ization. Figure 3 shows the t-SNE plot for 6000 latent representations. As can be seen in the plot, the learned representations for visits and medications are nicely clustered and well separated, which partly explains the good performance of this architecture.

6 Conclusion

AdaptiveNet provides a flexible architecture to deal with multiple variable-sized lists of clinical events of different types, such as clinical visits or medication adjustments. The flexibility of the architecture allows to exploit and integrate *all available data* into the decision making process in a unified and compact manner. Compared to classical approaches, AdaptiveNet is more robust in disease prediction, avoiding missing values and handling irregular visits by a recurrent unit. AdaptiveNet can be applied for various ML applications based on clinical datasets like EMRs, registries or trials. We are convinced that flexible and integrative deep learning systems such as AdaptiveNet can boost personalized medicine by considering as much data as possible to create high-quality machine-learned evidence.

Acknowledgements A list of rheumatology offices and hospitals that are contributing to the SCQM registries can be found on www.scqm.ch/institutions. The SCQM is financially supported by pharmaceutical industries and donors. A list of financial supporters can be found on www.scqm.ch/sponsor. We gratefully thank all patients and doctors for their support.

References

1. Baytas, I.M., Xiao, C., Zhang, X., Wang, F., Jain, A.K., Zhou, J.: Patient subtyping via time-aware LSTM networks. In: Proceedings of the 23rd ACM SIGKDD International Conference on Knowledge Discovery and Data Mining, KDD 2017, pp. 65–74. ACM, New York (2017). https://doi.org/10.1145/3097983.3097997
2. Breiman, L.: Random forests. Mach. Learn. **45**, 5–32 (2001)
3. Che, Z., Purushotham, S., Cho, K., Sontag, D., Liu, Y.: Recurrent neural networks for multivariate time series with missing values. Sci. Rep. **8**, 1–12 (2016). https://doi.org/10.1038/s41598-018-24271-9
4. Choi, E., Xu, Z., Li, Y., Dusenberry, M.W., Flores, G., Xue, Y., Dai, A.M.: Graph convolutional transformer: learning the graphical structure of electronic health records. CoRR abs/1906.04716 (2019). arXiv:1906.04716
5. Hoang, K.H., Ho, T.B.: Learning and recommending treatments using electronic medical records. Knowl.-Based Syst. **181**, 104788 (2019). https://doi.org/10.1016/j.knosys.2019.05.031. http://www.sciencedirect.com/science/article/pii/S0950705119302436
6. Hochreiter, S., Schmidhuber, J.: Long short-term memory. Neural Comput. **9**(8), 1735–1780 (1997). https://doi.org/10.1162/neco.1997.9.8.1735
7. Hügle, M., Kalweit, G., Mirchevska, B., Werling, M., Boedecker, J.: Dynamic input for deep reinforcement learning in autonomous driving. CoRR abs/1907.10994 (2019). arXiv:1907.01099
8. Kingma, D.P., Ba, J.: Adam: a method for stochastic optimization. CoRR abs/1412.6980 (2014). arXiv:1412.6980
9. Komorowski, M., Celi, L., Badawi, O., Gordon, A., Faisal, A.: The artificial intelligence clinician learns optimal treatment strategies for sepsis in intensive care. Nat. Med. **24**, 1716–1720 (2018). https://doi.org/10.1038/s41591-018-0213-5
10. Kourou, K., Exarchos, T.P., Exarchos, K.P., Karamouzis, M.V., Fotiadis, D.I.: Machine learning applications in cancer prognosis and prediction. Comput. Struct. Biotech. J. **13**, 8–17 (2015). https://doi.org/10.1016/j.csbj.2014.11.005. http://www.sciencedirect.com/science/article/pii/S2001037014000464
11. LeCun, Y., Bengio, Y., Hinton, G.: Deep learning. Nature **521**, 436–44 (2015). https://doi.org/10.1038/nature14539
12. Lee, G., Nho, K., Kang, B., Sohn, K.A., Kim, D.: Predicting alzheimer's disease progression using multi-modal deep learning approach. Sci. Rep. **9**, 1952 (2019). https://doi.org/10.1038/s41598-018-37769-z
13. Li, Y., Rao, S., Solares, J.R.A., Hassaïne, A., Canoy, D., Zhu, Y., Rahimi, K., Khorshidi, G.S.: BEHRT: transformer for electronic health records. CoRR abs/1907.09538 (2019). arxiv:1907.09538
14. Lin, C., Karlson, E.W., Canhao, H., Miller, T.A., Dligach, D., Chen, P.J., Perez, R.N.G., Shen, Y., Weinblatt, M.E., Shadick, N.A., Plenge, R.M., Savova, G.K.: Automatic prediction of rheumatoid arthritis disease activity from the electronic medical records. PLOS ONE **8**(8), 1–10 (2013). https://doi.org/10.1371/journal.pone.0069932
15. Lipton, Z.C., Kale, D.C., Elkan, C., Wetzel, R.C.: Learning to diagnose with LSTM recurrent neural networks. CoRR arxiv:1511.03677 (2015)
16. Liu, L., Li, H., Hu, Z., Shi, H., Wang, Z., Tang, J., Zhang, M.: Learning hierarchical representations of electronic health records for clinical outcome prediction. CoRR abs/1903.08652 (2019). arxiv:1903.08652
17. Liu, L., Shen, J., Zhang, M., Wang, Z., Tang, J.: Learning the joint representation of heterogeneous temporal events for clinical endpoint prediction. CoRR abs/1803.04837 (2018). arxiv:1803.04837
18. van der Maaten, L., Hinton, G.: Visualizing data using t-SNE. J. Mach. Learn. Res. **9**, 2579–2605 (2008). http://www.jmlr.org/papers/v9/vandermaaten08a.html
19. Maragatham, G., Devi, S.: LSTM model for prediction of heart failure in big data. J. Med. Syst. **43**(5), 1–13 (2019). https://doi.org/10.1007/s10916-019-1243-3

20. Miotto, R., Li, L., Kidd, B.A., Dudley, J.T.: Deep patient: an unsupervised representation to predict the future of patients from the electronic health records. Sci. Rep. **6**, 1–10 (2016)
21. Nancy, J.Y., Khanna, N.H., Arputharaj, K.: Imputing missing values in unevenly spaced clinical time series data to build an effective temporal classification framework. Comput. Stat. Data Anal. **112**(C), 63–79 (2017). https://doi.org/10.1016/j.csda.2017.02.012
22. Nguyen, P., Tran, T., Wickramasinghe, N., Venkatesh, S.: Deepr: a convolutional net for medical records (2016)
23. Pham, T., Tran, T., Phung, D., Venkatesh, S.: Predicting healthcare trajectories from medical records: a deep learning approach. J. Biomed. Inform. **69**, 218–229 (2017). https://doi.org/10.1016/j.jbi.2017.04.001. http://www.sciencedirect.com/science/article/pii/S1532046417300710
24. Rajkomar, A., Oren, E., Chen, K., Dai, A., Hajaj, N., Liu, P., Liu, X., Sun, M., Sundberg, P., Yee, H., Zhang, K., Duggan, G., Flores, G., Hardt, M., Irvine, J., Le, Q., Litsch, K., Marcus, J., Mossin, A., Dean, J.: Scalable and accurate deep learning for electronic health records. NPJ Digit. Med. **1**, 18 (2018). https://doi.org/10.1038/s41746-018-0029-1
25. Rubin, D.B.: Inference and missing data. Biometrika **63**(3), 581–592 (1976)
26. Schafer, J.L., Graham, J.W.: Missing data: our view of the state of the art. Psychol. Methods **7**(2), 147–177 (2002)
27. Shickel, B., Tighe, P., Bihorac, A., Rashidi, P.: Deep EHR: a survey of recent advances on deep learning techniques for electronic health record (EHR) analysis. CoRR abs/1706.03446 (2017). arxiv:1706.03446
28. Shiezadeh, Z., Sajedi, H., Aflakie, E.: Diagnosis of rheumatoid arthritis using an ensemble learning approach, pp. 139–148 (2015). https://doi.org/10.5121/csit.2015.51512
29. Uitz, E., Fransen, J., Langenegger, T., Stucki, P.D.M.G.: Clinical quality management in rheumatoid arthritis: putting theory into practice. Rheumatology **39**, 542–549 (2000). https://doi.org/10.1093/rheumatology/39.5.542. swiss clinical quality management in rheumatoid arthritis
30. Vodencarevic, A., Goes, M., Medina, O., de Groot, M., Haitjema, S., Solinge, W., Hoefer, I., Peelen, L., Laar, J., Zimmermann-Rittereiser, M., Hamans, B., Welsing, P.: Predicting flare probability in rheumatoid arthritis using machine learning methods, pp. 187–192 (2018). https://doi.org/10.5220/0006930501870192
31. Zaheer, M., Kottur, S., Ravanbakhsh, S., Poczos, B., Salakhutdinov, R.R., Smola, A.J.: Deep sets. In: Guyon, I., Luxburg, U.V., Bengio, S., Wallach, H., Fergus, R., Vishwanathan, S., Garnett, R. (eds.) Advances in Neural Information Processing Systems, vol. 30, pp. 3391–3401. Curran Associates, Inc. (2017). http://papers.nips.cc/paper/6931-deep-sets.pdf
32. Zhou, J., Cui, G., Zhang, Z., Yang, C., Liu, Z., Sun, M.: Graph neural networks: a review of methods and applications. CoRR abs/1812.08434 (2018). arxiv:1812.08434

DeStress: Deep Learning for Unsupervised Identification of Mental Stress in Firefighters from Heart-Rate Variability (HRV) Data

Ali Oskooei, Sophie Mai Chau, Jonas Weiss, Arvind Sridhar, María Rodríguez Martínez, and Bruno Michel

Abstract In this work we perform a study of various unsupervised methods to identify mental stress in firefighter trainees based on unlabeled heart rate variability data. We collect RR interval time series data from nearly 100 firefighter trainees that participated in a drill. We explore and compare three methods in order to perform unsupervised stress detection: (1) traditional K-Means clustering with engineered time and frequency domain features (2) convolutional autoencoders and (3) long short-term memory (LSTM) autoencoders, both trained on the raw RR data combined with DBSCAN clustering and K-Nearest-Neighbors classification. We demonstrate that K-Means combined with engineered features is unable to capture meaningful structure within the data. On the other hand, convolutional and LSTM autoencoders tend to extract varying structure from the data pointing to different clusters with different sizes of clusters. We attempt at identifying the true stressed and normal clusters using the HRV markers of mental stress reported in the literature. We demonstrate that the clusters produced by the convolutional autoencoders consistently and successfully stratify stressed versus normal samples, as validated by several established physiological stress markers such as RMSSD, Max-HR, Mean-HR and LF-HF ratio.

1 Introduction

It is known that when an individual is exposed to a stressor, the autonomic nervous (ANS) system is triggered resulting in the suppression of the parasympathetic nervous system and the activation of sympathetic nervous system [30]. This reaction which is known as the fight-or-flight response can involve physiological manifestations such as: vasoconstriction of blood vessels, increased blood pressure, increased muscle

A. Oskooei (✉) · S. M. Chau · J. Weiss · A. Sridhar · M. R. Martínez · B. Michel
IBM Research–Zurich, Säumerstrasse 4, 8803 Rüschlikon, Switzerland
e-mail: ali.oskooei@gmail.com

B. Michel
e-mail: bmi@zurich.ibm.com

© The Editor(s) (if applicable) and The Author(s), under exclusive license to Springer Nature Switzerland AG 2021

A. Shaban-Nejad et al. (eds.), *Explainable AI in Healthcare and Medicine*, Studies in Computational Intelligence 914,
https://doi.org/10.1007/978-3-030-53352-6_9

tension and a change in heart rate (HR) and heart rate variability (HRV) [25]. Among these, HRV has become a standard metric for the assessment of the state of body and mind, with multiple markers derived from HRV being routinely used for identifying mental stress or lack thereof [25]. HRV is a time series of the variation of the heart rate over time and is determined by calculating the difference in time between two consecutive occurrences of QRS-complexes, also known as the RR interval (RRI) [31]. An optimal HRV points to healthy physiological function, adaptability and resilience [19]. Increased HRV (beyond normal) may point to a disease or abnormality [19]. Reduced HRV, on the other hand, points to an impaired regulatory capacity and is known to be a sign of stress, anxiety and a number of other health problems [19].

Identifying stress has been the focus of much research as an increasing body of evidence suggests a rising prevalence of stress-related health conditions associated with the stressful contemporary lifestyle [13]. A significant portion of the contemporary stress is due to the occupational stress. Occupational stress can not only result in chronic health conditions such as heart disease [16] but can also have more immediate catastrophic effects such as accidents, injury and even death [23, 28]. Firefighters and smoke divers, in particular, are susceptible to acute stress due to the sensitive nature of their work. It is imperative that mental stress in firefighters is monitored to prevent injury to personnel or the public [2, 10]. In this work, our objective is to leverage HRV data and unsupervised machine learning methods, in order to detect mental stress in firefighters.

With the rise of modern machine learning and deep learning methods, these methods have been applied in the study of heart rate variability. Machine learning and deep learning methods have previously been used with HRV and electrocardiography (ECG) data for various applications such as: fatigue and stress detection [3, 12, 26, 27], congestive heart failure detection [32], cardiac arrhythmia classification [9]. The vast majority of prior arts, however, are supervised or based on labeled public datasets as opposed to unlabeled real-world data. There have been few previous attempts at unsupervised detection of mental stress, from ECG data, using for instance, self-organizing maps (SOMs) [11] or traditional clustering algorithms [20]. To the best of our knowledge, deep learning based unsupervised detection of mental stress in firefighters using short-term HRV data has not been studied before. In this paper, we propose an unsupervised approach using autoencoders and density-based clustering, combined with prior knowledge, in order to cluster and label raw HRV data collected from firefighters.

In this work, we collect RR interval time series data from 100 fire fighter trainees. We break down the collected time series data into 30 measurement windows prior to further processing or modeling. Successful calculation of stress markers from short-term HRV measurements (~10 s) have been previously demonstrated [7, 17, 32]. We then use the HRV samples with both classical clustering methods as well as convolutional and long short-term memory (LSTM) autoencoders to find meaningful clusters in the context of mental stress detection.

2 Results and Discussion

2.1 *Classical Feature Engineering and K-Means*

As a first attempt at unsupervised classification of HRV data, we employed K-means clustering on 18 engineered features (see supplementary table S1). The K-means clustering results for k = 2 are shown in Fig. 1C, plotted in two dimensions for mean heartrate (MeanHR) and root mean square of successive differences (RMSSD), two of the top reported biomarkers of stress in the literature [4, 29]. As shown in the figure, the identified clusters appear synthetic without a clear separation in the data. This is in part due to the high dimensionality of the data [1] as well as the intrinsic tendency of K-means clustering algorithm to cluster samples based on the Euclidean distance from the centroids of clusters, regardless of true separation within the data [22].

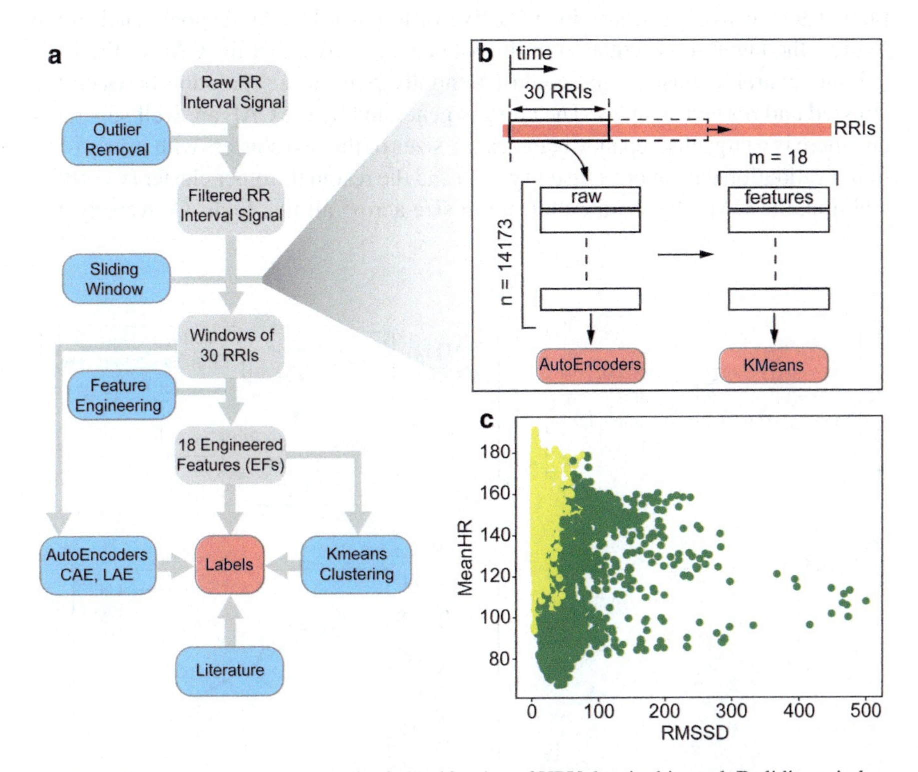

Fig. 1 **A** The workflow for unsupervised classification of HRV data in this work **B** sliding window transformation of the RR intervals into non-overlapping windows of 30 measurements and **C** the results of K-means clustering with $k = 2$ on the samples using the 18 engineered features. As seen in the plot K-means does not produce distinct well-separated clusters within the data

2.2 Convolutional and LSTM AutoEncoders

Having observed the inability of K-means clustering in finding meaningful structure in the data, we explored autoencoders to compress the data and find meaningful structures that we can leverage to identify mental stress. We trained and evaluated two autoencoder architectures: a convolutional autoencoder (CAE) and a LSTM autoencoder (LAE). The architectures of the two models are shown in Figs. 2A and 2B.

We adopted a 5-fold cross-validation scheme and trained five models for each autoencoder. The reconstruction errors across all fold for both CAE and LAE are shown in Fig. 2C. The LAE has slightly lower reconstruction error and results in a smoother reconstructed signal. A reconstructed validation sample using both CAE and LAE is presented in Fig. 2D.

Figures 3 and 4 show the training and validation results for the CAE and LAE models. On the top left corner of the figures, the validation error during training is plotted for all five models trained for five different folds. As demonstrated in the figures, the latent representation of the data encoded by either the CAE or the LAE exhibit separable clusters that could potentially point to a separation between the stressed and normal samples. The clusters generated by the CAE are well-separated and there is a large discrepancy between the sizes of the two clusters with nearly ~8% of the validation data in one cluster (y = 1) and the rest in the other cluster (y = 0). In addition, the CAE clusters are uniform in size across all five folds. Conversely, the

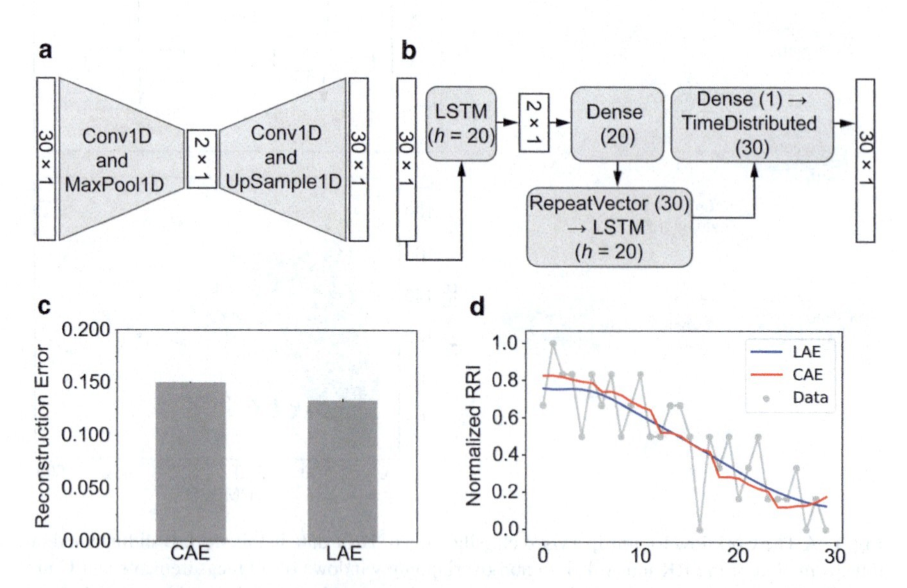

Fig. 2 The architecture and comparison of convolutional and LSTM autoencoders. **A** The convolutional autoencoder with a 2D bottleneck consisting of 1D convolutions, maxpooling and upsampling layers. **B** The LSTM autoencoder with a 2D bottleneck consisting of LSTM cells hidden dimension of 20. **C** The reconstruction error (i.e., MAE) on the validation datasets across five folds for the CAE and the LAE. **D** Reconstruction of a sample from the validation dataset by both the CAE and LAE

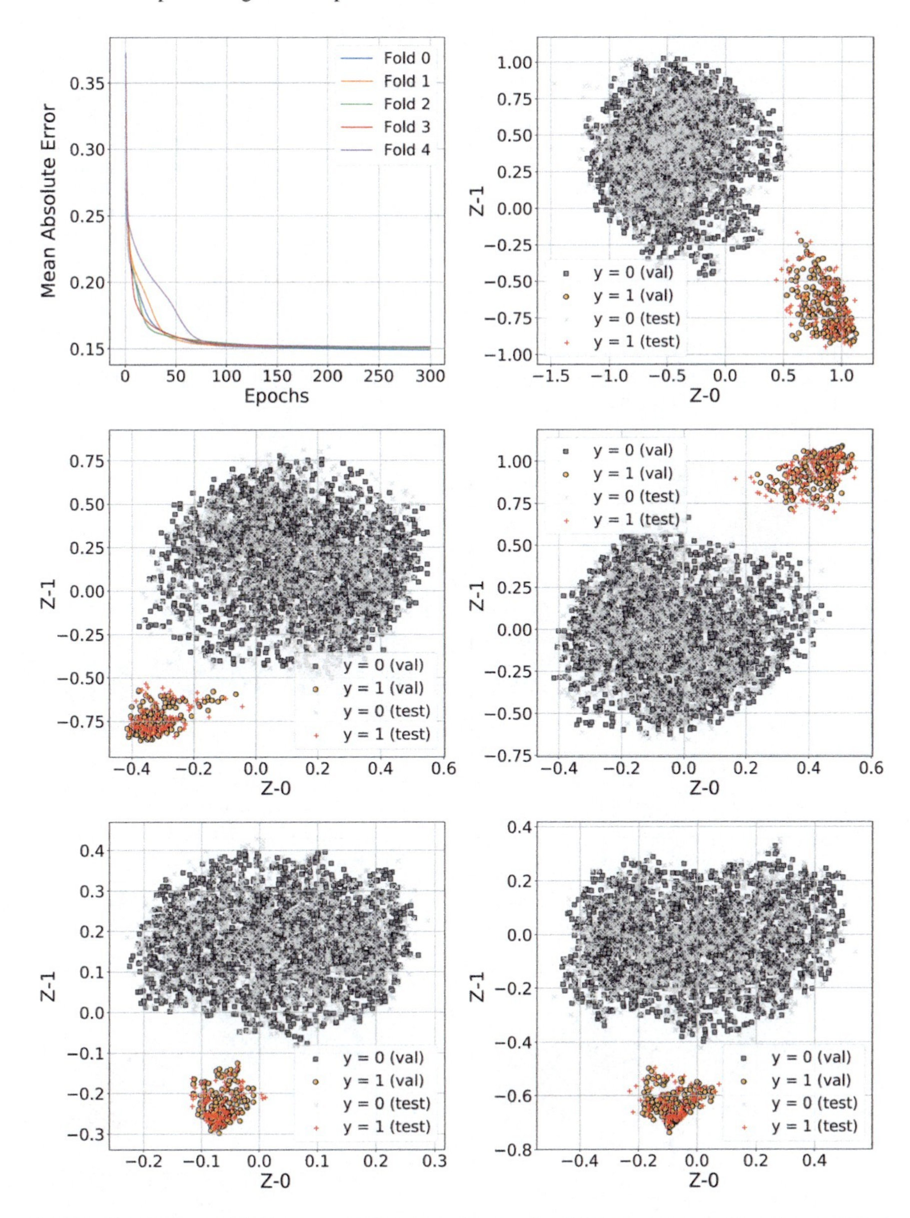

Fig. 3 Training and 5-fold cross-validation results for the CAE model as well as the predicted labels for the test dataset. The plot at the top left illustrates the validation error during the training of CAE in each fold. The scatter plots show the two clusters identified using DBSCAN clustering for the encoded validation data. We arbitrarily label the clusters "0" and "1". Cluster "0" contains the majority of the data (~90%) while the rest belong to cluster "1". The encoded test data and their KNN-predicted labels are superimposed with the validation clusters. As shown in the scatter plots, for all five folds, the encoded test data follow the same pattern as the validation data

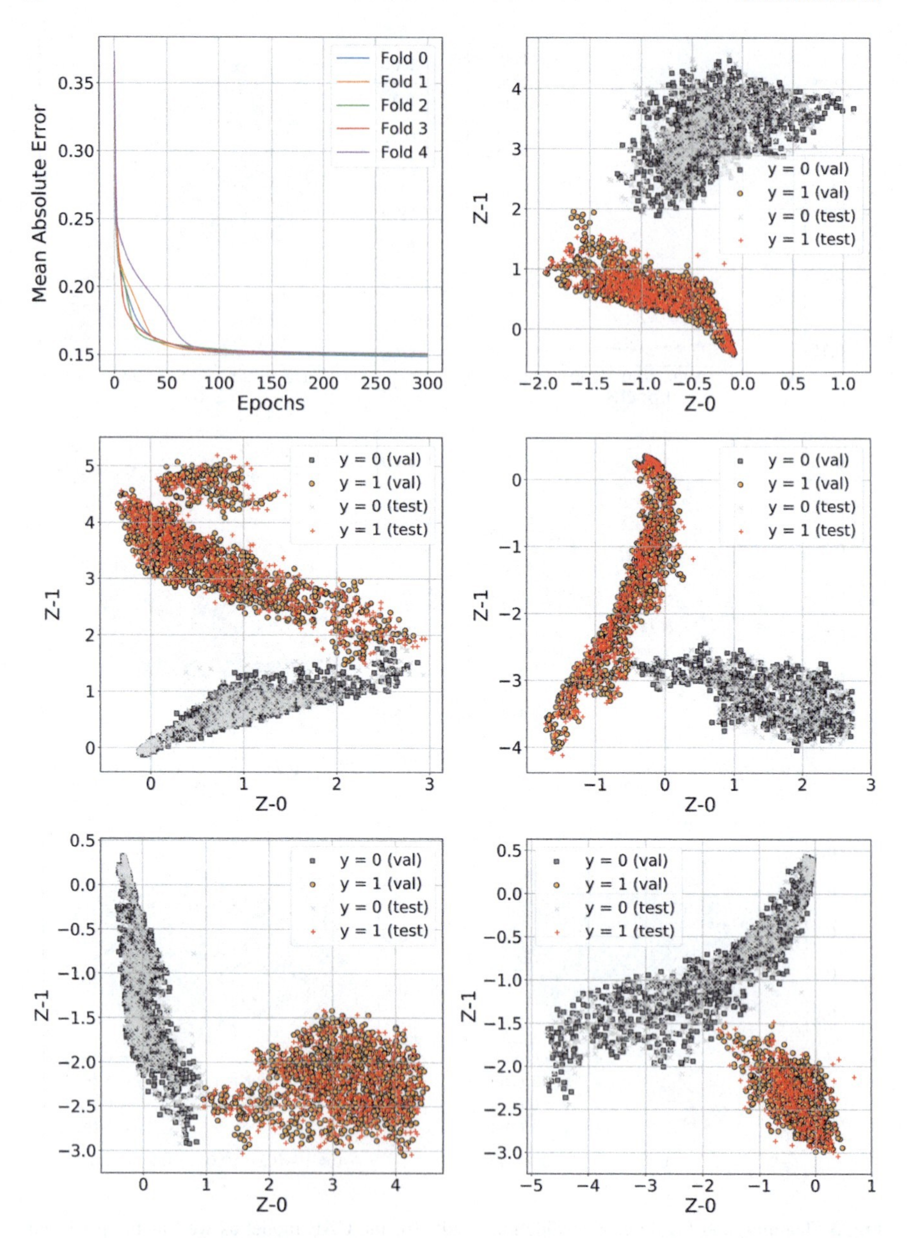

Fig. 4 Training and 5-fold cross-validation results for the LAE model as well as the predicted labels for the test dataset. The plot at the top left illustrates the validation error during the training of LAE in each fold. The scatter plots show the two clusters identified using DBSCAN clustering for the encoded validation data. We arbitrarily labeled the clusters "0" and "1". In comparison with the CAE results, the LAE-encoded clusters exhibit poor separation. In addition, the cluster sizes are more balanced with significant variability across different folds. The encoded test data and their KNN-predicted labels are superimposed with the validation clusters. As shown in the scatter plots, for all five folds, the encoded test data follow the same pattern as the validation data

LAE clusters are more balanced in size and the separation between the clusters is not significant. In addition, the sizes of LAE clusters are variable across different folds. The poor separation combined with the variable cluster size suggests that the LAE clusters do not represent a reproducible underlying structure within the data. We will investigate the meaningfulness of the clusters produced by each model shortly.

As described in Sect. 4, in order to classify unseen data, we used a KNN classifier with the clusters and labels obtained from the validation results. The label prediction results on the unseen test samples are shown in Figs. 3 and 4 for the CAE and LAE models respectively. As shown in the Fig. 3, the test samples encoded by the CAE follow the same cluster pattern as the validation data with the majority of the samples in cluster "0" and the remaining samples in cluster "1".

Similarly, as seen in Fig. 4, the test samples encoded by the LAE follow the same pattern as the validation clusters. Having the test clusters in hand, the question was whether any of the CAE or LAE encoded clusters are meaningful in the context of mental stress detection. To address this question, we resorted to established HRV markers of mental stress and calculated and compared them across the two clusters for each of the CAE and LAE models. The objective was to determine whether samples in each cluster share specific characteristics pertaining mental stress that separate them from the samples in the other cluster.

The mean RMSSD, a reported biomarker for mental stress [4, 13], is compared across the two models and clusters in Fig. 5. As shown in the figure, there is a significant (two-sided t-test, p-value $= 2e-6$) difference between the mean RMSSD for cluster "0" and cluster "1" of the test data encoded by the CAE. Cluster "0" (the grey bar) demonstrates a significantly lower RMSSD compared with cluster "1". The low RMSSD may signify low vagal tone and mental stress in cluster "0", as reported in the literature [5]. Conversely, cluster "1" has high RMSSD, and as such a higher vagal tone indicating normal parasympathetic function [18].

On the other hand, the difference in mean RMSSD between the two LAE clusters are insignificant (two-sided t-test, p-value $= 0.22$), as shown in Fig. 5. The low discrepancy between the RMSSD of the LAE clusters further solidifies the hypothesis that the LAE-encoded clusters are not meaningful in the context of mental stress detection while the CAE-encoded clusters successfully stratify stressed versus normal samples. To further verify this hypothesis, we calculated and compared other HRV markers of stress across different clusters and models.

The remaining barplots in Fig. 5 demonstrate a comparison between three other crucial HRV markers across the clusters given by the CAE and LAE model. Three features namely, maximum heart rate (Max-HR), mean RR interval (Mean-RR) and low frequency (LF) to high frequency (HF) ratio (LF-HF Ratio) were selected based on their reported importance in detecting mental stress from HRV data [4, 17, 29]. As shown in Fig. 5, the CAE values show a statistically significant discrepancy between the two clusters for all three features: Max-HR (p-value $= 7.2e-11$), Mean-RR (p-value $= 3.7e-8$) and LF-HF Ratio (p-value $= 4.8e-10$). We observed that Max-HR is higher in CAE cluster "1" versus cluster "0" which is consistent with the reported correlation between Max-HR and RMSSD in the literature [25]. Interestingly, Mean-RR is lower in cluster "1" compared to cluster "0". This is while Mean-RR has been

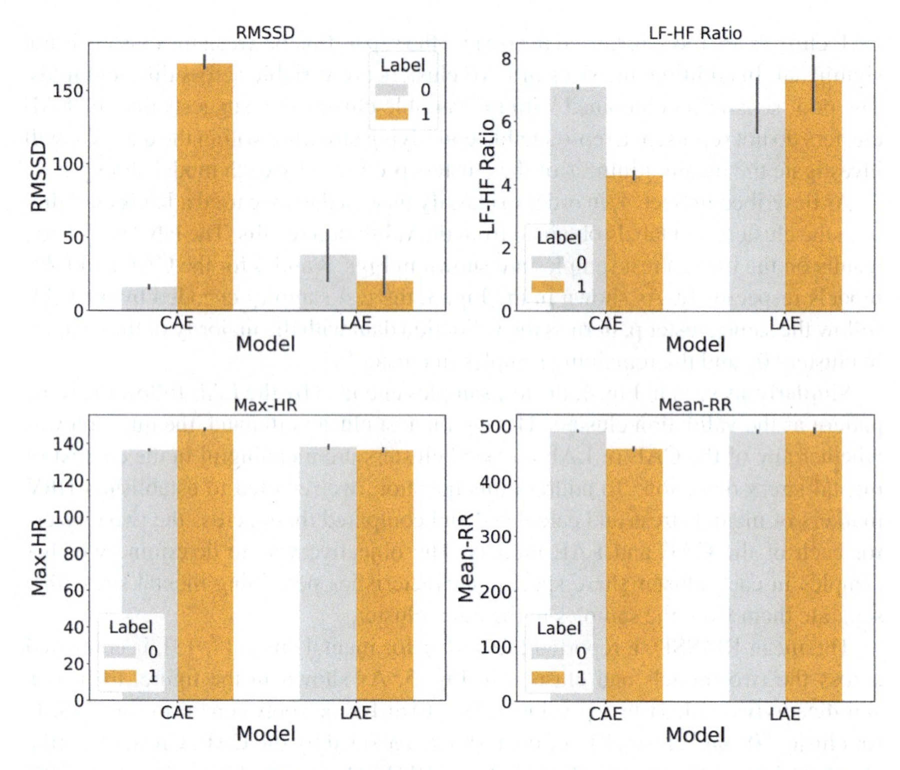

Fig. 5 Comparison of four HRV markers of stress across the clusters given by the CAE and LAE models. The CAE clusters show significant discrepancy in the HRV markers between the two clusters while the LAE clusters exhibit insignificant difference in the four HRV markers. These results further confirm our hypothesis that the CAE clusters are reproducible and relevant in the context of mental stress detection while the LAE clusters fail to stratify the samples in a meaningful manner

reported to decrease with the induction of mental stress compared with the resting state [30]. However, it should be noted that in our experiment, none of the subjects are in the resting state as they are all actively participating in a drill. LF-HF ratio is higher in cluster 0 than in cluster "1". Increased LF-HF ratio has been linked to mental stress in the literature [31]. In summary, two of the three HRV markers of stress suggest impaired vagal tone in cluster "0" and normal vagal tone for cluster "1". This is in agreement with the RMSSD results discussed earlier. Based on the results in Figs. 3 and 5, we postulate that the CAE cluster "1" corresponds to individuals that are physically stressed, as indicated by a high Max-HR and a slightly lower Mean-RR, but mentally relaxed, as indicated by a much higher RMSSD and a lower LF-HF ratio. CAE cluster "0", on the other hand, corresponds to samples experiencing both physical (i.e. high Max-HR and low Mean-RR) and mental stress (low RMSSD and high LF-HF Ratio).

For LAE clusters, on the other hand, HRV markers of stress do not exhibit significant discrepancy between the clusters: Max-HR (p-value = 0.05), Mean-RR (p-value

= 0.66) and LF-HF Ratio (p-value = 0.22). As shown in Fig. 5, the barplots corresponding to the LAE clusters and their associated p-values, do not demonstrate a meaningful stratification of test samples based on the evaluated markers of mental stress. The results of Fig. 5 for the LAE clusters are in line with our earlier observations and further confirm our hypothesis that LAE clusters do not appear meaningful in the context of mental stress detection.

3 Data Acquisition

The RR intervals were collected from 100 firefighters and smoke divers who volunteered to participate in this study. This study was conducted in March 2018 at the cantonal civil emergency services training center of the Canton Aargau at Eiken AG in Switzerland where firefighters participated in a weeklong training exercise to be certified as smoke diving team leaders. Among the various training exercises they participated in was the "Hotpot", a darkened chamber with a 3D maze made of cages, which simulated a building structure on fire. The exercise subjected the participants to a period (10–20 min) of intense stress, both physical and mental. The participants were given heart rate measuring chest belts (Polar H10) and "communication hub" smartphones to carry before they entered the maze. The RR interval time series were extracted from the ECG records. No ground truth regarding the mental stress status of the firefighters was given for the collected RR intervals. The outliers in the data were replaced with their nearest normal neighbors (i.e. Winsorized). The cleaned time-series data were then divided into non-overlapping windows of size 30 and shuffled as shown in Fig. 1B. 10% of the data was randomly selected and held out as the unseen test dataset and the remaining 90% was split into training and validation datasets according to a 5-fold cross-validation scheme. To enhance optimization and convergence of AEs, each raw RR interval window of size 30 was scaled between 0 and 1 using min-max scaling. Time and frequency domain HRV features were extracted using hrv-analysis 1.0.3 python library. The time and frequency domain features extracted and used in this work were: HFms, HFnu, HFpeak, HFrel, LF_HF Ratio, LFms, LFnu, LFpeak, LFrel, MaxRR, MeanHR, MeanRR, MinRR, NN50, RMSSD, SDNN, VLFms, pNN50. A detailed description of time and frequency domain HRV features can be found elsewhere [25].

4 Methods

4.1 K-Means Clustering

The traditional K-means clustering algorithm [14] was used to cluster the data transformed into 18 engineered features. We employed the K-means clustering algorithm implemented in Scikit-Learn Python library [21] with k = 2 and all other parameters set as default. However, common distance metrics, such as the Euclidean distance used in K-means, are not useful in finding similarity in high-dimensional data [1]. As a result, we explored autoencoders (AEs) as way to compress the raw samples into a lower dimensional latent space (2D in our case), and then search for patterns or clusters within the compressed (or encoded) data. We discuss autoencoders in detail in the following subsection.

4.2 AutoEncoders

Autoencoders are neural nets that ingest a sample, x, and attempt to reconstruct the sample at the output. When the autoencoder involves a hidden layer, h, that is of lower dimension than x, it is called an undercomplete autoencoder. The idea is to encode the data into a lower dimensional, h, which contains the most salient features of the data. The learning process of an autoencoder involves minimizing a loss function, J:

$$J = L(g(f(x)) \; where \; h = f(x) \tag{1}$$

where f is the encoder, g is the decoder and L is a loss function that penalizes g(h) for being dissimilar to x [8]. We explored both mean squared error (MSE) and mean absolute error (MAE) as the loss function, L, and found mean absolute error to offer better convergence and lower reconstruction error compared with MSE.

As shown in Fig. 2A, in our convolutional autoencoder (CAE), both the encoder and the decoder consisted of four 1-dimensional convolution layers (kernel size = 2) and two maxpooling layers for the encoder and two upsampling layers for the decoder. Relu activation was used in all convolutional layers. The total number of trainable weights for the CAE was 710.

The architecture of the LSTM autoencoder (LAE) is shown in Fig. 2B. The encoder consists of a LSTM layer with hidden dimension of 20 followed by a linear transformation to the 2D bottleneck. The decoder consists of a dense transformation of the 2D bottleneck to 20 dimensions followed by vector repetitions (30 times) and a LSTM layer followed by a dense layer that reconstructs the input sample. Elu activation [6] was used in both the encoder and the decoder of the LAE. The total number of trainable weights for the LAE was 5163 which is an order of magnitude higher than the CAE.

4.3 DBSCAN Clustering

We employed DBSCAN clustering algorithm [24] to identify clusters in the latent representation of HRV data given by the AEs. DBSCAN is a density-based algorithm that clusters densely-packed samples together while disregarding samples in low-density areas as outliers.

4.4 Training and Evaluation

The AEs were implemented using tensorflow 1.12.0 (tf.keras) deep learning library. Adam optimizer [15] with a learning rate of $1e-4$ and a batch size of 64 was used to train the AEs. 5-fold cross validation scheme was used to train the models and tune the hyperparamters. Both CAE and LAE were trained for 300 epochs. CAE loss plateaued after nearly 150 epochs while LAE plateaued much later at about 290 epochs. Both models were trained using a virtual machine with a 12-core CPU and 24 GB of RAM.

5 Conclusions

We presented a new approach for unsupervised detection of mental stress from raw HRV data using autoencoders. We demonstrated that classical K-means clustering combined with time and frequency domain features was not suitable for identifying mental stress. We then explored two different architectures of autoencoders to encode the data and find underlying patterns that may enable us to detect mental stress in an unsupervised manner. We trained convolutional and LSTM autoencoders and demonstrated that despite being more powerful and producing lower reconstruction error, LSTM autoencoders failed to identify useful patterns within the data. On the other hand, the convolutional autoencoders with their much fewer trainable weights, produced clusters that were verifiably distinct and pointed to different levels of mental stress according to the reported markers of mental stress. Based on the results given by the convolutional autoencoder, more than 90% of the samples collected from firefighter trainees during a drill were mentally stressed while less than 10% had normal HRV. As a future direction, it is imperative that the observed results in this work are thoroughly validated via new experiments.

References

1. Aggarwal, C.C., Hinneburg, A., Keim, D.A.: On the surprising behavior of distance metrics in high dimensional space. Presented at the International Conference on Database Theory, pp. 420–434. Springer (2001)
2. Beaton, R., Murphy, S., Pike, K., Jarrett, M.: Stress-symptom factors in firefighters and paramedics (1995)
3. Bhardwaj, R., Natrajan, P., Balasubramanian, V.: Study to determine the effectiveness of deep learning classifiers for ECG based driver fatigue classification. Presented at the 2018 IEEE

13th International Conference on Industrial and Information Systems (ICIIS), pp. 98–102. IEEE (2018)
4. Blásquez, J.C.C., Font, G.R., Ortís, L.C.: Heart-rate variability and precompetitive anxiety in swimmers. Psicothema **21**, 531–536 (2009)
5. Camm, A.J., Malik, M., Bigger, J.T., Breithardt, G., Cerutti, S., Cohen, R.J., Coumel, P., Fallen, E.L., Kennedy, H.L., Kleiger, R.: Heart rate variability: standards of measurement, physiological interpretation and clinical use. Task Force of the European Society of Cardiology and the North American Society of Pacing and Electrophysiology (1996)
6. Clevert, D.A., Unterthiner, T., Hochreiter, S.: Fast and accurate deep network learning by exponential linear units (elus). arXiv preprint arXiv:1511.07289 (2015)
7. Giannakakis, G., Grigoriadis, D., Giannakaki, K., Simantiraki, O., Roniotis, A., Tsiknakis, M.: Review on psychological stress detection using biosignals. IEEE Trans. Affect. Comput., 1 (2019)
8. Goodfellow, I., Bengio, Y., Courville, A.: Deep Learning. MIT Press, Cambridge (2016)
9. Hannun, A.Y., Rajpurkar, P., Haghpanahi, M., Tison, G.H., Bourn, C., Turakhia, M.P., Ng, A.Y.: Cardiologist-level arrhythmia detection and classification in ambulatory electrocardiograms using a deep neural network. Nat. Med. **25**, 65 (2019)
10. Harris, M.B., Baloğlu, M., Stacks, J.R.: Mental health of trauma-exposed firefighters and critical incident stress debriefing. J. Loss Trauma **7**, 223–238 (2002)
11. Huysmans, D., Smets, E., De Raedt, W., Van Hoof, C., Bogaerts, K., Van Diest, I., Helic, D.: Unsupervised learning for mental stress detection. Presented at the Proceedings of the 11th International Joint Conference on Biomedical Engineering Systems and Technologies, pp. 26–35 (2018)
12. Hwang, B., You, J., Vaessen, T., Myin-Germeys, I., Park, C., Zhang, B.-T.: Deep ECGNet: an optimal deep learning framework for monitoring mental stress using ultra short-term ECG signals. Telemed. E-Health **24**, 753–772 (2018)
13. Järvelin-Pasanen, S., Sinikallio, S., Tarvainen, M.P.: Heart rate variability and occupational stress-systematic review. Ind. Health **56**, 500–511 (2018)
14. Kanungo, T., Mount, D.M., Netanyahu, N.S., Piatko, C.D., Silverman, R., Wu, A.Y.: An efficient k-means clustering algorithm: analysis and implementation. IEEE Trans. Pattern Anal. Mach. Intell. **24**, 881–892 (2002)
15. Kingma, D.P., Ba, J.: Adam: a method for stochastic optimization. arXiv preprint arXiv:1412.6980 (2014)
16. Kivimäki, M., Leino-Arjas, P., Luukkonen, R., Riihimäi, H., Vahtera, J., Kirjonen, J.: Work stress and risk of cardiovascular mortality: prospective cohort study of industrial employees. BMJ **325**, 857 (2002)
17. Salahuddin, L., Cho, J., Jeong, M.G., Kim, D.: Ultra short term analysis of heart rate variability for monitoring mental stress in mobile settings. In: 2007 29th Annual International Conference of the IEEE Engineering in Medicine and Biology Society. Presented at the 2007 29th Annual International Conference of the IEEE Engineering in Medicine and Biology Society, pp. 4656–4659 (2007)
18. Laborde, S., Mosley, E., Thayer, J.F.: Heart rate variability and cardiac vagal tone in psychophysiological research-recommendations for experiment planning, data analysis, and data reporting. Front. Psychol. **8**, 213 (2017)
19. McCraty, R., Shaffer, F.: Heart rate variability: new perspectives on physiological mechanisms, assessment of self-regulatory capacity, and health risk. Glob. Adv. Health Med. **4**, 46–61 (2015)
20. Medina, L.: Identification of stress states from ECG signals using unsupervised learning methods. Presented at the Portuguese Conference on Pattern Recognition-RecPad (2009)
21. Pedregosa, F., Varoquaux, G., Gramfort, A., Michel, V., Thirion, B., Grisel, O., Blondel, M., Prettenhofer, P., Weiss, R., Dubourg, V.: Scikit-learn: machine learning in Python. J. Mach. Learn. Res. **12**, 2825–2830 (2011)
22. Raykov, Y.P., Boukouvalas, A., Baig, F., Little, M.A.: What to do when k-means clustering fails: a simple yet principled alternative algorithm. PLoS ONE **11**, e0162259 (2016). https://doi.org/10.1371/journal.pone.0162259

23. Salminen, S., Kivimäki, M., Elovainio, M., Vahtera, J.: Stress factors predicting injuries of hospital personnel. Am. J. Ind. Med. **44**, 32–36 (2003)
24. Schubert, E., Sander, J., Ester, M., Kriegel, H.P., Xu, X.: DBSCAN revisited, revisited: why and how you should (still) use DBSCAN. ACM Trans. Database Syst. (TODS) **42**, 19 (2017)
25. Shaffer, F., Ginsberg, J.P.: An overview of heart rate variability metrics and norms. Front. Public Health **5**, 258 (2017)
26. Smets, E., Casale, P., Großekathöfer, U., Lamichhane, B., De Raedt, W., Bogaerts, K., Van Diest, I., Van Hoof, C.: Comparison of machine learning techniques for psychophysiological stress detection. Presented at the International Symposium on Pervasive Computing Paradigms for Mental Health, pp. 13–22. Springer (2015)
27. Song, S.H., Kim, D.K.: Development of a stress classification model using deep belief networks for stress monitoring. Healthc. Inform. Res. **23**, 285–292 (2017)
28. Soori, H., Rahimi, M., Mohseni, H.: Occupational stress and work-related unintentional injuries among Iranian car manufacturing workers (2008)
29. Sun, F.T., Kuo, C., Cheng, H.T., Buthpitiya, S., Collins, P., Griss, M.: Activity-aware mental stress detection using physiological sensors. In: Gris, M., Yang, G. (eds.) Mobile Computing, Applications, and Services, pp. 282–301. Springer, Heidelberg (2012)
30. Taelman, J., Vandeput, S., Spaepen, A., Van Huffel, S.: Influence of mental stress on heart rate and heart rate variability. Presented at the 4th European Conference of the International Federation for Medical and Biological Engineering, pp. 1366–1369. Springer (2009)
31. von Rosenberg, W., Chanwimalueang, T., Adjei, T., Jaffer, U., Goverdovsky, V., Mandic, D.P.: Resolving ambiguities in the LF/HF ratio: LF-HF scatter plots for the categorization of mental and physical stress from HRV. Front. Physiol. **8**, 360 (2017). https://doi.org/10.3389/fphys.2017.00360
32. Wang, L., Zhou, X.: Detection of congestive heart failure based on LSTM-based deep network via short-term RR intervals. Sensors **19**, 1502 (2019)

A Deep Learning Approach for Classifying Nonalcoholic Steatohepatitis Patients from Nonalcoholic Fatty Liver Disease Patients Using Electronic Medical Records

Pradyumna Byappanahalli Suresha, Yunlong Wang, Cao Xiao, Lucas Glass, Yilian Yuan, and Gari D. Clifford

Abstract Nonalcoholic Steatohepatitis (NASH), an advanced stage of Nonalcoholic Fatty Liver Disease (NAFLD) causes liver inflammation and can lead to cirrhosis. In this paper, we present a deep learning approach to identify patients at risk of developing NASH, given that they are suffering from NAFLD. For this, we created two sub cohorts within NASH (NASH *suspected* (NASH-S) and NASH *biopsy-confirmed* (NASH-B)) based on the availability of liver biopsy tests. We utilized medical codes from patient electronic medical records and augmented it with patient demographics to build a long short-term memory based NASH vs. NAFLD classifier. The model was trained and tested using five-fold cross-validation and compared with baseline models including XGBoost, random forest and logistic regression. An out-of-sample area under the precision-recall curve (AUPRC) of 0.61 was achieved in classifying NASH patients from NAFLD. When the same model was used to classify out-of-sample NASH-B cohort from NAFLD patients, a highest AUPRC of 0.53 was achieved which was better than other baseline methods.

Keywords Nonalcoholic Steatohepatitis · Nonalcoholic Fatty Liver Disease · Electronic medical records · Deep learning

---

P. B. Suresha (✉) · G. D. Clifford
Georgia Institute of Technology, Atlanta, GA 30332, USA
e-mail: pradyumna.suresha@gmail.com

Y. Wang · L. Glass · Y. Yuan
IQVIA, 1 IMS Drive, Plymouth Meeting, PA 19462, USA

C. Xiao
IQVIA, Boston, MA 02210, USA

G. D. Clifford
Emory University, Atlanta, GA 30322, USA

© The Editor(s) (if applicable) and The Author(s), under exclusive license to Springer Nature Switzerland AG 2021

A. Shaban-Nejad et al. (eds.), *Explainable AI in Healthcare and Medicine*, Studies in Computational Intelligence 914,
https://doi.org/10.1007/978-3-030-53352-6_10

1 Introduction

Nonalcoholic Steatohepatitis or simply NASH is a liver condition in which fat buildup in the liver causes inflammation and liver damage [6, 13]. An advanced stage in the Nonalcoholic Fatty Liver disease (NAFLD) spectrum, NASH occurs in people with no alcohol abuse history and can worsen to cause scarring of liver, leading to cirrhosis [4]. Typically patient medical history and physical exams including blood tests, imaging and liver biopsy are used for diagnosing NASH [3]. In the United States (US) NASH affects somewhere between 1.5–6.45% of the population whereas NAFLD has a prevalence of 24% [14]. Recently, much research is being done to discover new drugs that can cure them [8, 9, 12] and clinical trials to confirm the efficacy of these discovered drugs in curing the ailment is on the rise. For this, screening for valid patients suffering from the ailment is paramount. As manual selection of these patients is strenuous and not scalable, automated patient screening tools are need of the hour. In this regard, we attempt to develop a patient screening tool for NASH given a list of patients already suffering from NAFLD. Specifically, we used the electronic medical record (EMR) data of patients to train a classifier that yields the probability of a patient developing NASH in the next month, given that the patient is already suffering from NAFLD. Much has been done to develop machine learning techniques to deal with EMR data [1, 2, 10, 11]. Similar to these works, we used a deep learning approach based on long short-term memory (LSTM) [5] to train the NASH classifier. In conjunction to modeling longitudinal EMR data via LSTM, we utilized patient demographics information to train a 'demographics network' (see Fig. 1) which aided classification.

2 Data and Method

In this work, we had three separate cohorts. The first cohort was the set of patients who had diagnosis codes corresponding to NAFLD but did not have diagnosis codes corresponding to NASH in their database. In all our experiments, this cohort was the negative cohort and is referred as NAFLD cohort. The second cohort was the group of patients who had a diagnosis code for NASH and NAFLD but did not have a record for liver biopsy in their database. We call this the NASH *suspected* or NASH-S cohort. The final set was the group of patients who not only had diagnosis codes for NASH and NAFLD but also had a positive liver biopsy test within six months of their NASH diagnosis. We call this as the NASH *biopsy-confirmed* or the NASH-B cohort. The NASH-B and NASH-S cohort when combined and considered as a single cohort, is referred to as NASH cohort. Note that all NASH patients always had a NAFLD diagnosis code as NASH is an advanced stage in the NAFLD spectrum. For model development, a patient list was created by applying the following rules sequentially. First, we included all patients whose database had an 'International Classification of Diseases' diagnosis code for NASH or NAFLD. Next, we discarded all patients

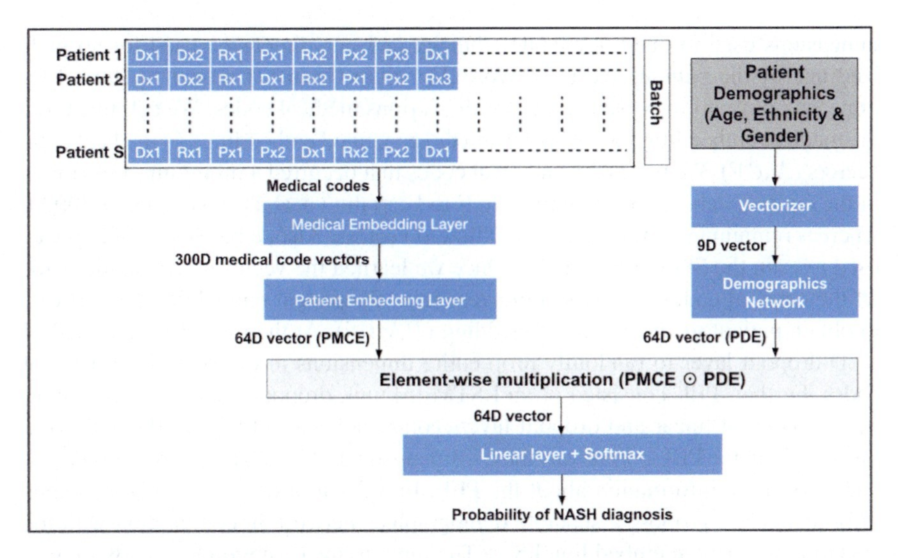

Fig. 1 Deep learning architecture for NASH vs. NAFLD classification. $\mathcal{PMCE}$ stands for patient medical code embedding and $\mathcal{PDE}$ stands for patient demographics embedding. Symbol '$\odot$' signifies element-wise multiplication. A number 'N' followed by the letter 'D' is used to represent 'N-dimensional'. Longitudinal patient EMRs are fed through the medical embedding layer and patient embedding layer to obtain the 64-dimensional vector $\mathcal{PMCE}$. Parallelly, patient demographics is fed through the demographics network to obtain the 64-dimensional $\mathcal{PDE}$. An element-wise multiplication of these two vectors, when passed via a linear layer with softmax non-linearity, yields the probability of the given patient belonging to NASH class

with less than five years of EMR data before the disease diagnosis. In the end, we obtained 172 NASH-B patients and 15284 NASH-S patients giving us a total of 15456 NASH patients. In the pool of NAFLD patients obtained, we randomly sampled 61824 patients to maintain the following proportion–NASH : NAFLD :: 1: 4. This proportion conforms with the prevalence rate of the two diseases in the US. For all these patients, we pulled diagnosis (Dx), procedure (Px) and drug (Rx) codes (together called medical codes) from the EMR tables. To prevent possible data-leakage, we only considered patient medical codes at least a month before the diagnosis of their respective ailment and manually deleted all medical codes corresponding to the diagnosis of NAFLD, NASH or liver biopsy test.

3 Proposed Model Framework

We propose a recurrent neural network based approach to perform classification. For a given patient $\mathcal{S}$, along with the longitudinal medical codes say $\mathcal{L} =$ [Dx1, Dx2, Rx1, Px1, Rx2, Px2, ...] we had a feature set that contained the patient demographics information $\mathcal{D}$. Specifically, patient demographics was a nine-dimensional vector with two dimensions used to one-hot encode the gender, six dimensions used to one-hot encode ethnicity and one-dimension used for normalized age of the patient. We used Word2Vec Skip-Gram [7] model to learn 300-dimensional vector representations for the various medical codes. We call this layer the medical embedding layer (MEL) and the generated embeddings as medical code vectors ($\mathcal{MCV}$). We used those medical codes that occurred a minimum of 30 times in the entire database for learning $\mathcal{MCV}$ to keep the vocabulary size under 10000, whereas remaining codes were set to all-zero vectors. Also, a batch size of 256 was used to train the Skip-Gram model. Once we learned the vector representations for all the medical codes, we passed the vectors to the patient embedding layer (PEL) to obtain patient medical code embedding ($\mathcal{PMCE}$). At the top, PEL consisted of a 2D-dropout layer to randomly drop entire dimensions in the vectorized medical codes. Further, PEL consisted of an LSTM [5] unit, dropout layer, Max Pool unit, and a couple of linear and dropout layers connected sequentially. In the end, for a given patient, the PEL output was the 64-dimensional $\mathcal{PMCE}$. Please refer to Figs. 1 and 2 for more information about the PEL. In order to leverage the demographics information, we propose a separate demographics network in conjunction with the LSTM based patient embedding layer. The input to the demographics network was the nine-dimensional patient demographics vector discussed earlier. The output of the demographics network was a 64-dimensional vector called the patient demographics embedding ($\mathcal{PDE}$) which matched the dimensions of $\mathcal{PMCE}$. In Fig. 2 the block diagram of demographics network is shown with more details. An element-wise multiplication between $\mathcal{PMCE}$ and $\mathcal{PDE}$ was performed to combine demograph-

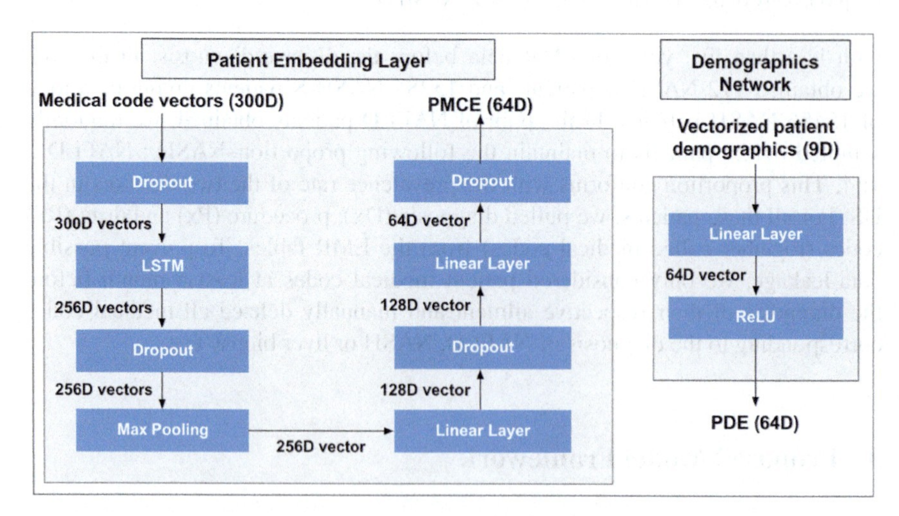

Fig. 2 A close-up view of the patient embedding layer and the demographics network. Notations- 'D' stands for dimensional, $\mathcal{PMCE}$ stands for patient medical code embedding and $\mathcal{PDE}$ stands for patient demographics embedding. The patient embedding layer takes in a 300-dimensional medical code vector via the different linear and non-linear units manipulating the vector size along the way to yield the 64-dimensional $\mathcal{PMCE}$. Demographics network takes in the 9-dimensional vectorized patient demographics and converts it into the 64-dimensional $\mathcal{PDE}$ vector via a fully-connected linear layer and a rectified linear unit

ics information with the longitudinal medical codes. We fed this vector to a linear layer with a softmax non-linearity to obtain the probability of the given patient being diagnosed with NASH in a month.

4 Experiment

In the two experiments performed, we used five-fold cross-validation and computed out-of-sample area under the precision-recall curve (AUPRC) and out-of-sample area under the receiver operating characteristic (AUROC) to assess performance. We compared the performance of the proposed model with three baseline models, namely XGBoost (XGB), random forest and logistic regression. For all the baseline models, we used bag-of-words of the patient medical codes augmented with nine-dimensional demographics vector as features. Further, hyperparameter tuning was done using a subset of the training data. In the training phase, NASH (NASH-S+NASH-B) was the positive cohort and NAFLD was the negative cohort. In the testing phase, we had two settings. First, we compared the performances of different models in classifying NASH (NASH-S+NASH-B) from the NAFLD class. In terms of AUPRC, XGB performed the best with an AUPRC of 0.64. The proposed deep learning model was a close second with an AUPRC of 0.61. Next, we compared the performance of classifying NASH-B from the NAFLD cohort. In this experiment, the proposed model performed better than the baseline models with the highest AUPRC = 0.53. This experiment revealed model's ability to identify biopsy-confirmed NASH patients. Figure 3 illustrates the precision-recall curves and receiver operating characteristic curves for both experiments.

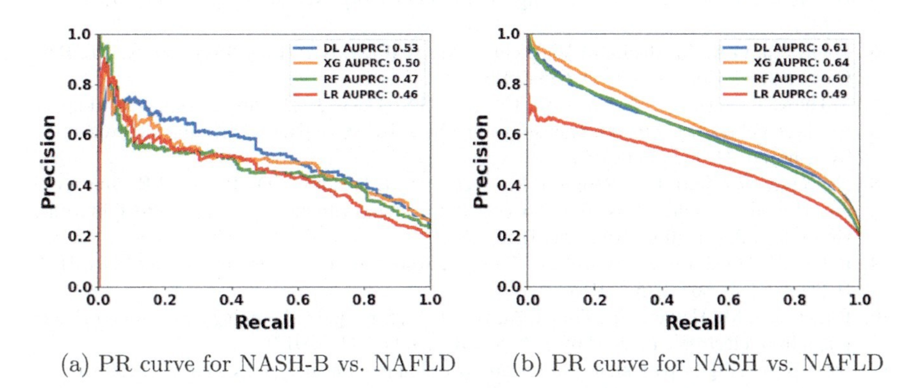

(a) PR curve for NASH-B vs. NAFLD (b) PR curve for NASH vs. NAFLD

Fig. 3 Performance curves for identifying patients (suffering from NAFLD) with risk for NASH. DL stands for deep learning which is the proposed model. XG is short for XGBoost, RF is short for random forest, LR is short for logistic regression. XG, RL and LR are baseline models. Subfigure (**a**) shows different model performance in classifying NASH-B cohort from NAFLD. Subfigure (**b**) shows different model performance in classifying all of NASH cohort from NAFLD

5 Conclusion

In this work, we presented a deep learning approach to build a screening tool for identifying patients at high risk to develop NASH given a pool of NAFLD patients. For this, we designed an architecture that can deal with sequential time-series features as well as "static" features. The deep learning model had a better performance than XGBoost, random forest and logistic regression when used to identify patients who had a positive liver biopsy test. It had comparable performance with XGBoost when used to identify NASH patients whose liver biopsy reports were unavailable. Future work includes improving the deep learning architecture to achieve a better classification performance, performing patient subtyping using $\mathcal{PMCE}$ and $\mathcal{PDE}$ vectors and training interpretable deep learning models to identify effect of each medical codes on the decision of the classifier.

References

1. Ching, T., Himmelstein, D.S., Beaulieu-Jones, B.K., Kalinin, A.A., Do, B.T., Way, G.P., Ferrero, E., Agapow, P.M., Zietz, M., Hoffman, M.M., et al.: Opportunities and obstacles for deep learning in biology and medicine. J. R. Soc. Interface **15**(141), 20170387 (2018)
2. Choi, E., Bahadori, M.T., Schuetz, A., Stewart, W.F., Sun, J.: Doctor AI: predicting clinical events via recurrent neural networks. In: Machine Learning for Healthcare Conference, pp. 301–318 (2016)
3. Dyson, J.K., Anstee, Q.M., McPherson, S.: Non-alcoholic fatty liver disease: a practical approach to diagnosis and staging. Frontline Gastroenterol. **5**(3), 211–218 (2014)
4. Farrell, G.C., Larter, C.Z.: Nonalcoholic fatty liver disease: from steatosis to cirrhosis. Hepatology **43**(S1), S99–S112 (2006)
5. Hochreiter, S., Schmidhuber, J.: Long short-term memory. Neural Comput. **9**(8), 1735–1780 (1997)
6. Michelotti, G.A., Machado, M.V., Diehl, A.M.: NAFLD, NASH and liver cancer. Nat. Rev. Gastroenterol. Hepatol. **10**(11), 656 (2013)
7. Mikolov, T., Sutskever, I., Chen, K., Corrado, G.S., Dean, J.: Distributed representations of words and phrases and their compositionality. In: Advances in Neural Information Processing Systems, pp. 3111–3119 (2013)
8. Neuschwander-Tetri, B.A., Brunt, E.M., Wehmeier, K.R., Oliver, D., Bacon, B.R.: Improved nonalcoholic steatohepatitis after 48 weeks of treatment with the ppar-γ ligand rosiglitazone. Hepatology **38**(4), 1008–1017 (2003)
9. Ratziu, V., Goodman, Z., Sanyal, A.: Current efforts and trends in the treatment of NASH. J. Hepatol. **62**(1), S65–S75 (2015)
10. Rotmensch, M., Halpern, Y., Tlimat, A., Horng, S., Sontag, D.: Learning a health knowledge graph from electronic medical records. Sci. Rep. **7**(1), 5994 (2017)
11. Saria, S.: The digital patient: machine learning techniques for analyzing electronic health record data. Ph.D. thesis, Stanford University (2011)
12. Sumida, Y., Yoneda, M.: Current and future pharmacological therapies for NAFLD/NASH. J. Gastroenterol. **53**(3), 362–376 (2018)

13. Wree, A., Broderick, L., Canbay, A., Hoffman, H.M., Feldstein, A.E.: From NAFLD to NASH to cirrhosis-new insights into disease mechanisms. Nat. Rev. Gastroenterol. Hepatol. **10**(11), 627 (2013)
14. Younossi, Z., Anstee, Q.M., Marietti, M., Hardy, T., Henry, L., Eslam, M., George, J., Bugianesi, E.: Global burden of NAFLD and NASH: trends, predictions, risk factors and prevention. Nat. Rev. Gastroenterol. Hepatol. **15**(1), 11 (2018)

Visualization of Deep Models on Nursing Notes and Physiological Data for Predicting Health Outcomes Through Temporal Sliding Windows

Jienan Yao, Yuyang Liu, Brenna Li, Stephen Gou, Chloe Pou-Prom, Joshua Murray, Amol Verma, Muhammad Mamdani, and Marzyeh Ghassemi

Abstract When it comes to assessing General Internal Medicine (GIM) patients' state, physicians often rely on structured, time series physiological data because it's more efficient and requires less effort to review than unstructured nursing notes. However, these text-based notes can have important information in predicting a patient's outcome. Therefore, in this paper we train two convolutional neural networks (CNN) on in-house hospital nursing notes and physiological data with temporally segmented sliding windows to understand the differences. And we visualize the process in which deep models generate the outcome prediction through interpretable gradient-based visualization techniques. We find that the notes model provides overall better predictions results and it is capable of sending warnings for crashing patients in a more timely manner. Also, to illustrate the different focal points of the models, we identified the top contributing factors each deep model utilizes to make predictions.

Keywords CNN · EHR · Clinical notes · Early warning system · Visualization

---

J. Yao · Y. Liu (✉) · B. Li · S. Gou · M. Ghassemi
Department of Computer Science, University of Toronto, Toronto, Canada
e-mail: yuyang@cs.toronto.edu

J. Yao
e-mail: jnyao@cs.toronto.edu

B. Li
e-mail: brli@cs.toronto.edu

S. Gou
e-mail: gouzhen1@cs.toronto.edu

C. Pou-Prom · J. Murray · A. Verma · M. Mamdani
Unity Health Toronto, Toronto, Canada

© The Editor(s) (if applicable) and The Author(s), under exclusive license to Springer Nature Switzerland AG 2021
A. Shaban-Nejad et al. (eds.), *Explainable AI in Healthcare and Medicine*, Studies in Computational Intelligence 914,
https://doi.org/10.1007/978-3-030-53352-6_11

1 Introduction

With recent investments in clinical digitization, many health organizations are interested in developing Clinical Early Warnings Systems (CEWSs) with physiological data and clinicians' notes data, to assist in prioritizing treatment and patient care [4, 18]. However, little is known how these two data types compare relative to clinical predictions. The problem is further complicated because physiological data and textual note data are collected differently and have different temporal segmentation and sparsity [3]. Therefore, we employ the use of sliding windows on both datasets to normalize variability in sparsity and segmentation for more effective comparison. The purpose of this work is to investigate if unstructured textual data can provide comparably valuable information to the structured physiological data (heart rates, oxygen level, lab test results, etc.) in the context of predicting whether a patient in the General Internal Medicine (GIM) ward would crash.[1] In order to evaluate this, we train machine learning models using either physiological data ("**tabular**" data) or clinical nursing notes ("**notes**" data). We implement convolutional neural net (CNN) models which take as input either of the feature sets (e.g., *tabular* or *notes*), as previous research has shown the effectiveness of CNN architectures in healthcare problems [9, 11, 19]. We then design the following research questions to help us understand how nursing notes and tabular data can be used to predict patient health outcome in GIM: (1) How do our models trained on tabular data compare to models trained on nurse notes data. (2) How early can our deep models make the correct prediction. (3) What information are the deep models using to make the predictions. (4) What are the top contributing factors of our deep models.

2 Related Work

2.1 Clinical Early Warning Systems

Traditionally, CEWSs take in a set of physiological parameters such as, blood glucose level, oxygen saturation, temperature, etc., with corresponding ranges to calculate a score to determine the patient's health outcomes [16, 18]. However, these traditional CEWSs are dependant on the parameters selected and the method of calculation used, which limits their robustness and predictive power for individual patient encounters [2]. Thus lately, the trend has been to use neural network models in CEWS to account for personalized patient encounter predictions.

[1] We define a "crashing" patient as someone who is experiencing one of the following outcomes: death on the GIM ward, transfer to intensive care unit (ICU), or transfer to palliative unit.

2.2 Neural Network Approaches

With increasingly publicly available large-scaled health records, such as, the MIMIC II/III dataset [6], opportunities and interests to apply deep learning neural methods to health outcome predictions are rising [4]. Various studies have shown promising results using raw physiological data in predicting patient mortality, ICU transfers, and health deterioration trends [2, 3, 18]. And recently, through advancements in natural language processing, we are also seeing a greater focus on textual data, such as, clinical notes and discharge summaries [5, 8]. As an example, Waudby *et al.* demonstrated that sentiments inferences extracted from the MIMIC III clinical notes can be used as predictors for patient mortality [17]. And other works comparing neural models have found that convolutional neural networks are more suitable for long, unstructured notes [7]. And from these findings, we built our own text extraction CNN model that is fast and easy to tune.

2.3 Interpreting Prediction Outcomes

A caveat with using deep neural models is the opaqueness in understanding how the machine generates these predictions, especially for clinical notes. It's often unclear which sentences, or words were used and the weight they were given [15]. To visualize the computation behind CNN models on clinical data, [10] applied attention-based deconstruction to emphasize the words contributing to the prediction. However, nursing notes as we know can be very long and heterogeneous, therefore a larger scope on text is needed to make sense behind the reasoning of why it was selected. Therefore, we adapted Grad-CAM, a CNN based visualization technique traditionally on imaging data to our clinical notes [15]. And when applied, with our sliding window approach, as defined in the Method section, allows us to visualize on a time scale the changes in sentences that contribute to the prediction.

3 Methods

3.1 Data

We use de-identified clinical data from the General Internal Medicine (GIM) ward of a hospital in North America,[2] and include patients that have completed visits between December 2014 and April 2019. The data consist of the following sets of features:

[2] A Hospital that is part of Unity Health Toronto.

- **Tabular data:** This consists of routinely-collected lab results (e.g., blood glucose) and vitals (e.g., hear rate). This data also includes clinical orders, such as diet change orders (e.g., if a patient is moving to a *nil per os*[3] diet) and imaging orders (e.g., chest X-ray).
- **Notes data:** This includes de-identified nursing notes.

We train different models on each set of features (i.e., tabular data vs. notes data) and the models each take as input a window consisting of 3× 8-h intervals (i.e., 24 h of data).

3.1.1 Notes Data Pre-processing

To fully utilize the temporal data and provide timely predictions, we concatenate the notes data through sliding windows consisting of 24 h of data (i.e., 3× 8-h intervals of notes). We also notice that sometimes only one note is recorded and appeared in several windows as we slide over the time point when outcome turns 1 from 0, making all those windows share the same content but their corresponding outcomes may vary, thus creating conflicting training pairs with the same input corresponding to different output labels. To deal with this issue caused by the sparsity in the clinical notes, we append a notes sparsity indicator vector of length 3 (i.e., the window size) indicating whether notes were recorded in the corresponding interval, as seen in the Notes Sparsity Indicator in Fig. 1b.

Then, for text pre-processing, we remove of white-space, numerical values and standard English stop-words using the `Gensim` library [14]. From the pre-processed notes, we then extract LDA topic probabilities and get the GloVe word representation (more details in the *Models* section).

3.1.2 Tabular Data Pre-processing

8-h intervals of the tabular data are created by taking the mean average whenever there are multiple measures within an interval. Missing data are imputed with last-observation-carried-forward and then filled in by the mean. For each measure, we then create two additional variables:

[3]*Nil per os* is Latin for "nothing by mouth" and is used when a patient cannot receive food orally.

Original Table

rowIndex	Encounter	Window	Timestamp	Temp (F)	Heart rate	02 count	~300 measures	Outcome
1	1111	1	00:00	98	77	89	...	0
2	1111	2	08:00	98	83	89	...	0
3	1111	3	16:00	98	86	87	...	0
4	1111	4	00:00	98	80	85	...	1
5	2222	1	00:00	101	88	99	...	0
6	2222	2	08:00	101	90	99	...	0
7	2222	3	16:00	101	87	99	...	0
8	3333	1	00:00	98	103	97	...	0
9	3333	2	08:00	98	114	97	...	0
10	3333	3	16:00	99	116	98	...	0
11	3333	4	00:00	99	130	98	...	0
12	3333	5	08:00	103	141	96	...	1
...	...	...	...	...	...	...	...	...

After Sliding Window

rowIndex	Encounter	Window	Timestamp	Temp (F)	Heart rate	02 count	~300 measures	Outcome
1	1111	[1, 2, 3]	[00:00, 08:00, 16:00]	[98,98,98]	[77,83,86]	[89,89,87]	...	0
2	1111	[2, 3, 4]	[08:00, 16:00, 00:00]	[98,98,98]	[83,86,80]	[89,87,85]	...	1
3	2222	[1, 2, 3]	[00:00, 08:00, 16:00]	[101,101,101]	[88,90,87]	[99,99,99]	...	0
4	3333	[1, 2, 3]	[00:00, 08:00, 16:00]	[98,98,99]	[103,114,116]	[97,97,98]	...	0
5	3333	[2, 3, 4]	[08:00, 16:00, 00:00]	[98,99, 99]	[114,116,130]	[97,98,98]	...	0
6	3333	[3, 4, 5,]	[16:00, 00:00, 08:00]	[99,99,103]	[116,130,141]	[98,98,96]	...	1
...	...	...	...	...	...	...	...	...

(a) Sliding windows example for tabular data

Original Table

rowIndex	Encounter	Window	Timestamp	Notes	Outcome
1	1111	1	00:00	'	0
2	1111	2	08:00	the fox in the box	0
3	1111	3	16:00	said hello to the vox	0
4	1111	4	00:00	'	1
5	2222	1	00:00	the cat said hi	0
6	2222	2	08:00	'	0
7	2222	3	16:00	'	0
8	3333	1	00:00	'	0
9	3333	2	08:00	'	0
10	3333	3	16:00	'	0
11	3333	4	00:00	then the cat said bye	0
12	3333	5	08:00	'	1
...	...	...	...	...	...

After Sliding Window

rowIndex	Encounter	Window	Timestamp	Notes	Notes Sparsity Indicator	Outcome
1	1111	[1, 2, 3]	[00:00, 08:00, 16:00]	the fox in the box said hello to the vox	011	0
2	1111	[2, 3, 4]	[08:00, 16:00, 00:00]	the fox in the box said hello to the vox	110	1
3	2222	[1, 2, 3]	[00:00, 08:00, 16:00]	the cat said hi	100	0
4	3333	[1, 2, 3]	[00:00, 08:00, 16:00]	'	000	0
5	3333	[2, 3, 4]	[08:00, 16:00, 00:00]	then the cat said bye	001	0
6	3333	[3, 4, 5,]	[16:00, 00:00, 08:00]	then the cat said bye	010	1
...	...	...	...	...	...	...

(b) Sliding windows example for nurse notes data

Fig. 1 **a** and **b** represent an example of corresponding tabular and notes data with 24 h sliding window (length 3). Each rowIndex is an unique 8 h entry, that belongs to a given Encounter (a patient encounter number), Window length, Timestamp and Outcome (maximum of the original binary Outcome). In **a** around 300 physiological measures were averaged across the 24 h time interval. And in **b** the Notes in the original table were concatenated within each 24 h window, with the notes sparsity indicator capturing which window the note came from

- an *indicator variable* for each feature to differentiate between measured and imputed values (i.e., 1 if the value is measured in that interval, 0 if it's imputed); and
- a *time elapsed variable* for keeping track of the number of hours since the previous measure.

We also use the sliding window technique on the tabular data, we construct the data for each sliding window by concatenating the three observations in a window. (see Fig. 1a).

To ensure a fair comparison, we remove windows of tabular data where the corresponding notes data window has empty notes. We end up with **133,848**, **22,393**, and **8,631** windows of data for the training, validation, and test sets respectively.

3.2 *Models*

3.2.1 Notes Data Model

We build a CNN model using Global Vectors for Word Representation (GloVe) word embeddings trained on the notes data [12]. The CNN aims to predict patient health outcomes from the nursing notes, which are largely unstructured, free-hand and heterogeneous, with spontaneous entries. GloVe consists of a co-occurrence matrix method based on the idea that words with similar distributions have similar meanings [12]. In early experiments, we found that training GloVe on our own notes yielded better performance.

Each 24-h interval of nursing notes is treated as a document represented by an array of word embedding vectors. This is given as input to the CNN with 1D convolutions. This is to capture the localized context in the document. We then pair this with the doc2vec rerpesentation of the note to provide more global context on the document level. doc2vec [13] is a document embedding that provides fixed-length dense vector representation for text of variable length and encodes information such as the topic of the paragraph. Overall, our CNN is trained conditioned on both the fixed-length doc2vec vectors and notes sparsity indicators (e.g., [0, 1, 0] indicates notes were recorded only at the second window) to form an ensemble model. See Fig. 2 for the architecture of the CNN architecture.

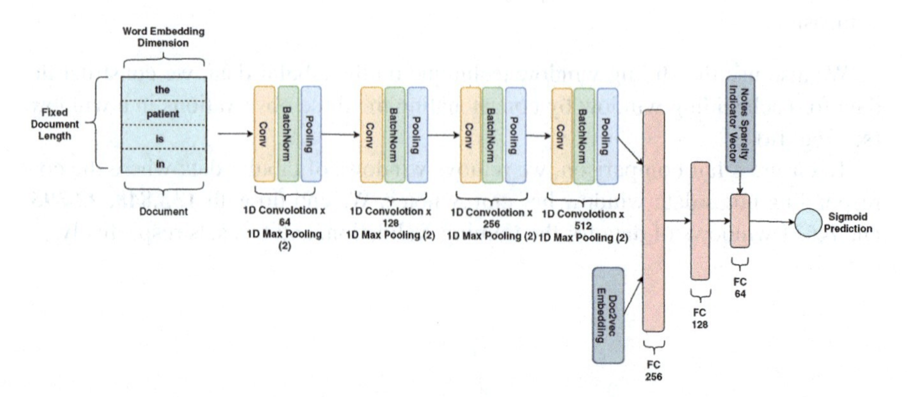

Fig. 2 Model architecture of CNN for unstructured notes data, the inputs of word embedding have fixed length of 2,000 tokens, shorter notes are padded with zeros and longer notes are truncated. Each document consists of notes concatenated from a sliding window. There are four convolution steps before the fully connected layers taking into considerations of the notes level embedding and notes sparsity indicators

For our GloVe embedding input, we cap the maximum number of tokens to 2,000, and pad with 0's when necessary. This number is determined by the mean average document length in our data. The CNN consists of four 1D convolutional layers, each with double the number of filters as the previous layer. All filters are of size 3, and each convolutional layer is followed by a max pooling layer. We train the doc2vec model (Distributed Memory Model of Paragraph Vectors, PV-DM) on the nurse notes data with 150-dimensional embeddings. We then concatenate the output of the last layer in CNN with the doc2vec representation for the input document to obtain the ensemble model. The concatenated vector is given as input to two densely connected layers, together with notes sparsity indicators.

To reduce overfitting, we apply dropout (45%) before each fully connected layer, and use early stopping when the validation error begins to increase. These parameters are optimized by informal grid search.

3.2.2 Tabular Data Model

In each sliding window of size **3**, there are **375** lab/vital features with **3** temporal observations. In order to leverage the patterns hidden in each feature, we structure the data to have a dimension of $\mathbf{375} \times \mathbf{3}$ so that each convolutional kernel of size $\mathbf{1} \times \mathbf{3}$ can focus on a specific feature. During the training processes, the kernels will learn the temporal pattern change across every physiological signal within each sliding window. Extracted features are flattened and fed to two densely connected layers with ReLU activation, followed by a sigmoid layer.

3.2.3 Logistic Regression Models

The logistic regression nurse notes model takes as input the topic probabilities from a 50-topic Latent Dirichlet Allocation (LDA) representation of the notes [1]. LDA generates an efficient and low-dimensional representation and offers decent interpretability. The logistic regression tabular data model is trained on $\mathbf{375} \times \mathbf{3}$ windows of data.

4 Results

4.1 How Do Our Models Trained on Tabular Data Compare to Models Trained on Nurse Notes Data?

For Table 1, the threshold is selected by a grid search to achieve a recall score close to 0.8 that is suggested practically by physicians, which results in a threshold of 0.695 for both the CNN notes model and tabular model.

The metrics for the CNN models in Table 1 reveal that although both CNN models share the same recall of 0.862, the nursing note CNN model has a higher precision of 0.163 compared to 0.156 from the tabular CNN model, which suggests the former model is more capable of identifying crashing patients, at the cost of lower accuracy, 0.840 as opposed to 0.861. The nursing note data are equally as important, if not more significant, as the tabular data, evidenced by the higher ROC-AUC values on the notes data model. And we observed similar results in the logistic regression models, that the model trained on notes data (ROC-AUC of 0.826) is performing equally as good as the model trained on tabular data (ROC-AUC of 0.820).

Table 1 Overall comparisons between notes model and tabular model and detailed comparisons among ICU transfer, Death and Palliative transfer for the CNN models. The number outside the parenthesis is from the test set whereas the number inside the parenthesis is from the validation set

		OVERALL	ICU	DEATH	PAL
Notes	Accuracy	0.840 (0.831)	0.857 (0.846)	0.859 (0.861)	0.845 (0.850)
	Precision	0.163 (0.195)	0.069 (0.107)	0.015 (0.017)	0.095 (0.094)
	Recall	0.862 (0.807)	0.731 (0.765)	1.000 (0.786)	0.964 (0.867)
	ROC-AUC	**0.858** (0.802)	0.780 (0.774)	**0.984** (0.922)	**0.925** (0.848)
Tabular	Accuracy	0.861 (0.837)	0.882 (0.854)	0.887 (0.869)	0.873 (0.856)
	Precision	0.156 (0.181)	0.069 (0.101)	0.011 (0.019)	0.091 (0.082)
	Recall	0.862 (0.807)	0.769 (0.786)	0.750 (0.929)	0.964 (0.813)
	ROC-AUC	0.839 (0.816)	**0.787** (0.800)	0.908 (0.943)	0.895 (0.843)

Based on Table 1, among all three individual crashing types (i.e., ICU transfer, palliative transfer, death), ICU transfer has the worst performance. We speculate this is because ICU transfers are inherently highly unpredictable and very spontaneous. We also notice a shift in the precision and recall between the validation and test data, whereby we observed higher recall and lower precision in the test data. We speculate this might be because the distributions in the test and validation data might be different. However, overall, the performance results from the notes and tabular deep models are very similar, with notes performing better on death and palliative conditions.

We also investigate how our deep models capture the true crashing patients. In Fig. 3 there are 58 patients who crashed (i.e., who experienced an ICU transfer, a

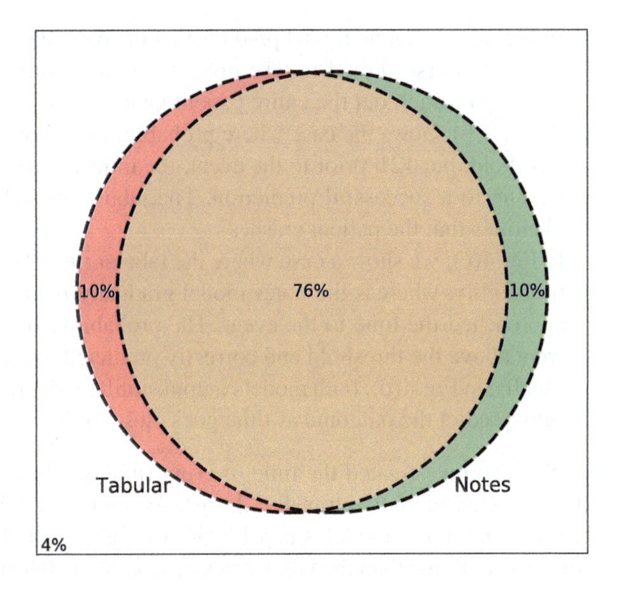

Fig. 3 The pink circle represents the correctly classified patient visits by our tabular model, and the green circle stands for the correctly classified patient visits by our notes model. There are 2 positive patient visits in the test data that are correctly classified by neither of our model, and both of which are ICU transfers patient visits

palliative transfer, or death) in our test dataset, out of which, 44 are correctly classified by both of our deep models. And 6 patients are correctly predicted by our notes model only, and another 6 patients are captured by tabular model only. There are only 2 crashing patients who had been missed by both of the deep models.

4.2 How Early Can Our Deep Models Make the Correct Prediction?

While our deep models are capable of predicting the patients crashing events, we would like to investigate how early before the patient crashing could our deep model signal the warning. The most valuable warning will be those timely before the actual event time. The model can bring assistance to the medical team as resources can be allocated just before the event happens and intervention can be initiated within proper time interval.

For each patient visit, we calculate the time-to-event (in days) from the warning when the model first passes the threshold to the recorded timestamp for the patient crashing. See Fig. 4 for examples of predicted risks of crashing by both tabular model and notes model along patient trajectories. The trajectories show how the probability of crashing (y-axis), as predicted by the notes or tabular model, changes over time (x-axis). In the examples, all visits eventually experience the outcome. We highlight the following scenarios:

- In Fig. 4(a), the note model predicts that the patient will crash from the beginning when the nurse started to take notes while the tabular model fails to trigger the warning through out the entire patient visit.
- Figure 4(b) shows the case where probabilities of both models start off below the threshold, but 12 h prior to the event, the trajectory of the notes model increases, leading to a successful prediction. The tabular model, on the other hand, never identifies that the patient crashes.
- In Fig. 4(c), we show a case where the tabular model decreased its risk prediction probability, whereas the notes model gradually increased its risk prediction as it approached the time to the event. The probability of the notes model eventually went above the threshold and correctly predicted that the crashing event.
- Finally, in Fig. 4(d), both models output similar risk prediction scores and successfully predict the outcome as time goes close to the crashing event.

We also investigated the time-to-event (in days) for each specific outcome (i.e., ICU transfer, palliative transfer, death). As shown in Table 2, on all three types of crashing, the notes model is capable of making correct classifications with a tighter time to event time than the tabular model, except the ICU transfer at the first quartile.

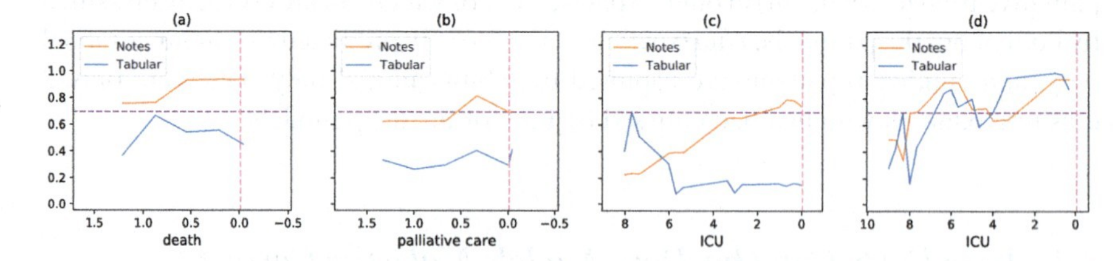

Fig. 4 There are two trajectories in each subplot, one is for the notes model and other one is for the tabular model. The x-axis is right-aligned to the time of event in days. The y-axis is the predictions made by both models. The horizontal line is the threshold used to predict the outcomes. And the vertical line marks the point in which the event happened

Table 2 The time to event (days) for crashing patients visits at three quartiles on the test data set at threshold 0.695

	Quartile	CRASH	ICU	DEATH	PAL
Notes	0.25	1.102	0.539	2.914	1.413
	0.5	3.536	2.381	11.079	3.817
	0.75	8.203	5.011	19.448	9.391
Tabular	0.25	1.499	0.503	11.913	1.748
	0.5	4.668	3.330	20.347	7.416
	0.75	11.333	6.112	22.049	11.705

4.3 *What Information Are the Deep Models Using to Make the Predictions?*

To understand how the deep models make decisions, we adopt the guided grad-CAM technique [15] on both of our models (i.e., tabular model and notes model). This approach calculates the gradients of the risk score with respect to the input: higher gradient magnitude indicates higher activation, which can be used as a measure of how important a given word token is to the prediction. The magnitude of the activation provides a visualization in the form of a heatmap indicating the important tokens contributing to the final prediction.

Guided grad-CAM in the case of 2D image that can be visualized as a 3-channel RGB image. In the case of word embedding with 150 dimensions, it is hard to visualize the guide grad-CAM heatmap. Hence, we take the sum of absolute values along the embedding dimension to represent the magnitude of the activation. This is based on the observation that gradients of the words having high activation will vary dramatically along the embedding dimension. The similar technique can be used in the tabular model to locate important physiological data. We present some interesting cases using the same visits as in Fig. 4 in which the notes model provides timely prediction based on meaningful information.

For the case in Fig. 4(a), we show an example heatmap in Fig. 5a, that the model focuses on keywords such as **ccrt (critical care response team)** and ventolin that are terminologies related to treatment, and **wheezy** and **amber urine** which are symptoms that can give us insights on why the patient died.

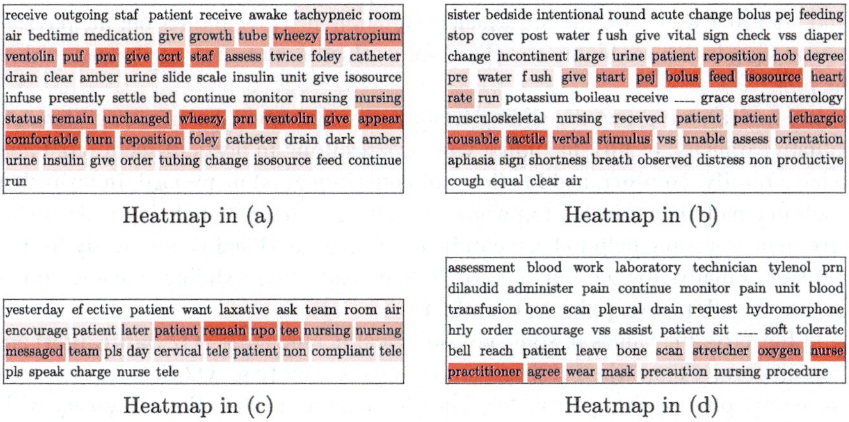

receive outgoing staf patient receive awake tachypneic room air bedtime medication give growth tube wheezy ipratropium ventolin puf prn give ccrt staf assess twice foley catheter drain clear amber urine slide scale insulin unit give isosource infuse presently settle bed continue monitor nursing nursing status remain unchanged wheezy prn ventolin give appear comfortable turn reposition foley catheter drain dark amber urine insulin give order tubing change isosource feed continue run

Heatmap in (a)

sister bedside intentional round acute change bolus pej feeding stop cover post water f ush give vital sign check vss diaper change incontinent large urine patient reposition hob degree pre water f ush give start pej bolus feed isosource heart rate run potassium boileau receive ___ grace gastroenterology musculoskeletal nursing received patient patient lethargic rousable tactile verbal stimulus vss unable assess orientation aphasia sign shortness breath observed distress non productive cough equal clear air

Heatmap in (b)

yesterday ef ective patient want laxative ask team room air encourage patient later patient remain npo tee nursing nursing messaged team pls day cervical tele patient non compliant tele pls speak charge nurse tele

Heatmap in (c)

assessment blood work laboratory technician tylenol prn dilaudid administer pain continue monitor pain unit blood transfusion bone scan pleural drain request hydromorphone hrly order encourage vss assist patient sit ___ soft tolerate bell reach patient leave bone scan stretcher oxygen nurse practitioner agree wear mask precaution nursing procedure

Heatmap in (d)

Fig. 5 Heatmap activation over four notes

Next, we looked at an example of a visit whose probabilities followed different trajectories in the tabular model and in the notes model (i.e., Fig. 4(b)). We report the corresponding heatmap in Fig. 5b, and find that the notes model focuses on words such as **rousable, bolus feed** which are typical actions of end-of-life care. This particular visit eventually experienced the palliative transfer outcome. Meanwhile, the tabular model fails to identify the outcome based on the physiological signals.

Then, we look at a case where the tabular data fails to capture the outcome while the notes model does (i.e., Fig. 4)(c). In this case, the patient eventually ends up transferring to the ICU. The words with high activation shown in Fig. 5c indicate that the patient is under the state of NPO (Latin for "nothing by mouth") and waiting for a TEE test (i.e., a transesophageal echocardiography).

Finally, in Fig. 4(d), we show an example where the notes model's prediction is in sync with that of tabular model from the beginning of encounter to the crashing time. The notes model identifies the outcome ahead of the tabular model. From the heatmap (see Fig. 5d), we find the model is focusing on words like **oxygen**, **mask**, and **precaution**. These words suggest that the patient was undergoing a preemptive act to prevent potential consequence at least noticed by the nurse.

4.4 What Are the Top Contributing Factors of Our Deep Models?

We include a matrix representation of the important words and tabular features from both deep models in Fig. 6. Each column represents a sliding window in the test set. In notes model for each word in a sliding window, we add up its magnitude of activation and then normalize it within each sliding window. We then follow the same process to get the matrix representation and manually select top predictive words to be included in the plot.

And for tabular model, we apply the similar process to get top predictive features. And after that we utilize the Spearman correlation analysis to filter out the physiological signals that are highly correlated for a selection of features to be included in the plot.

In Fig. 6(a), we group the sliding windows by outcome type (i.e., ICU transfer, palliative transfer, death). In sliding windows resulting in death, it seems common to have **family**, **comfort**, and **loc (level of consciousness)** mentioned. In visits that result in a palliative transfer, top words are related to the patient being under **measurements**, needing help to keep **comfortable (comfort)** and getting ready for the **transfer**. Finally, for the outcome of ICU transfer, these sliding windows often involves **nasal**, **prong**, **pain**, and **comfort**.

In Fig. 6(b), the values of some physiological features (e.g., **O2 Saturation**) are more important than whether they are measured or not (e.g., **O2 Saturation measured**). We observe the opposite trend in other features such as **Pain Intensity with Movement**, where the fact that this was measured is more important than the actual

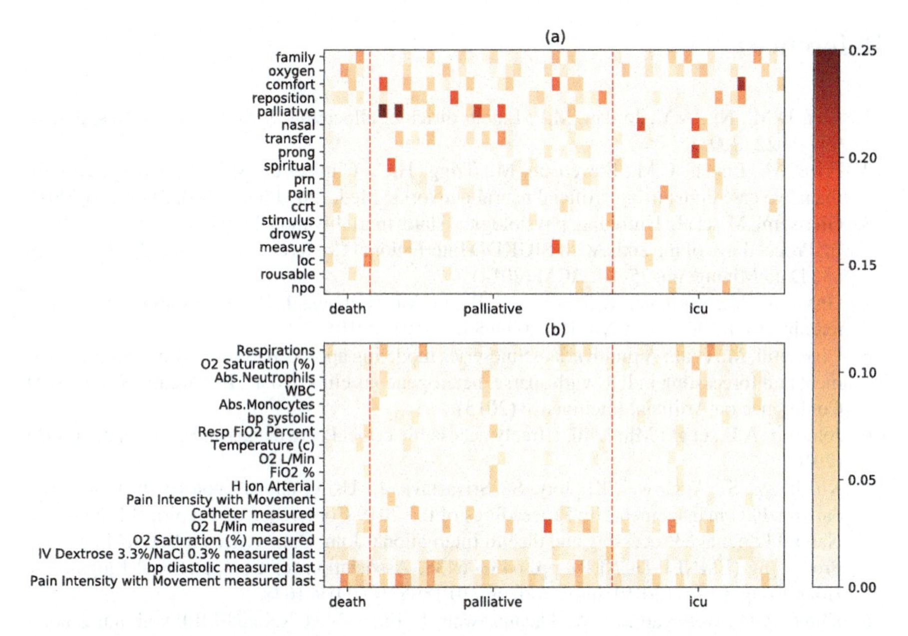

Fig. 6 Important words/tabular features among nursing notes and physiological measurements. x-axis represents sliding windows during the visits of crashing patients (grouped by death, palliative and ICU outcomes), y-axis represents subset of the top indicative words/tabular data as well as those picked from spearman correlation analysis. A physiological data name ending with 'measured' indicates whether the entry is measured or imputed, and 'measured last' indicates the number of hours since last measurement

value. We also notice that the time since last measurements of certain physiological data usually contributes the most to our model.

5 Conclusion

In summary, we have shown that, although nursing notes are inherently noisy, there is still value in the notes for predicting health outcomes. Deep models trained on nursing notes data can achieve similar performance as those of tabular data. In addition, the notes model is capable of predicting patient crashing closer to the event, ICU transfer, death, or palliative care. When predicting, notes model and tabular model make prediction based on different perspectives. Considering the limitations of the models and potential improvements, we discarded sliding windows if there are no nursing notes data when we build the notes model. To have a fair comparison, we also remove these sliding windows when building the tabular model even if there are other tabular data inside these sliding windows. Future work includes removing high-frequency words in the dataset in the preprocessing step and experimenting with various sliding window sizes or different combination of word embeddings.

References

1. Blei, D.M., Ng, A.Y., Jordan, M.I.: Latent dirichlet allocation. J. Mach. Learn. Res. **3**(Jan), 993–1022 (2003)
2. Frize, M., Ennett, C.M., Stevenson, M., Trigg, H.C.: Clinical decision support systems for intensive care units: using artificial neural networks. Med. Eng. Phys. **23**(3), 217–225 (2001)
3. Ghassemi, M., et al.: Unfolding physiological state: mortality modelling in intensive care units. In: Proceedings of the 20th ACM SIGKDD International Conference on Knowledge Discovery and Data Mining, pp. 75–84. ACM (2014)
4. Ghassemi, M., Naumann, T., Schulam, P., Beam, A.L., Ranganath, R.: Opportunities in machine learning for healthcare. arXiv:1806.00388 [cs, stat] (2018)
5. Ghassemi, M., et al.: A multivariate timeseries modeling approach to severity of illness assessment and forecasting in ICU with sparse, heterogeneous clinical data. In: Twenty-Ninth AAAI Conference on Artificial Intelligence (2015)
6. Johnson, A.E., et al.: Mimic-iii, a freely accessible critical care database. Sci. Data **3**, 160035 (2016)
7. Khadanga, S., Aggarwal, K., Joty, S., Srivastava, J.: Using clinical notes with time series data for ICU management. In: Proceedings of the 2019 Conference on Empirical Methods in Natural Language Processing and the 9th International Joint Conference on Natural Language Processing (EMNLP-IJCNLP), pp. 6433–6438. Association for Computational Linguistics, Hong Kong, China (2019). https://doi.org/10.18653/v1/D19-1678
8. Khine, A.H., Wettayaprasit, W., Duangsuwan, J.: Ensemble CNN and MLP with nurse notes for intensive care unit mortality. In: 2019 16th International Joint Conference on Computer Science and Software Engineering (JCSSE), pp. 236–241. IEEE (2019)
9. Kim, Y.: Convolutional neural networks for sentence classification. In: Proceedings of the 2014 Conference on Empirical Methods in Natural Language Processing (EMNLP), pp. 1746–1751. Association for Computational Linguistics, Doha, Qatar (2014). https://doi.org/10.3115/v1/D14-1181
10. Lovelace, J.R., Hurley, N.C., Haimovich, A.D., Mortazavi, B.J.: Explainable prediction of adverse outcomes using clinical notes. arXiv preprint arXiv:1910.14095 (2019)
11. Nguyen, P., Tran, T., Wickramasinghe, N., Venkatesh, S.: Deepr: a convolutional net for medical records. IEEE J. Biomed. Health Inform. **21**(1), 22–30 (2017). https://doi.org/10.1109/JBHI.2016.2633963
12. Pennington, J., Socher, R., Manning, C.: Glove: global vectors for word representation. In: Proceedings of the 2014 Conference on Empirical Methods in Natural Language Processing (EMNLP), pp. 1532–1543. Association for Computational Linguistics, Doha, Qatar (2014). https://doi.org/10.3115/v1/D14-1162
13. Quoc, L., Tomas, M.: Distributed representations of sentences and documents. In: International Conference on Machine Learning, pp. 1188–1196 (2014)
14. Řehůřek, R., Sojka, P.: Software framework for topic modelling with large corpora. In: Proceedings of the LREC 2010 Workshop on New Challenges for NLP Frameworks, pp. 45–50. ELRA, Valletta (2010)
15. Selvaraju, R.R., Cogswell, M., Das, A., Vedantam, R., Parikh, D., Batra, D.: Grad-CAM: Visual explanations from deep networks via gradient-based localization. In: Proceedings of the IEEE International Conference on Computer Vision, pp. 618–626 (2017)
16. Smith, G.B., Prytherch, D.R., Meredith, P., Schmidt, P.E., Featherstone, P.I.: The ability of the national early warning score (news) to discriminate patients at risk of early cardiac arrest, unanticipated intensive care unit admission, and death. Resuscitation **84**(4), 465–470 (2013)
17. Waudby-Smith, I.E., Tran, N., Dubin, J.A., Lee, J.: Sentiment in nursing notes as an indicator of out-of-hospital mortality in intensive care patients. PloS one **13**(6), e0198687 (2018). Public Library of Science

18. Wellner, B., et al.: Predicting unplanned transfers to the Intensive Care Unit: a machine learning approach leveraging diverse clinical elements. JMIR Med. Inf. **5**(4) (2017). https://doi.org/10.2196/medinform.8680
19. Wu, Y., Jiang, M., Xu, J., Zhi, D., Xu, H.: Clinical named entity recognition using deep learning models. In: AMIA Annual Symposium Proceedings, vol. 2017, p. 1812. American Medical Informatics Association (2017)

Constructing Artificial Data for Fine-Tuning for Low-Resource Biomedical Text Tagging with Applications in PICO Annotation

Gaurav Singh, Zahra Sabet, John Shawe-Taylor, and James Thomas

Abstract Biomedical text tagging systems are plagued by the dearth of labeled training data. There have been recent attempts at using pre-trained encoders to deal with this issue. Pre-trained encoder provides representation of the input text which is then fed to task-specific layers for classification. The entire network is fine-tuned on the labeled data from the target task. Unfortunately, a low-resource biomedical task often has too few labeled instances for satisfactory fine-tuning. Also, if the label space is large, it contains few or no labeled instances for majority of the labels. Most biomedical tagging systems treat labels as indexes, ignoring the fact that these labels are often concepts expressed in natural language e.g. 'Appearance of lesion on brain imaging'. To address these issues, we propose constructing extra labeled instances using label-text (i.e. label's name) as input for the corresponding label-index (i.e. label's index). In fact, we propose a number of strategies for manufacturing multiple artificial labeled instances from a single label. The network is then fine-tuned on a combination of real and these newly constructed artificial labeled instances. We evaluate the proposed approach on an important low-resource biomedical task called *PICO annotation*, which requires tagging raw text describing clinical trials with labels corresponding to different aspects of the trial i.e. PICO (Population, Intervention/Control, Outcome) characteristics of the trial. Our empirical results

G. Singh (✉) · J. Shawe-Taylor · J. Thomas
University College London, London, UK
e-mail: g.singh@cs.ucl.ac.uk

J. Shawe-Taylor
e-mail: j.shawe-taylor@ucl.ac.uk

J. Thomas
e-mail: james.thomas@ucl.ac.uk

Z. Sabet
AIG, London, UK
e-mail: zahra.sabetsarvestani@aig.com

© The Editor(s) (if applicable) and The Author(s), under exclusive license to Springer Nature Switzerland AG 2021

A. Shaban-Nejad et al. (eds.), *Explainable AI in Healthcare and Medicine*, Studies in Computational Intelligence 914,
https://doi.org/10.1007/978-3-030-53352-6_12

show that the proposed method achieves a new state-of-the-art performance for PICO annotation with very significant improvements over competitive baselines.

Keywords Biomedical text tagging · PICO annotation · Artificial data · Transfer learning

1 Introduction

Biomedical text is often associated to terms that are drawn from a pre-defined vocabulary and indicate the information contained in the text. As these terms are useful for extracting relevant literature, biomedical text tagging has remained an important problem in BioNLP [4, 28] and continues to be an active area of research [20, 21]. Currently, a standard text tagging model consists of an encoder that generates a context-aware vector representation for the input text [1, 23, 26], which is then fed to a 2-layered neural network or multi-layered perceptron that classifies the given text into different labels. The encoder is generally based on CNN [10, 11, 23], RNN [11], or (more recently) multi-head self-attention network e.g. Transformer [1]. This works well for tasks that have a lot of labeled data to train these models (Fig. 1).

Unfortunately, lack of labeled data is a common issue with many biomedical tagging tasks. We focus on one such low-resource biomedical text tagging task called *PICO* annotation [20]. It requires applying distinct sets of concepts that describe complementary clinically salient aspects of the underlying trial: the population enrolled, the interventions administered and the outcomes measured, i.e. the PICO elements. Currently, these PICO annotations are obtained using manual effort, and it can take years to get just a few thousand documents annotated. Not surprisingly, there are not enough PICO annotations available to train a sophisticated classification model based on deep learning.

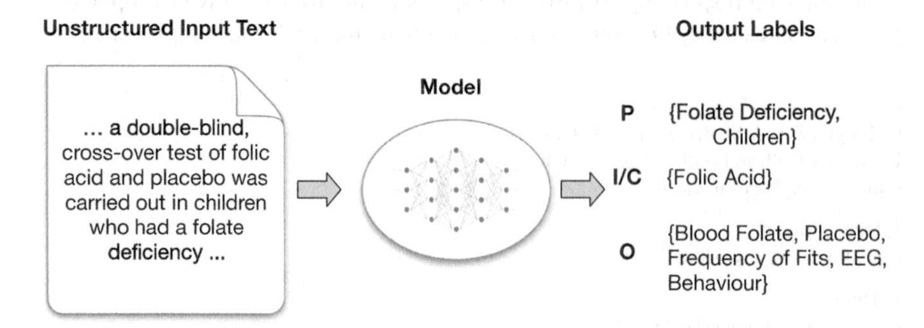

Fig. 1 An illustration of the PICO annotation task. The model receives input text describing a clinical trial and applies concepts from a pre-defined vocabulary corresponding to Population, Intervention and Outcome characteristics of the trial

Recently, there have been attempts to address this issue of low-resource biomedical text tagging using transfer learning [19, 25]. These methods begin by training a (large) deep neural network model on a high-resource biomedical task followed by fine-tuning on the low-resource target task. One such work from Zeng et al. [25] proposed training a deep CNN on the source task of applying MeSH tags to biomedical abstracts, and then used this pre-trained network for applying ICD-9 codes to clinical text after fine-tuning. They used a freely available repository of biomedical abstracts on Pubmed, which are annotated with medical concepts called MeSH tags (i.e. Medical Subject Heading) for source task. In a similar work, Rios et al. [19] focused on a different target task and experimented with ablations of network architectures. There have also been attempts in the past to solve PICO annotation using concept embeddings pre-trained using deepwalk on the vocabulary graph [20].

One drawback of above methods is that they treat labels as indexes, despite the fact that these labels are concepts expressed in natural language e.g. 'Finding Relating To Institutionalization'. After training on the source task, the encoder is retained, while the task specific deeper layers are replaced with new layers, and the entire network is fine-tuned. Consequently, task-specific layers have to be re-trained from the very beginning on a low-resource biomedical task. This poses problems, not only because there is insufficient labeled data for fine-tuning, but there are few or no labeled instances for a large number of labels.

To address this, we propose constructing additional artificial instances for fine-tuning by using the label-text as an input for the corresponding label-index. We propose various strategies to construct artificial instances using label-text, which are then used in combination with real instances for fine-tuning. This is based on the hypothesis that an encoder pre-trained on a related biomedical source task can be used to extract good quality representations for the label-text, where these label-texts are concepts expressed in natural language.

To summarize the contribution of this work:

- To address the scarcity of labeled data for PICO annotation, we construct additional artificial labeled instances for fine-tuning using label-texts as inputs for label-indexes.
- In order to obtain rich encoder representations for label-text, we pre-train the encoder on a related source task i.e. MeSH annotation.

2 Related Works

We focus on the problem of low-resource biomedical multi-label text classification using transfer learning. As such there are three broad areas pertinent to our work: (1) multi-label text classification; (2) biomedical text annotation; (3) transfer learning for classification. We will give an overview of relevant literature in these three areas.

2.1 Multi-label Text Classification

The task of PICO annotation can be seen as an instance of multilabel classification, which has been an active area of research for a long time, and therefore has a rich body of work [8, 9, 13, 18]. One of the earliest works [13] in multi-label text classification represented multiple labels comprising a document with a mixture model. They use expectation maximization to estimate the contribution of a label in generating a word in the document. The task is then to identify the most likely mixture of labels required to generate the document. Unfortunately, when the set of labels is prohibitively large, these label mixture models are not practical. Then, a kernel based approach for multilabel classification based on using a large margin ranking system was proposed [7]. In particular, it suggests a SVM like learning system for direct multi-label classification, as opposed to breaking down these into many binary classification problems.

More recently, there has been a shift towards using deep learning based methods for multilabel classification [12, 15, 24]. One such work [15] proposes using a simple neural network approach, which consists of two modules: a neural network that produces label scores, and a label predictor that converts labels scores into classification using threshold. Later, [12] focused on the problem of assigning multiple labels to each document, where the labels are drawn from an extremely large collection. They propose using a CNN with dynamic max-pooling to obtain document representation, which is then fed to an output layer of the size of label vocabulary, with sigmoid activation and binary cross-entropy loss. In [24], they propose learning a joint feature and label embedding space by combining DNN architectures of canonical correlation analysis and autoencoder, coupled with correlation aware loss function. This neural network model is referred to as Canonical Correlated AutoEncoder (C2AE).

2.2 Biomedical Text Annotation

Biomedical text annotation is a very broad and active area of research [4, 19, 20, 25, 28], e.g. PICO annotations and MeSH tagging. The authors in Singh *et al.* [20] focused on the problem of PICO annotation under the constraint of sparse labelled data. Their approach is based on generating a list of candidate concepts and then filtering this list using a deep neural network. The candidates concepts for a given abstract are generated using a biomedical software called Metamap, which uses various heuristic, rules and IR-based techniques to generate a list of concepts that might be contained in a piece of text.

Another work [21] focused on the similar task of MeSH tagging such that the tags are drawn from a tree-structured vocabulary. They treat biomedical text tagging as an instance of seq-to-seq modelling so that the input text sequence is encoded into a vector representation, which is then decoded into a sequence corresponding to the path of a label from the root of the Ontology. A similar work [14] on applying ICD (diagnostic) codes to clinical text broke down the multilabel classification problem into many binary classification problems. They use a CNN to obtain feature maps of the input text, and then apply per-label attention to obtain a contextual representation, which is then fed to a single-neuron output layer with sigmoid activation. They focus on explaining predictions using the attention distribution over the input text and share the feature maps generated using CNN across different labels.

2.3 Transfer Learning for Text Classification

There have been various attempts to use transfer learning for text classification in the past [2, 3, 6, 17], but with very limited success. Some of these earlier methods were based on inferring parameters of the classification model from dataset statistics [6], thus eliminating the need to train on every task. While this is an interesting approach, these statistics are often unable to transfer anything more than general characteristics of the model that are available e.g. mean or variance. In a different line of work [16], authors proposed an algorithm based on projecting both source and target domain onto the same latent space before classification.

Lately, the popularity of deep learning models for solving various NLP tasks has multiplied manifolds. As a matter of fact, the two works closest to us from Rios *et al.* [19] and Zeng *et al.* [25] are based on CNNs. Both the methods begin by pre-training a CNN on tagging abstracts with MeSH terms, which is then used for applying ICD-9 (disease codes) codes to unstructured EMRs. More recently, significant success has been achieved on variety of NLP tasks [1, 19, 25] by using pre-trained bidirectional encoders. Hence, instead of CNN, we use one such pre-trained bidirectional encoder in this work.

3 Method

In this section, we begin with a brief description of the model architecture used in PICO annotation. This includes a description of the encoder architecture used for both, source and target task. Then, we give details of the source task (i.e. MeSH annotation) used for pre-training the encoder. This is followed by a discussion on the target task (i.e. PICO annotation) where we provide motivations for constructing artificial instances. We also describe our proposed strategies for constructing artificial instances to be used in fine-tuning. Finally, we briefly mention our fine-tuning strategy that uses a mixture of real and artificial data.

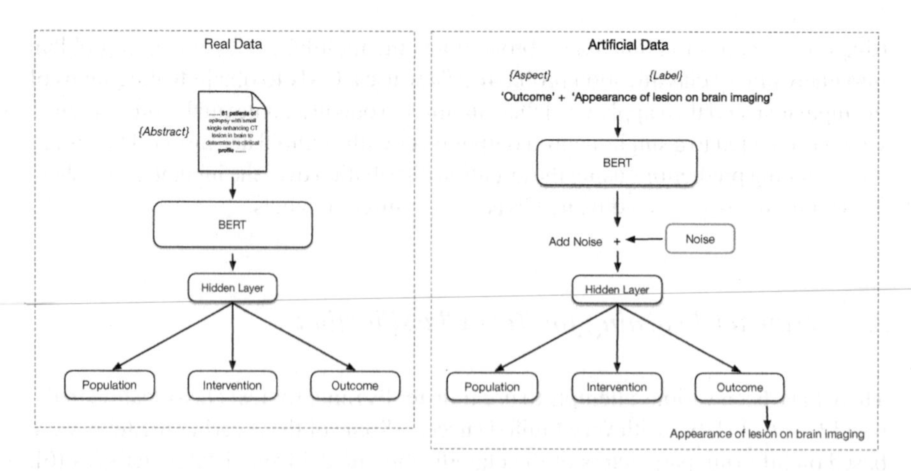

Fig. 2 A pictorial representation of the model architecture and training process. **Left:** The model takes as input an unstructured piece of text e.g. abstract, and applies the assigned labels. **Right:** To address scarcity of labeled data for fine-tuning, we construct artificial inputs for labels using the label-text as input for the label-index

3.1 Model

We provide an overview of the model architecture used for PICO annotation in Fig. 2. The model consists of a bidirectional encoder (i.e. Transformer encoder), followed by a shared hidden layer and three task-specific output layers—one each for concepts corresponding to aspects P, I/C and O.

The encoder provides contextual representation of the input text (e.g. abstract), which is then passed through the shared hidden layers, and then to the task-specific output layers. We append a [CLS] token to the beginning of the input text, and the vector representation of this token is taken to be the vector representation of the input text. The task specific output layers then classify the text into pre-defined vocabulary of PICO concepts. The shared hidden layer is used to leverage correlations between different P, I/C and O concepts, for example, if the trial Population is '*Male*' then Outcome will not generally be '*Breast Cancer*'.

3.1.1 Encoder

We use the encoder of the Transformer [22] network to generate contextual representations of input text. It consists of repeating blocks and each block consists of two sublayers: multi-head self-attention layer and position-wise convolution layer. There are residual connections and layer normalization after every sublayer. The input is a sequence of word embeddings added with positional embeddings before being fed to the encoder.

Recently, it has been shown that pre-trained transformer encoder i.e. BERT [5], can generate rich representations for input text, leading to superior results over a number of low-resource target NLP tasks. Hence, we pre-trained BERT on a biomedical task similar to PICO annotation, this task applies MeSH terms to biomedical abstracts available on PubMed.[1]

3.2 *Source Task: MeSH Annotation*

We extract over 1 million abstracts from Pubmed along with the annotated MeSH tags, where these MeSH terms correspond to specific characteristics of a biomedical text. The extracted dataset covered the entire vocabulary of MeSH terms with ≈29K unique terms. These terms are arranged in the from of a tree-structured hierarchy e.g. *Disease* → *Virus Diseases* → *Central Nervous System Viral Diseases* → *Slow Virus Diseases*. These MeSH terms are similar in functionality to the PICO tags, but they are more coarse and not based on PICO characteristics. The model consisted of BERT as the encoder, followed by a hidden layer and an output layer of the size of MeSH vocabulary. Afterwards, the hidden layer and the output layer are replaced with new layers. Please note that the vocabulary of PICO and MeSH is completely different. We trained the model using binary cross-entropy loss for over 45 days on a single GPU (2080) with a minibatch size of 4.

3.3 *Target Task: PICO Annotation*

We use the BERT encoder trained on the MeSH annotation task for PICO annotations. BERT generates vector representation of a document, these representations are then passed through a new hidden layer followed by three separate output layers—one each corresponding to P, I/C and O aspects of the trial—and the entire model is fine-tuned on the PICO annotation task.

As the task-specific layers are completely replaced, these layers have to be re-trained from scratch on the PICO annotation task. This would not be an issue if there was sufficient labeled data available for effectively fine-tuning these task-specific layers. Unfortunately, biomedical tasks—such as PICO annotation—have scarce labeled data available for fine-tuning. Also, if the label space is large then a number of labels never appear in the fine-tuning set.

Interestingly, labels in PICO annotation are often self-explanatory biomedical concepts expressed in natural language e.g. 'Finding Relating To Institutionalization'. Since the encoder has been pre-trained on a large corpora of biomedical abstracts, we hypothesize that it can be used to obtain rich contextual representations for these PICO labels. A label-text can then be used as an input for the label-index;

[1] An online repository of biomedical abstracts.

in other words, we are constructing artificial instances for labels using the label-text as input and label-index as output. We propose different strategies for constructing artificial labeled instances in this fashion.

3.3.1 Artificial Instance Construction

In this section we describe our proposed strategies for constructing artificial inputs for label-indexes. These strategies are based on using label-text to construct artificial inputs. These newly created artificial instances can then be used in combination with real instances for fine-tuning. Here, we describe these strategies in details:

- **LI:** We refer to the first strategy as LI which stands for *Label* as *Input*. In this strategy, we feed the network with the label-text (e.g. 'Finding Relating To Institutionalization') as an input for the label-index. We also append the [CLS] token to the beginning of the resulting string. The encoder generated representation for the label-text is then passed on to the shared hidden layer and then to the (three) output layers for classification. This way we have generated extra labeled instances for labels, which helps in effective fine-tuning of the task-specific layers, especially for labels that are rare.
- **LIS:** LIS stands for LI + Synonyms. We randomly replace different words in the label-text with their synonyms. More specifically, we randomly select one of the words in the label-text and replace it with its synonym, we then select another word from the label-text and repeat the same process over to generate multiple instances from a single label-text. As an example, 'Finding Relating To Institutionalization' → 'Conclusion Relating To Institutionalization'. We only replace one word at a time to ensure that semantics of the label-text do not venture too far from the original meaning. These synonyms are extracted from the WordNet corpus.
- **LISA:** LISA stands for *LIS + Aspect*, where Aspect is the P, I/C and O characteristic of the trial. In this strategy, we append one of the aspects to the label-text e.g. 'Outcome'+ 'Finding Relating To Institutionalization', which is then classified as the label in the output layer corresponding to that specific aspect—we have three output layers corresponding to P, I/C and O aspects.
- **LISAAS:** LISA+AS stands for *LIS + Aspect + Auxiliary Sentences*. As opposed to the previous strategy of simply appending the aspect to the label-text, we construct artificial sentences by prepending a pre-defined fixed aspect-based sentence to the label-text. For example, 'The population of the trial consists of patients with' + 'Diabetes' as an input text for the label 'Diabetes', where the first part is the sentence containing the aspect of the label and last part is the label-text. We prepare a fixed set of three aspect-sentences, one corresponding to each P, I/C and O aspect.
- **LISAAS + Noise:** While *LISAAS* is closer to a real abstract than other strategies, it can still bias the model towards expecting the exact sentence structure at test time. We can not possibly construct all possible input texts in this manner to train the network. To address this issue, we add random Gaussian noise to the vector representations of the input text generated by the encoder. The encoder

should be able to attend to important bits of information at test time but due to the differences in sentence structure the vector representation might be slightly different. By adding noise we wanted to replicate that scenario during training. We apply layer normalization after adding noise to ensure the norm of the signal remains similar to the real input text.

3.3.2 Training

We randomly choose to construct a minibatch from either all real or all artificial instances for the first 50% epochs during fine-tuning. Afterwards, we continue with only real data until the end. This training process is followed with all the different strategies for constructing artificial instances.

4 Experimental Setup

In this section, we describe in details the dataset used for experiments, the baseline models used for comparison and the evaluation setup and the metrics used to assess the performance.

4.1 Dataset

We use a real-world dataset provided by Cochrane[2] which consists of manual annotations applied to biomedical abstracts. More specifically, as we have outlined throughout this paper, trained human annotators have applied tags from a subset of the Unified Medical Language System (UMLS) to free-text summaries of biomedical articles, corresponding to the PICO aspects. We should recall that PICO stands for Population, Intervention/Comparator and Outcomes. Population refers to the characteristics of clinical trial participants (e.g., diabetic males). Interventions are the active treatments being studied (e.g., aspirin); Comparators are baseline or alternative treatments to which these are compared (e.g., placebo)—the distinction is arbitrary, and hence we collapse I and C. While the Outcomes are the variables measured to assess the efficacy of treatments (e.g., headache severity).

These annotations are performed by trained human annotators, who attached concept terms corresponding to each PICO aspect of individual trial summaries.

[2]Cochrane is an international organization that focusses on improving healthcare decisions through evidence: http://www.cochrane.org/.

4.2 Baselines

We use three straightforward baselines for comparison, these are: (1) training a CNN-based multitask classifier that directly predicts the output labels based on the input text, (2) BERT pre-trained on *Wikipedia* and *BookCorpus*, and, (3) BERT pre-trained on MeSH tagging task.

The CNN-based model consists of an embeddings layer, which is followed by three parallel convolution layers with filter sizes 1, 3 and 5. Each of these convolution layers are followed by a ReLU activation and max-pooling layer, after which the outputs from these 3 layers are concatenated. We then apply: a dropout layer, dense layer, ReLU activation layer; in that order. Finally, we pass the hidden representation to three different output layers for classification.

The other obvious choice for a baseline is to use BERT pre-trained on BookCorpus [27] and English Wikipedia, which are the same datasets used in the original paper [5]. The encoder consists of 12 consecutive blocks where each such block consists of 12 multihead attention heads and position-wise convolution layers. The encoder representations are then passed through three separate dense layers where the output of each layer is passed to one of the three output layers. These three output layers classify the input text into corresponding P, I/C and O elements.

There have been works in the past using CNNs pre-trained on MeSH tagging for separate biomedical annotation tasks. Hence, we decided to pre-train BERT on a dataset of over one million abstracts tagged with MeSH terms. The architecture of the model exactly resembles the one described above. All the baselines are only trained on real data.

4.3 Evaluation Details

We divided the dataset into 80/20 for train/test split. We had ground truth annotations for all instances corresponding to PICO aspects of the trial i.e. all texts have been annotated by domain experts with labels from the UMLS vocabulary. The texts here are summaries of each element extracted for previous reviews; we therefore concatenate these summaries to form continuous piece of text that consists of text spans describing respective elements of the trials. The exhaustive version of vocabulary used by Cochrane consists of 366,772 concepts, but we restricted the vocabulary to labels that are present in the dataset. We summarize some of these statistics in Table 1.

We performed all of the hyper-parameter tuning via nested-validation. More specifically, we separated out 20% of training data for validation, and kept it aside for tuning over hyper-parameters, which included iteratively experimenting with various values for different hyper-parameters and also the structure of the network. The values for dropout were fixed for all dropout layers to reduce the workload of optimization, and that value was optimized over 10 equidistant steps in the range of

Table 1 Dataset statistics

Samples (clinical trials)	10137
Distinct population concepts	1832
Distinct intervention concepts	1946
Distinct outcome concepts	2556
Population concepts	15769
Intervention concepts	10537
Outcome concepts	19547

[0, 1]. The threshold of binary classification in the output layer of all the networks was also tuned over 10 equidistant values in the range [0, 1]. We trained for a maximum of 150 epochs and used early stopping to get the optimal set of parameters that performed the best on the nested-validation set. We optimized for the highest micro-f1 score for all of the tuning.

4.4 Metrics

We evaluated our approach on 3 standard metrics (1) Precision, (2) Recall and (3) F1 score. We computed macro and micro versions of these three metrics for each of the three aspects i.e. P, I/C and O separately. Finally, we compute the average of Micro-F1 scores for all the three aspects, and this was the ultimate metric used for tuning hyper-parameters and to decide for early stopping while training.

5 Results

We report results for all the models using various metrics in Table 2. CNN refers to a standard CNN-based multitask classifier that directly maps an input text into the output labels. Also, we use two separate variants of BERT, one that was trained on the BookCorpus and Wikipedia—as described in the original paper on BERT, and the other one referred to as BERT-MeSH was pre-trained on a subset of MeSH-tagged biomedical abstracts. We compare our proposed method with these competitive baselines (Fig. 3).

Our proposed artificial instance based approaches uniformly beat the three competitive baselines in terms of macro and micro values of precision, recall and f1 score, as can be seen in Table 2. This means that the model not only correctly predicts frequent labels, but also improves performance on infrequent labels. Please note that for biomedical annotation tasks that have a large proportion of rare labels even slightly improving macro-f1 score can be very challenging, therefore, these

Table 2 Precisions, recalls and f1 measures realized by different models on the respective PICO elements. Best result for each element and metric is **bolded**. CNN refers to the CNN-based multi-task classifier, BERT refers to the transformer encoder pre-trained using BookCorpus and English Wikipedia in the original paper [5], while BERT-MeSH refers to the transformer encoder pre-trained on MeSH tagging task. Please note that all methods used the same subword tokenizer provided by BERT. Also, all baselines were trained on only real data

Category	Model	Macro-pr	Macro-re	Macro-f1	Micro-pr	Micro-re	Micro-f1
Population	CNN	0.031	0.027	0.027	0.614	0.506	0.555
	BERT	0.040	0.036	0.037	0.714	0.572	0.635
	BERT-MeSH	0.043	0.038	0.038	0.719	0.560	0.629
	LI	0.048	0.044	0.044	0.726	0.604	0.659
	LIS	0.046	0.042	0.043	0.733	0.596	0.658
	LISA	0.046	0.042	0.042	**0.757**	0.606	**0.673**
	LISAAS	0.048	0.044	0.045	0.740	0.603	0.664
	+ Noise	**0.049**	**0.046**	**0.046**	0.737	**0.613**	0.670
Interventions/Comparator	CNN	0.027	0.023	0.024	0.671	0.418	0.515
	BERT	0.032	0.028	0.029	0.717	0.465	0.564
	BERT-MeSH	0.035	0.032	0.032	0.729	0.483	0.581
	LI	0.043	0.041	0.041	0.700	0.549	0.616
	LIS	0.044	0.040	0.041	**0.753**	0.528	0.621
	LISA	0.042	0.037	0.038	0.743	0.526	0.616
	LISAAS	0.044	0.041	0.041	0.748	0.540	0.627
	+ Noise	**0.045**	**0.043**	**0.042**	0.740	**0.550**	**0.631**
Outcomes	CNN	0.037	0.030	0.030	0.530	0.333	0.409
	BERT	0.043	0.035	0.036	0.603	0.398	0.480
	BERT-MeSH	0.048	0.039	0.041	0.612	0.412	0.492
	LI	0.052	0.046	0.046	0.611	**0.468**	**0.530**
	LIS	0.052	0.044	0.045	0.621	0.442	0.517
	LISA	0.050	0.042	0.043	**0.629**	0.442	0.519
	LISAAS	**0.054**	0.046	0.047	0.620	0.453	0.523
	+ Noise	0.053	**0.047**	**0.047**	0.620	0.455	0.525

results are encouraging. In addition, we also compute the average of micro-f1 score and macro-f1 score for all three PICO aspects in Table 3. Our proposed approach ***LISAAS+Noise*** beats all the other baselines, especially in terms of macro metrics. Implying, the strategy of adding noise to encoder representations helps the model classify rare labels more accurately.

Another interesting insight is that pre-training BERT on biomedical abstracts (i.e. MeSH-tagged Pubmed abstracts) performs better in comparison to simply using the original BERT pre-trained on English Wikipedia. Therefore, we can safely infer that pre-training encoders on biomedical abstracts can lead to slightly improved performance over target biomedical tasks.

We have described the motivations for adding noise to encoder representations in the Section on Methodology. We draw the noise vector from a standard normal distribution and then scale its norm to a certain fraction of the norm of the encoder representation. In order to find out what level of noise leads to best results, we plot

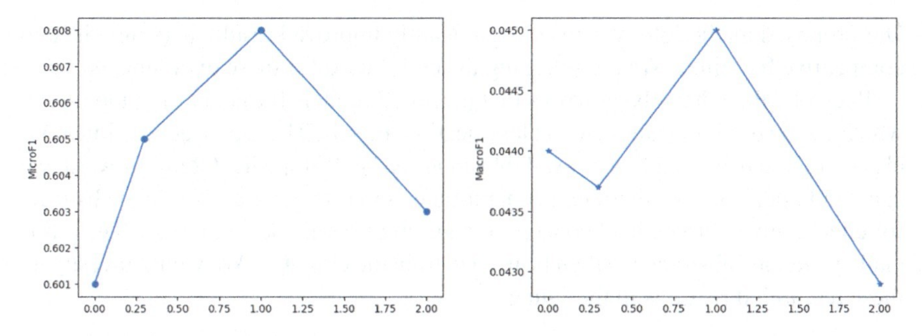

Fig. 3 Performance in terms of average micro and macro-f1 scores versus the norm of the noise added to encoder representations. We scale the norm of the random noise prior to adding it. Hence, the x-axis represents the norm of the noise as a fraction of the norm of the encoder representation

Table 3 Average of macro-f1/micro-f1 scores for all three aspects of a clinical trial i.e. Population, Intervention/Control and Outcome. Best result for each metric is **bolded**

Model	Avg. micro-f1	Avg. macro-f1
CNN	0.493	0.027
BERT	0.559	0.034
BERT-MeSH	0.567	0.037
LI	0.601	0.044
LIS	0.599	0.043
LISA	0.603	0.041
LISAAS	0.605	0.044
+ Noise	**0.609**	**0.045**

the average micro-f1 and macro-f1 scores for various values of this fraction. We tune this fraction over the validation set and obtain a value of 1.0, i.e. the noise has the same norm as the encoder vector representation. We were expecting the noise to be much lower for best results, but contrary to our expectations larger noise was needed by the model to handle artificial data.

6 Conclusion

We proposed a new method for biomedical text annotation that can work with limited training data. More specifically, our model uses pre-trained bidirectional encoders for mapping clinical texts to output concepts. Generally, such models require fine-tuning on the target task, but there is often insufficient labeled data for fine-tuning. To address this we proposed constructing extra artificial instances for fine-tuning using the labels themselves. We describe various strategies for constructing these artificial instances.

Our proposed methodology leads to significantly improved results in comparison to competitive baselines. We are releasing the code[3] used for these experiments.

Recently, there have been works that proposed models for generating input-text, which can be used for training text classification models. These text-generation models are trained using publically available datasets e.g. Wikipedia. Going forward, we can build upon some of these text generation methods to generate artificial abstracts for either a given biomedical concept or a group of biomedical concepts. The artificially generated abstract can then be used to train the classifier. We would investigate these research directions in the future.

Acknowledgements We would like to thank Cochrane for providing us with PICO annotation data.

References

1. Adhikari, A., Ram, A., Tang, R., Lin, J.: Docbert: bert for document classification. arXiv preprint arXiv:1904.08398 (2019)
2. Dai, W., Chen, Y., Xue, G.R., Yang, Q., Yu, Y.: Translated learning: transfer learning across different feature spaces. In: Advances in Neural Information Processing Systems, pp. 353–360 (2009)
3. Dai, W., Xue, G.R., Yang, Q., Yu, Y.: Transferring naive bayes classifiers for text classification. AAAI **7**, 540–545 (2007)
4. Demner-Fushman, D., Elhadad, N., et al.: Aspiring to unintended consequences of natural language processing: a review of recent developments in clinical and consumer-generated text processing. IMIA Yearbook (2016)
5. Devlin, J., Chang, M.W., Lee, K., Toutanova, K.: Bert: pre-training of deep bidirectional transformers for language understanding. arXiv preprint arXiv:1810.04805 (2018)
6. Do, C.B., Ng, A.Y.: Transfer learning for text classification. In: Advances in Neural Information Processing Systems, pp. 299–306 (2006)
7. Elisseeff, A., Weston, J.: A kernel method for multi-labelled classification. In: Advances in Neural Information Processing Systems, pp. 681–687 (2002)
8. Elisseeff, A., Weston, J., et al.: A kernel method for multi-labelled classification. NIPS **14**, 681–687 (2001)
9. Fürnkranz, J., Hüllermeier, E., Loza Mencía, E., Brinker, K.: Multilabel classification via calibrated label ranking. Mach. Learn. **73**(2), 133–153 (2008)
10. Lai, S., Xu, L., Liu, K., Zhao, J.: Recurrent convolutional neural networks for text classification. In: Twenty-Ninth AAAI Conference on Artificial Intelligence (2015)
11. Lee, J.Y., Dernoncourt, F.: Sequential short-text classification with recurrent and convolutional neural networks. arXiv preprint arXiv:1603.03827 (2016)
12. Liu, J., Chang, W.C., Wu, Y., Yang, Y.: Deep learning for extreme multi-label text classification. In: Proceedings of the 40th International ACM SIGIR Conference on Research and Development in Information Retrieval, pp. 115–124. ACM (2017)
13. McCallum, A.: Multi-label text classification with a mixture model trained by EM. In: AAAI Workshop on Text Learning, pp. 1–7 (1999)
14. Mullenbach, J., Wiegreffe, S., Duke, J., Sun, J., Eisenstein, J.: Explainable prediction of medical codes from clinical text. arXiv preprint arXiv:1802.05695 (2018)
15. Nam, J., Kim, J., Mencía, E.L., Gurevych, I., Fürnkranz, J.: Large-scale multi-label text classification–revisiting neural networks. In: Joint European Conference on Machine Learning and Knowledge Discovery in Databases, pp. 437–452. Springer (2014)

[3]https://github.com/gauravsc/pico-tagging.

16. Pan, S.J., Kwok, J.T., Yang, Q., et al.: Transfer learning via dimensionality reduction. AAAI **8**, 677–682 (2008)
17. Pan, S.J., Yang, Q.: A survey on transfer learning. IEEE Trans. Knowl. Data Eng. **22**(10), 1345–1359 (2009)
18. Read, J., Pfahringer, B., Holmes, G., Frank, E.: Classifier chains for multi-label classification. In: Machine Learning and Knowledge Discovery in Databases, pp. 254–269 (2009)
19. Rios, A., Kavuluru, R.: Neural transfer learning for assigning diagnosis codes to EMRs. Artif. Intell. Med. **96**, 116–122 (2019)
20. Singh, G., Marshall, I.J., Thomas, J., Shawe-Taylor, J., Wallace, B.C.: A neural candidate-selector architecture for automatic structured clinical text annotation. In: Proceedings of the 2017 ACM on Conference on Information and Knowledge Management, pp. 1519–1528. ACM (2017)
21. Singh, G., Thomas, J., Marshall, I., Shawe-Taylor, J., Wallace, B.C.: Structured multi-label biomedical text tagging via attentive neural tree decoding. In: Proceedings of the 2018 Conference on Empirical Methods in Natural Language Processing, pp. 2837–2842 (2018)
22. Vaswani, A., et al.: Attention is all you need. In: Advances in Neural Information Processing Systems, pp. 5998–6008 (2017)
23. Yang, Z., Yang, D., Dyer, C., He, X., Smola, A., Hovy, E.: Hierarchical attention networks for document classification. In: Proceedings of the 2016 Conference of the North American Chapter of the Association for Computational Linguistics: Human Language Technologies, pp. 1480–1489 (2016)
24. Yeh, C.K., Wu, W.C., Ko, W.J., Wang, Y.C.F.: Learning deep latent space for multi-label classification. In: Thirty-First AAAI Conference on Artificial Intelligence (2017)
25. Zeng, M., Li, M., Fei, Z., Yu, Y., Pan, Y., Wang, J.: Automatic ICD-9 coding via deep transfer learning. Neurocomputing **324**, 43–50 (2019)
26. Zhou, C., Sun, C., Liu, Z., Lau, F.: A C-LSTM neural network for text classification. arXiv preprint arXiv:1511.08630 (2015)
27. Zhu, Y., et al.: Aligning books and movies: towards story-like visual explanations by watching movies and reading books. In: Proceedings of the IEEE ICCV, pp. 19–27 (2015)
28. Zweigenbaum, P., Demner-Fushman, D., Yu, H., Cohen, K.B.: Frontiers of biomedical text mining: current progress. Briefings Bioinform. **8**(5), 358–375 (2007)

Character-Level Japanese Text Generation with Attention Mechanism for Chest Radiography Diagnosis

Kenya Sakka, Kotaro Nakayama, Nisei Kimura, Taiki Inoue, Yusuke Iwasawa, Ryohei Yamaguchi, Yoshimasa Kawazoe, Kazuhiko Ohe, and Yutaka Matsuo

Abstract Chest radiography is a general method for diagnosing a patient's condition and identifying important information. Therefore, a large amount of chest radiographs have been taken. In order to reduce the burden on medical professionals, methods for generating findings have been proposed. However, the study of generating chest radiograph findings has primarily focused on the English language, and

K. Sakka (✉) · K. Nakayama · N. Kimura · T. Inoue · Y. Iwasawa · R. Yamaguchi · Y. Kawazoe · K. Ohe · Y. Matsuo
The University of Tokyo, Tokyo, Japan
e-mail: ksakka0309@gmail.com

K. Nakayama
e-mail: nakayama@weblab.t.u-tokyo.ac.jp

N. Kimura
e-mail: kimura@weblab.t.u-tokyo.ac.jp

T. Inoue
e-mail: taiki-inoue@g.ecc.u-tokyo.ac.jp

Y. Iwasawa
e-mail: iwasawa@weblab.t.u-tokyo.ac.jp

R. Yamaguchi
e-mail: r-yamagu@m.u-tokyo.ac.jp

Y. Kawazoe
e-mail: kawazoe@hcc.h.u-tokyo.ac.jp

K. Ohe
e-mail: kohe@hcc.h.u-tokyo.ac.jp

Y. Matsuo
e-mail: matsuo@weblab.t.u-tokyo.ac.jp

K. Nakayama
NABLAS Inc., Bunkyo City, Japan

© The Editor(s) (if applicable) and The Author(s), under exclusive license to Springer Nature Switzerland AG 2021
A. Shaban-Nejad et al. (eds.), *Explainable AI in Healthcare and Medicine*, Studies in Computational Intelligence 914,
https://doi.org/10.1007/978-3-030-53352-6_13

to the best of our knowledge, no studies have studied Japanese data on this subject. The difficult points of the Japanese language are that the boundaries of words are not clear and that there are numerous orthographic variants. For deal with two problems, we proposed an end-to-end attention-based model that generates Japanese findings at the character-level from chest radiographs. We evaluated the method using a public dataset of Japanese chest radiograph findings. Furthermore, we confirmed via visual inspection that the attention mechanism captures the features and positional information of radiographs.

Keywords Medical image captioning · Character level approach · Attention mechanism · Japanese text

1 Introduction

Chest radiography is a general method for determining the patient's condition and for identifying important information. Thus, chest radiography is performed for a large number of patients in several situations. Such as emergency medical care and medical checkup. However, a high level of expertise is required for interpreting chest radiographs [6]. Thus, a large workforce of medical specialists is necessary. As a result, medical workers including radiologists are over-burdened with work, and solutions are required to resolve this issue. Therefore, the study has been actively conducted to automatically findings generation [2, 5, 14, 19]. However, the study of generating medical findings primarily focused on the English language.

The Japanese Social of Radiological Technology (JSRT) Dataset [18] created by JSRT consists of Japanese chest radiographs (Fig. 1). Several study using the JSRT dataset has been conducted [3, 9]. However, to the best of our knowledge, no study has been conducted to target findings generation in the Japanese language. There are two challenges involved in generating findings in the Japanese language. The first challenge is that word splitting is difficult (Fig. 1). For example, "Upper-left lung abnormal shadow" is expressed as "左上肺野異常影" in the Japanese language. A sentence in the English language can be split into words based on the spaces. However, the boundaries of words are not clear in the Japanese language. The second challenge is that there are many orthographic variants of words in the Japanese language. In the case of findings on chest radiographs, when the abnormal area is on both sides of the lungs, there are multiple expressions, such as "両肺", "両側肺", "両肺部", and "両側肺部" (English: both sides of the lung). For overcoming two challenges, we proposed an end-to-end model that generates Japanese findings in character-level [8, 12] using an attention mechanism.

We evaluated our method using a public JSRT Dataset. The effectiveness of the method was confirmed using the BLEU score [13]. We found that our method could generate Japanese findings corresponding to the orthographic variants. Furthermore, we have confirmed via visual inspection that the attention mechanism captured the features and positional information of chest radiographs.

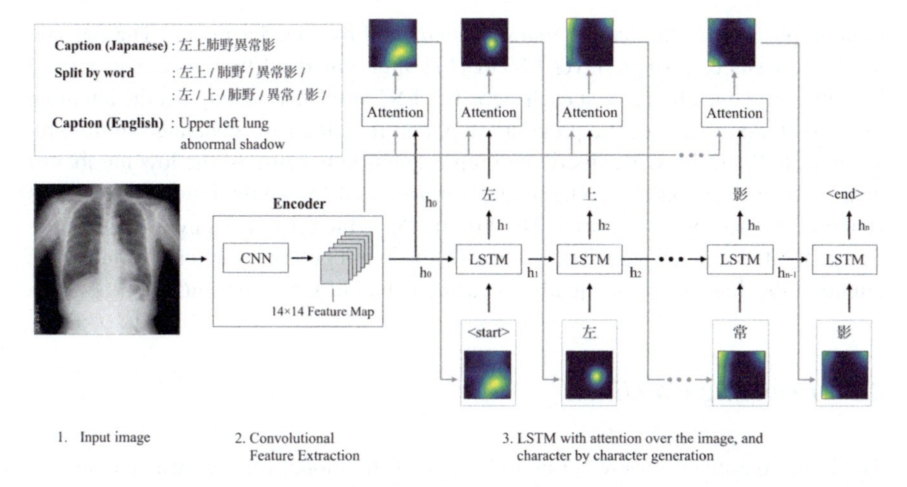

Fig. 1 Architecture of the proposed model. (1) Input the pre-processed chest radiograph into the encoder. (2) The feature maps for the input image are extracted using Convolutional Neural Network (CNN). (3) In the decoder, the findings are generated at the character-level using long short-term memory (LSTM) and the attention mechanism based on the extracted feature maps

2 Materials and Methods

2.1 Datasets

We prepared two datasets from JSRT Dataset based on the threshold for the appearance frequency of the medical findings. The data with the appearance frequency of finding lower than the threshold were excluded because the model cannot learn the features of data sufficiently [2, 5]. The thresholds were set at 5 and 30. A dataset with a threshold of 5 covered 94.79% of all data and included 164 findings composed of 118 characters. The minimum number of characters in the findings was 3 and the maximum number of characters was 9. The number of training data was 12,810, validation data was 1,602 and test data was 1,602. Furthermore, a dataset with a threshold of 30 covered 87.02% of all the data, and included 26 findings composed of 65 characters. The minimum number of characters in the findings was 3, and the maximum number of characters was 13. The number of training data was 11,752, validation data was 1,453 and test data was 1,453.

2.2 Overview of Methods

In this paper, we propose an Encoder-Decoder model that uses a attention mechanism to output the medical findings of the Japanese language at character-level. The input to the encoder model using pre-trained ResNet-151 [11] is a three-channel chest radiograph. Thereafter, feature maps are extracted by encoder model. The decoder model consists of a single layer LSTM [17], and a medical finding was generated by outputting one character at each step. The LSTM takes three inputs: the attention map, the hidden state of the previous step, and the character generated in the previous step. We did not use distributed representation because of the low number of characters. The encoder-decoder model is trained end-to-end by minimizing the loss function defined by two factors. The first factor is the cross-entropy of the characters generated at each step. The second factor is a regularization term to make the attention mechanism work equally on each channel of the feature map [12].

2.3 Learning Condition

The hyper parameters were set as follows, and the model training was conducted. Batch size was 16, dropout rate of the decoder is 0.5 [15], learning rate of the encoder was 0.0001, the learning rate of the decoder was 0.0004, and the optimization function was Adam [7]. When the validation data was not improved for 10 epochs continuously, the learning rate was adjusted by multiplying the learning rate of the encoder and the decoder by 0.8 [1]. The maximum number of epoch was 200, and early stopping [16] was conducted when the BLEU-4 score was not improved for 20 consecutive epochs of the validation data to avoid overfitting. The model trained by oversampling and undersampling [4, 10] because the balance of dataset was imbalanced.

2.4 Evaluation Metrics

Two types of datasets were prepared as test datasets. The first test dataset was the original distribution dataset. In the original distribution dataset, the distribution of each finding is actually obtained in the medical field. Most findings obtained in the actual medical field are normal. Therefore, "異常なし (English: normal)" accounted for approximately 75.58% of the total data in the original distribution dataset. The second test dataset contained only abnormal data. The only abnormal dataset is prepared by excluding the data annotated by "異常なし (English: normal)". By evaluating with the only abnormal dataset, it is possible to prevent a model biased toward "異常なし (English: normal)" from being highly evaluated. We evaluated the performance of the model with the BLEU 1–4 score.

3 Results and Discussion

The result of the attention mechanism and the generated findings were shown in Fig. 2. A chest radiograph of the patient with the implanted device confirmed that the weight distribution of the attention mechanism was high around the device (Fig. 2A). The attention mechanism works strongly at the position of the device throughout the generation of findings. In the case of "両側肺尖胸肥厚 (English: bilateral pulmonary apical chest thickening)", we confirmed that the model captured the positional information (Fig. 2B). In order to diagnose the normal condition from information on the chest radiograph, it is necessary to deny the possibility of any other abnormal findings. The abnormal part of each finding is distributed in various areas in the chest radiograph. Therefore, the entire image must be interpreted comprehensively. We confirmed the distribution of weights of the attention mechanism when the model generated a finding with "異常なし (English: normal)". The attention mechanism was indicated to function over the entire region of the upper body (Fig. 2C). Furthermore, we evaluated the accuracy of the data for "異常なし (English: normal)" with perfect match. The accuracies were 93.84% and 76.99% for the datasets with thresholds of 5 and 30, respectively.

The evaluation result of two test dataset was shown in Table 1. In the original distribution dataset, both oversampling and undersampling methods have shown a high BLEU score. In the dataset with only abnormal findings, undersampling scored slightly higher. In the medical field, the model is required to consider abnormal findings. Therefore, it is generally important to increase recall over precision. From that perspective, undersampling was superior to oversampling in our experiments.

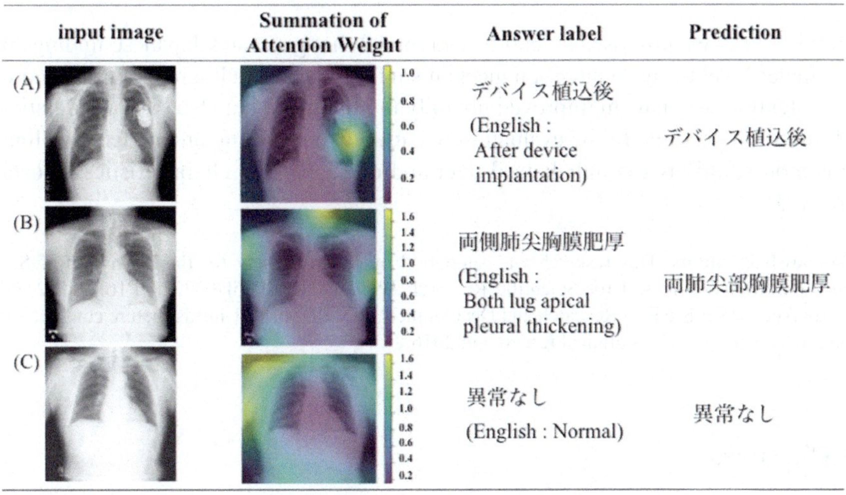

	input image	Summation of Attention Weight	Answer label	Prediction
(A)		1.0 0.8 0.6 0.4 0.2	デバイス植込後 (English : After device implantation)	デバイス植込後
(B)		1.6 1.4 1.2 1.0 0.8 0.6 0.4 0.2	両側肺尖胸膜肥厚 (English : Both lug apical pleural thickening)	両肺尖部胸膜肥厚
(C)		1.6 1.4 1.2 1.0 0.8 0.6 0.4 0.2	異常なし (English : Normal)	異常なし

Fig. 2 Visualization of the attention mechanism and the generated medical findings

Table 1 BLEU score of our methods. We evaluated the original distribution dataset and only abnormal dataset in oversampling and undersampling. Each dataset was filtered as per the threshold of the appearance frequency of the findings. BLEU-n indicates the BLEU score of the test data that uses up to n-grams. The last column indicates the number of findings generated with our method

Datasets	Methods	Threshold	BLEU-1	BLEU-2	BLEU-3	BLEU-4	Number of findings
Original distribution dataset	Oversampling	5	**0.7708**	**0.7668**	**0.7604**	**0.7535**	72
		30	**0.8322**	**0.8304**	**0.8261**	**0.8214**	22
	Undersampling	5	0.5472	0.5353	0.5091	0.4934	42
		30	0.6203	0.6142	0.5962	0.5873	26
Only abnormal dataset	Oversampling	5	0.2348	0.2193	0.1980	**0.1691**	61
		30	0.1980	0.1882	0.1664	0.1376	16
	Undersampling	5	**0.2612**	**0.2314**	**0.2093**	0.1439	33
		30	**0.2641**	**0.2468**	**0.2258**	**0.1713**	23

There are orthographic variants in the findings of the chest radiograph; thus, the existing evaluation metrics, such as the BLEU score are difficult to evaluate correctly. For example, "両側肺尖胸膜肥厚" and "両肺尖部胸膜肥厚" have the same meaning in terms of words (Fig. 2B). However, the similarity of the two findings is low because the existing evaluation metrics consider the position of each character and do not consider the orthographic variants. Therefore, in the case of the only abnormal dataset, the total BLEU score was lower Table 1.

4 Conclusion

In this paper, we proposed an end-to-end model that generates Japanese findings in character-level using an attention mechanism for the chest radiograph. Furthermore, the attention mechanism improved not only the accuracy, but also the interpretation ability of the results. In the evaluation, we used oversampling and undersamplling, common solutions for imbalanced data and discussed the characteristics of each method.

Acknowledgements This research was supported by Japan Society for the Promotion of Science (JSPS) Grant-in-Aid for Scientific Research JP25700032, JP15H05327, JP16H06562 and Japan Agency for Medical Research and Development (AMED) of ICT infrastructure construction research business such as clinical research in 2016.

References

1. Anders, K., John, A.H.: A simple weight decay can improve generalization. In: NIPS (1992)

2. Baoyu, J., Pengtao, X., Eric, P.X.: On the automatic generation of medical imaging reports. In: ACL (2017)
3. Candemir, S., et al.: Lung segmentation in chest radiographs using anatomical atlases with nonrigid registration. IEEE Trans. Med. Imaging **33**, 577–590 (2014)
4. Chawla, V.N., Bowyer, W.K., Hall, O.L., Kegelmeyer, P.W.: SMOTE: synthetic minority over-sampling technique. J. Artif. Intell. Res. **16**, 321–357 (2002)
5. Christy, Y.L., Xiaodan, L., Zhiting, H., Eric, P.X.: Knowledge-driven Encode. Retrieve, Paraphrase for Medical Image Report Generation. In: AAAI (2019)
6. Delrue, L., Gosselin, R., Ilsen, B., Landeghem, V.A., de Mey, J., Duyck, P.: Difficulties in the interpretation of chest radiography. In: Comparative Interpretation of CT and Standard Radiography of the Chest (2010)
7. Diederik, P.K., Jimmy, L.B.: Adam: a method for stochastic optimization. In: ICLR (2015)
8. Dzmitry, B., KyungHyun, C., Yoshua, B.: Neural machine translation by jointly learning to align and translate. In: ICLR (2015)
9. Hamada, A.A., Samir, B., Suziah, S.: A computer aided diagnosis system for lung cancer based on statistical and machine learning techniques. J. Comput. **9**, 425–431 (2014)
10. He, H., Edwardo, A.G.: Learning from imbalanced data. Trans. Knowl. Data Eng. **21**, 1263–1284 (2009)
11. Kaiming, H., Zhang, X., Ren, S., Sun, J.: Deep residual learning for image recognition. In: CVPR (2016)
12. Kelvin, X., et al.: Show, attend and tell: neural image caption generation with visual attention. In: PMLR (2015)
13. Kishore, P., Salim, R., Todd, W., Wei, J.Z.: BLEU: a method for automatic evaluation of machine translation. In: ACL (2002)
14. Li, Y.C., Liang, X., Hu, Z., Xing, P.E.: Hybrid retrieval-generation reinforced agent for medical image report generation. In: NIPS (2018)
15. Nitish, S., Geoffrey, H., Alex, K., Ilya, S., Ruslan, S.: Dropout: a simple way to prevent neural networks from overfitting. J. Mach. Learn. Res. **15**, 1929–1958 (2014)
16. Rich, C., Steve, L., Lee, G.: Overfitting in neural nets: backpropagation, conjugate gradient, and early stopping. In: NIPS (2001)
17. Sepp, H., Jürgen, S.: Long short-term memory. Neural Comput. **9**, 1735–1780 (1997)
18. Shiraishi, J., et al.: Development of a digital image database for chest radiographs with and without a lung nodule receiver operating characteristic analysis of radiologists' detection of pulmonary nodules. Am. J. Roentgenol. **174**, 71–74 (2000)
19. Wang, X., Peng, Y., Lu, L., Lu, Z., Summers, M.R.: TieNet: text-image embedding network for common thorax disease classification and reporting in chest X-rays. In: CVPR (2018)

Extracting Structured Data from Physician-Patient Conversations by Predicting Noteworthy Utterances

Kundan Krishna, Amy Pavel, Benjamin Schloss, Jeffrey P. Bigham, and Zachary C. Lipton

Abstract Despite concerted efforts to mine various modalities of medical data, the conversations between physicians and patients at the time of care remain an untapped resource of insights. In this paper, we explore the possibility of leveraging this data to extract structured information that might assist physicians with post-visit documentation in electronic health records, potentially lightening the clerical burden. In this exploratory study, we describe a new dataset consisting of conversation transcripts, post-visit summaries, corresponding supporting evidence (in the transcript), and structured labels. We focus on the tasks of recognizing relevant diagnoses and abnormalities in the review of organ systems (RoS). One methodological challenge is that the conversations are long (around 1500 words) making it difficult for modern deep-learning models to use them as input. To address this challenge, we extract *noteworthy* utterances—parts of the conversation likely to be cited as evidence supporting some summary sentence. We find that by first filtering for (predicted) noteworthy utterances, we can significantly boost predictive performance for recognizing both diagnoses and RoS abnormalities.

K. Krishna (✉) · A. Pavel · J. P. Bigham · Z. C. Lipton
Carnegie Mellon University, Pittsburgh, USA
e-mail: kundank@andrew.cmu.edu

A. Pavel
e-mail: apavel@andrew.cmu.edu

J. P. Bigham
e-mail: jbigham@andrew.cmu.edu

Z. C. Lipton
e-mail: zlipton@andrew.cmu.edu

B. Schloss
Abridge AI Inc., Pittsburgh, USA
e-mail: bschloss@abridge.ai

© The Editor(s) (if applicable) and The Author(s), under exclusive license to Springer Nature Switzerland AG 2021
A. Shaban-Nejad et al. (eds.), *Explainable AI in Healthcare and Medicine*, Studies in Computational Intelligence 914,
https://doi.org/10.1007/978-3-030-53352-6_14

1 Introduction

Medical institutions collect vast amounts of patient Electronic Health Record (EHR) data including family history, past surgeries, medications and more. Such EHR data helps physicians recall past visits, assess patient conditions over time, and learn crucial information (e.g., drug allergies) in emergency scenarios. However, EHR data is tedious and time consuming for physicians to produce. For every hour of visiting patients, physicians spend around 45 min of EHR documentation [21], and often need to complete documentation outside of work hours which contributes to burnout [8]. Physicians spend much of the EHR documentation time recalling and manually entering information discussed with the patient (e.g., reported symptoms). While transcribing physician-patient discussions could aid EHR documentation, such conversations are long (roughly 10 min or 1500 words in our dataset) and difficult to read due to redundancies and disfluencies typical of conversation.

To mitigate the burden of EHR documentation, we leverage transcribed physician-patient conversations to automatically extract structured data. As an initial investigation, we explore two prediction tasks using the physician-patient conversation as input: relevant diagnosis prediction, and organ system abnormality prediction. In the first task, we extract the set of diagnosis mentioned in the conversations that are relevant to the chief complaint of the patient (i.e. the purpose of the visit), omitting irrelevant diagnosis. For instance, a patient's diagnosis of hypercholestremia (high cholesterol) may be relevant if his visit is for hypertension but not relevant if the visit is for common cold. For the second task, we extract the organ systems for which the patient reported an abnormal symptom during a review. For instance, a patient with a chief complaint of diabetes might report fatigue (symptom) indicating a musculoskeletal (system) abonormality. Taken together, the relevant diagnosis and symptomatic organ systems can provide a high-level overview of patient status to aid physicians in post-visit EHR documentation.

We formulate our tasks as multi-label classification problems and evaluate task performance for a medical-entity-based string-matching baseline, traditional learning approaches (e.g., logistic regression) and state-of-the-art neural approaches (e.g., BERT). One challenge is that conversations are long, containing information irrelevant to our tasks (e.g., small talk). A crucial finding is that a filtering-based approach to pre-select important parts of the conversation (we call them "noteworthy" utterances/sentences) before feeding them into a classification model significantly improves the performance of our models, increasing micro-averaged F1 scores by 10 points for diagnosis prediction and 5 points for RoS abnormality prediction. We compare different ways of extracting noteworthy sentences, such as using a medical entity tagger and training a model to predict such utterances, using annotations present in our dataset. An oracle approach using ground truth noteworthy sentences annotated in the dataset, boosts performance of the downstream classifiers significantly and, remarkably, we are able to realize a significant fraction of that gain by using our learned filters.

We find that using sentences that are specifically noteworthy with respect to medical diagnoses works best for the diagnosis prediction task. In contrast, for the RoS abnormality prediction task, the best performance is achieved when using sentences extracted by a medical entity tagger along with sentences predicted to be noteworthy with respect to review of systems.

2 Related Work

Prior work has focused on qualitative and quantitative evaluation of conversations between physicians and patients, which has been surveyed by [17]. Researchers have analyzed patients' questions to characterize their effects on the quality of interaction [18], and tried to draw correlations between questioning style of physicians and the kind of information revealed by the patients [19]. Although research on extracting information from clinical conversations is scarce, there is significant work on extracting information from other forms of conversation such as summarizing email threads [14] and decisions in meetings [24].

Compared to patient-physician conversations, EHR data has been heavily leveraged for a variety of tasks, including event extraction [6], temporal prediction [3], and de-identification [5]. We point to [20] for an overview. Researchers have used patient admission notes to predict diagnoses [11]. Using content from certain specific sections of the note improves performance of diagnosis extraction models when compared to using the entire note [4]. In our work too, making diagnosis predictions on a smaller part of conversations consisting of filtered noteworthy sentences leads to better model performance. Leveraging extracted symptoms from clinical notes using Metamap [2] medical ontology improves performance on diagnosis prediction [9]. This shows the usefulness of incorporating domain knowledge for diagnosis prediction, which we have also leveraged for our tasks by using a medical entity tagging system. Beyond diagnosis prediction, EHR data has been used to extract other information such as medications and lab tests [25], including fine-grained information like dosage and frequency of medicines and severity of diseases [10].

The systems in all of this work are based on clinical notes in the EHR, which are abundant in datasets. The research on information extraction from medical conversations is scarce likely owed in part to the paucity of datasets containing both medical conversations and annotations. Creating such a dataset is difficult due to the medical expertise that is required to annotate medical conversations with tags such as medical diagnoses and lab test results. One notable work in this area extracts symptoms from patient-physician conversations [16]. Their model takes as input snippets of 5 consecutive utterances and predicts whether the snippet has a symptom mentioned and experienced by the patient, using a recurrent neural network. In contrast, we make predictions of diagnoses and RoS abnormalities from an entire conversation using a variety of models including modern techniques from deep NLP, and introduce an approach to aid this by filtering out noteworthy sentences from the conversation.

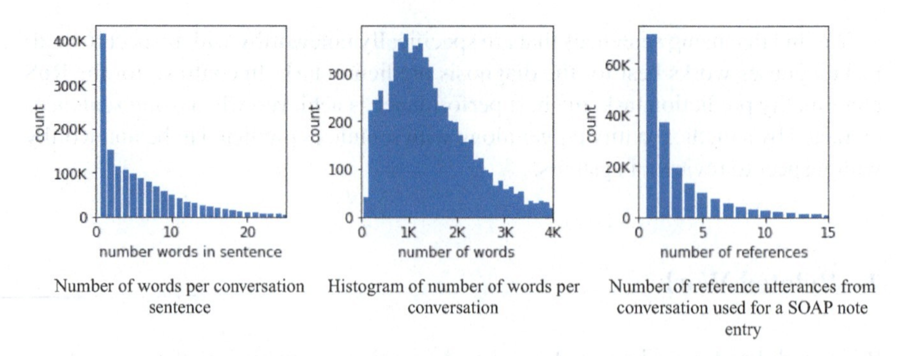

Number of words per conversation sentence

Histogram of number of words per conversation

Number of reference utterances from conversation used for a SOAP note entry

Fig. 1 Distribution of sentence lengths, number of words in physician-patient conversations of our dataset, and the number of evidence utterances in it referred by an entry of the corresponding SOAP note

3 Dataset

This paper addresses a dataset of human-transcribed physician-patient conversations. The dataset includes 2732 cardiologist visits, 2731 family medicine visits, 989 interventional cardiologist visits, and 410 internist visits. Each transcript consists of timestamped utterances with speaker labels. A typical conversation consists of 200–250 utterances. The median utterance is short (Fig. 1a), possibly due to the high frequency of back-chanelling (e.g., "umm-hmm", "okay", etc.). In total, each conversation contains around 1500 words (Fig. 1b).

In our dataset, the transcribed conversations are coupled with corresponding structured text summaries and summary annotations. The structured text summaries, or SOAP notes typically written by a physician to summarize a patient visit, and their annotations were constructed by expert clinical annotators who received task-specific training. The acronym SOAP in SOAP note stands for the four sections of the note: The (S)ubjective section contains a subjective accounting of the patient's current symptoms, and a history of the present illness, and miscellaneous details. The (O)bjective section contains objective information such as results from lab tests, and observations from a physical examination. The (A)ssessment and (P)lan sections contain the inferences made by the physician, including the differential diagnosis, and the plan for treatment, including further tests, planned changes to the patient's medications, other non-pharmaceutical therapeutics, and more.

In total, our dataset consists of 6862 datapoints (i.e., physician-patient conversation transcripts with corresponding annotated notes), which we have then divided into train and test sets with sizes 6270 and 592, respectively. To train our models, we set aside 500 points as a validation set for tuning hyperparmeters. The number of datapoints and the splits are the same for both the tasks.

In our dataset, each line in a SOAP note is classified into one of 12 total subsections within one of the high-level Subjective, Objective, Assessment, or Plan sections. For example, subsections for the Subjective section include *Subjective: Review Of Systems* and *Subjective:Past Medical History*. Each line in a SOAP note appears alongside structured categorical or numerical metadata. For instance, a SOAP note line about medication (e.g., "Take Aspirin once a day.") may be coupled with structured data for the medication name (e.g., "Aspirin") and the dosage (e.g., "daily"). Each SOAP note line is also associated with the lines in the transcript that were used as evidence by the annotator to create the line and its metadata. Each SOAP note line with its associated metadata, i.e. SOAP note entry, uses an average of 3.85 transcript lines as evidence (Fig. 1c). We take subsets of information from the dataset described above to design datasets for the relevant diagnosis prediction and review of systems abnormality prediction tasks.

3.1 Relevant Diagnosis Prediction

Given a physician-patient conversation, we aim to extract the mentioned past and present diagnoses of the patient that are relevant to the primary reason for the patient's visit (called the Chief Complaint). For each conversation, we create a list of the Chief Complaint and related medical problems by using categorical tags associated with the following subsections of the SOAP note:

1. The Chief Complaint of the patient from *Subjective: Chief Complaint* the subsection of the SOAP note.
2. All medical problems in the *Subjective: Past Medical History* subsection tagged with "HPI" (History of Present Illness) to signify that they are related to the Chief Complaint.
3. The medical problem tags present in the *Assessment and Plan: Assessment* subsection of the SOAP note.

We then simplified the medical problem tags by converting everything to lowercase, and removing elaborations given in parentheses. For example, we simplify "hypertension (moderate to severe)" to "hypertension". For each of the 20 most frequent tags retrieved after the previous simplifications, we searched among all medical problems and added the ones that had the original tag as a substring. For example, "systolic hypertension" was merged into "hypertension". After following the above procedure on the training and validation set, we take the 15 most frequent medical problem tags (Table 1) and restrict the task to predicting whether each of these medical problems were diagnosed for a patient or not.

3.2 Review of Systems (RoS) Abnormality Prediction

Given a physician-patient conversation we also predict the organ systems (e.g., respiratory system) for which the patient predicted a symptom (e.g., trouble breathing). During a patient's visit, the physician conducts a Review of Systems (RoS), where the physician reviews organ systems and potential associated symptoms and asks if

Table 1 Diagnoses and abnormal systems extracted from the train+validation split of the dataset with their number of occurrences

Diagnosis	Frequency
Hypertension	1573
Diabetes	1423
Atrial fibrillation	1335
Hypercholesterolemia	1023
Heart failure	584
Myocardial infarction	386
Arthritis	288
Cardiomyopathy	273
Coronary arteriosclerosis	257
Heart disease	240
Chronic obstructive lung disease	235
Dyspnea	228
Asthma	188
Sleep apnea	185
Depression	148

System	Frequency
Cardiovascular	2245
Musculoskeletal	1924
Respiratory	1401
Gastrointestinal	878
Skin	432
Head	418
Neurologic	385

the patient is experiencing each symptom. In our dataset SOAP notes, the *Subjective: Review of Systems* subsection contains annotated observations from the RoS, each containing a system, symptom and result. For instance, a system (e.g., "cardiovascular"), an associated symptom (e.g., "chest pain or discomfort") and a result based on patient feedback (e.g., "confirms", "denies"). To reduce sparsity in the data for system/symptom pairs, we consider only systems and whether or not each system contained a confirmed symptom. We also consider only the set of 7 systems for which more than 5% of patients reported abnormalities, for prediction (Table 1).

4 Methods

We use a single suite of models for both tasks.

4.1 Input-Agnostic Baseline

We establish the best value of each metric that can be achieved without using the input (i.e. an input-agnostic classifier). The behavior of the input-agnostic classifier depends on the metric. For example, to maximize accuracy, the classifier predicts the majority class (usually negative) for all diagnoses. On the other hand, to maximize

F1 and recall, the classifier predicts the positive class for all diagnoses. To maximize AUC and precision-at-1, the classifier assigns probabilities to each diagnosis according to their prevalence rates. For a detailed description of multilabel performance metrics, we point to [12].

4.2 Medical-Entity-Matching Baseline

This baseline uses a traditional string-matching tool. For extracting relevant diagnoses, for each diagnosis, we check to see whether it is mentioned in the conversation. Since a diagnosis can be expressed in different ways, e.g., "myocardial infarction" has the same meaning as the common term "heart attack", we use a system for tagging medical terms (QuickUMLS) that maps strings to medical entities with a unique ID. For example, "hypertension" and "high blood pressure" are both mapped to the same ID.

For predicting RoS abnormalities, our baseline predicts that the person has an abnormality in a system if any symptom related to the system is mentioned in the text as detected by QuickUMLS. The symptoms checked for each system are taken from the RoS tags in the dataset. For example, the cardiovascular system has symptoms like "chest pain or discomfort" and "palpitations, shortness of breath".

4.3 Learning Based Methods

We apply the following classical models: Logistic Regression, Support Vector Classifier, Multinomial Naive Bayes, Random Forest and Gradient Boosting. We use bag-of-words representation of conversations with unigrams and bigrams with TF-IDF transform on the features.

We also applied state of the art neural methods on the problem. We classified diagnoses and RoS abnormalities as present or not present using two BERT models with wordpiece [26] tokenization—one generic, pretrained BERT model, and one pretrained BERT model that is finetuned on clinical text [1]. Each of our BERT models are 12-layered with a hidden size of 768. The final hidden state of the [CLS] token is taken as the fixed-dimensional pooled representation of the input sequence. This is fed into a linear layer with sigmoid activation and output size equal to the number of prediction classes (15 for diagnosis prediction and 7 for the RoS abnormality prediction), thus giving us the probability for each class. Since the pretrained BERT models do not support a sequence length of more than 512 tokens, we break up individual conversations into chunks of 512 tokens, pass the chunks independently through BERT and mean-pool their [CLS] representations. Due to memory constraints we only feed the first 2040 tokens of a conversation into the model.

4.4 Hybrid Models

The long length of the input sequence makes the task difficult for the neural models. We tried a variety of strategies to pre-filter the contents of the conversation so that we only feed in sentences that are more relevant to the task. We call such sentences *noteworthy*. We have 3 ways for deciding if a sentence is noteworthy, which lead to 3 kinds of noteworthy sentences.

- **UMLS-noteworthy:** We designate a sentence as noteworthy if the QuickUMLS medical tagger finds an entity relevant to the task (e.g., a diagnosis or symptom) as defined in the medical-entity-matching baseline.
- **All-noteworthy:** We deem a sentence in the conversation noteworthy if it was used as evidence for any line in the annotated SOAP note. We train a classifier to predict the noteworthy sentences given a conversation.
- **Diagnosis/RoS-noteworthy:** In this we define noteworthy sentences as in the previous approach, the only difference being that here only those sentences are deemed noteworthy that were used as evidence for an entry containing the ground truth tags(diagnosis/RoS abnormality) that we are trying to predict.

In addition to trying out these individual filtering strategies, we also try their combinations as we shall discuss in the following section.

5 Results and Discussion

5.1 Metrics

We evaluate the performance of models using the following metrics: accuracy, area under the receiver-operator characteristics (AUC), F1 score, and precision-at-1. Because this is a multilabel classification task (e.g., predicting positive or negative occurrence of 15 diagnoses), reporting aggregate scores across labels requires some care. For both F1 and AUC, we aggregate scores using both micro- and macro-averaging [23] following the metrics for multilabel diagnosis prediction in [13]. Macro-averaging averages scores calculated separately on each label, while micro-averaging pools predictions across labels before calculating a single metric. We also compute precision-at-1 to capture the percentage of times that each model's most confident prediction is correct (i.e., the frequency with which the most confidently predicted diagnosis actually applies).

Table 2 Aggregate results for the medical diagnosis prediction task. AN: predicted noteworthy utterances, DN: utterances predicted to be noteworthy specifically concerning a summary passage discussing diagnoses, F2K: UMLS-extracted noteworthy utterances with added top predicted AN/DN utterances to get K total utterances, M-: macro average, m-: micro average

Model	Accuracy	M-AUC	M-F1	m-AUC	m-F1	Precision-at-1
Input agnostic baseline	0.9189	0.5000	0.1414	0.7434	0.3109	0.2027
UMLS Medical Entity Matching	0.9122	0.8147	0.5121	0.8420	0.5833	0.5034
Logistic Regression	0.9417	0.8930	0.2510	0.9317	0.5004	0.6064
LinearSVC	0.9395	0.8959	0.2113	0.9354	0.4603	0.6199
Multinomial NaiveBayes	0.9269	0.7171	0.0615	0.8296	0.1938	0.4848
Random Forest	0.9212	0.8868	0.0155	0.8795	0.0541	0.5304
Gradient Boosting Classifier	0.9467	0.9181	0.5024	0.9447	0.6514	0.5861
BERT	0.9452	0.8953	0.4413	0.9365	0.6009	0.6199
CLINICALBERT (CBERT)	0.9476	0.9040	0.4573	0.9413	0.6029	0.6300
AN+CBERT	0.9511	0.9222	0.4853	0.9532	0.6561	0.6470
DN+CBERT	0.9551	**0.9342**	**0.5655**	**0.9616**	0.7029	**0.6621**
UMLS+CBERT	0.9519	0.8615	0.5238	0.9290	0.6834	0.6030
UMLS-AN-CBERT	0.9541	0.9261	0.5317	0.9588	0.6803	**0.6621**
UMLS-DN-CBERT	0.9510	0.9359	0.5210	0.9593	0.6641	0.6368
UMLS-F2K-AN+CBERT	**0.9554**	0.9188	0.5599	0.9567	**0.7139**	0.6487
UMLS+F2K-DN+CBERT	0.9535	0.9354	0.5301	0.9610	0.6911	0.6486
ORACLE AN+CBERT	0.9509	0.9418	0.5500	0.9588	0.6789	0.6250
ORACLE DN+CBERT	0.9767	0.9771	0.7419	0.9838	0.8456	0.7162

5.2 *Results*

We evaluated the performance of all models aggregated across classes on the tasks of relevant diagnosis prediction (Table 2) and RoS abnormality prediction (Table 3). Predicting RoS abnormality proves to be a more difficult task than predicting relevant diagnoses as reflected by the lower values achieved on all metrics. We hypothesize that this is because of the variety of symptoms that can be checked for each system. The cardiovascular system has 152 symptoms in our dataset including 'pain in the ribs', 'palpitations', 'increased heart rate' and 'chest ache'. A learning-based model would have to learn to correlate all of these symptoms to the cardiovascular system in addition to predicting whether or not the patient experiences the symptom. For diagnosis prediction, we do not need to learn such a correlation.

For diagnosis prediction, medical-entity-matching baseline achieves better F1 scores than many of the classical models, which is due to its high recall at the cost of lower precision (0.76 and 0.47 respectively when micro-averaged). The high recall and low precision together demonstrate that if a diagnosis has been made for the patient, the diagnosis is often directly mentioned in the conversation but the converse

Table 3 Aggregate results for the RoS abnormality prediction task. AN: predicted noteworthy utterances, RN: utterances predicted to be noteworthy specifically concerning a summary passage discussing review of systems, F2K: UMLS-extracted noteworthy utterances with added top predicted AN/RN utterances to get K total utterances, M-: macro average, m-: micro average

Model	Accuracy	M-AUC	M-F1	m-AUC	m-F1	Precision-at-1
Input agnostic baseline	0.8677	0.5000	0.2235	0.7024	0.3453	0.3040
UMLS Medical Entity Matching	0.4532	0.7074	0.2797	0.7454	0.3079	0.3226
Logistic Regression	0.8819	0.8050	0.2102	0.8496	0.3506	0.3952
LinearSVC	0.8798	0.8093	0.1623	0.8516	0.3025	0.3986
Multinomial NaiveBayes	0.8687	0.6183	0.0369	0.7383	0.0653	0.3818
Gradient Boosting Classifier	0.8740	0.7949	0.2500	0.8405	0.3324	0.4020
Random Forest	0.8677	0.7210	0.0000	0.7670	0.0000	0.3412
BERT	0.8818	0.8240	0.3304	0.8620	0.4275	0.3986
CLINICALBERT (CBERT)	0.8784	0.8305	0.3878	0.8667	0.4857	0.4003
AN+CBERT	**0.8837**	0.8491	0.3560	0.8801	0.4761	0.4274
RN+CBERT	0.8861	0.8391	0.3720	0.8788	0.4925	0.4054
UMLS+CBERT	0.8769	0.8036	0.3421	0.8464	0.4457	0.3902
UMLS+AN+CBERT	0.8868	0.8252	0.3039	0.8626	0.4515	0.4139
UMLS+RN+CBERT	0.8810	0.8390	0.3122	0.8745	0.4152	0.3902
UMLS+F2K-AN+CBERT	0.8834	0.8169	0.2385	0.8585	0.3894	0.4189
UMLS+F2K-RN+CBERT	0.8827	**0.8595**	**0.3987**	**0.8895**	**0.5308**	**0.4291**
ORACLE AN+CBERT	0.8846	0.8535	0.3662	0.8841	0.5062	0.4375
ORACLE RN+CBERT	0.9454	0.9595	0.7235	0.9703	0.7847	0.4966

is not true. Among the BERT-based models, we see a modest improvement in F1 and precision-at-1, when using ClinicalBERT instead of the common BERT because it is fine-tuned on clinical text. Using predicted noteworthy sentences from the transcript instead of all of the transcript generally led to an improvement in performance. For diagnosis prediction, using a model that uses only predicted diagnosis-noteworthy sentences rather than all-noteworthy sentences performs the best for a majority of the metrics. For RoS abnormality prediction, the trend reverses and using predicted RoS-noteworthy sentences performs worse than using predicted AllNoteworthy sentences from the transcript. If we train on ORACLE noteworthy sentences, we achieve a precision-at-1 of 0.72 for diagnosis prediction and 0.50 for RoS abnormality prediction. Note that the maximum achievable precision-at-1 on the diagnosis prediction task is 0.7584 and for the RoS abnormality prediction task it is 0.5811, because the patients do not always have one of the diagnoses or RoS abnormalities that we are concerned with.

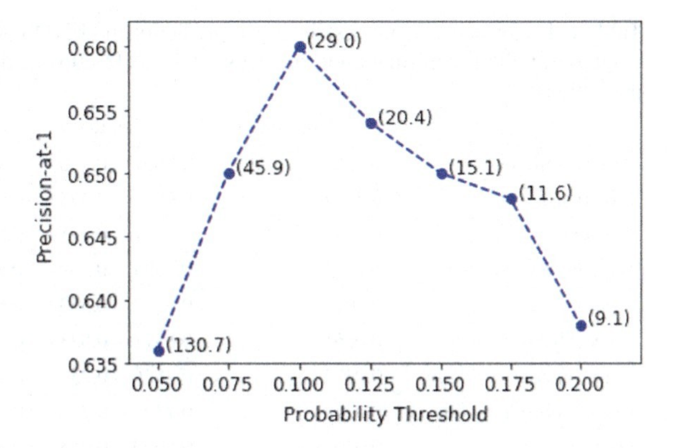

Fig. 2 Precision-at-1 at different thresholds of the diagnosis-noteworthy utterance classifier. Average number of noteworthy sentences extracted in parantheses

The average number of UMLS-noteworthy sentences extracted by QuickUMLS for diagnosis prediction and RoS abnormality prediction tasks is 4.42 and 5.51 respectively out of an average of 215.14 total sentences. We train BERT on only the UMLS-noteworthy sentences, on a union of UMLS-noteworthy sentences and predicted all/task-specific noteworthy sentences, and a FillUptoK (F2K) variant where we take the union but we only add the top-predicted all/task-specific noteworthy sentences until we reach a total of K sentences to be fed into BERT, where K is a hyperparameter. The last model achieves the best results for RoS abnormality prediction when we pool the UMLS-noteworthy sentences with the top-predicted RoS-noteworthy sentences. This is in contrast with results for disease prediction where only using predicted disease-noteworthy sentences performed best. This suggests that domain knowledge in the medical tagger may be more useful for RoS-abnormality prediction, because it gives an explicit listing of the large variety of symptoms pertaining to each organ system.

On both tasks, it is possible to use a small fraction of the transcript and still get performance comparable to models that use all of it. For the task of diagnosis prediction, the UMLS-noteworthy sentences only make up 2.1% of total sentences in a conversation on average, but using just them with the ClinicalBERT model still achieves higher F1 scores than all classical ML models which use the entire conversation. We carried out an experiment to observe the correlation between number of words input into the ClinicalBERT model and the performance achieved. To do this, we varied the threshold probability for the noteworthy utterance classifier in the Diagnosis-noteworthy+ClinicalBERT model. Fewer noteworthy sentences are extracted and passed to ClinicalBERT as the threshold goes up. The performance increases with a decrease in the number of filtered sentences and then goes down (Fig. 2). The best performance is achieved when we pass an average of 29 utterances for each transcript.

Table 4 Performance of our best diagnosis prediction model (DN+CBERT) at predicting individual diagnoses. CP@1: contribution to precision-at-1, the fraction of times a disease was a correct top prediction

Disease	Prevalence rate	Precision	Recall	F1	Accuracy	AUC	CP@1
Atrial fibrillation	0.2568	0.8667	0.9408	0.9022	0.9476	0.9773	0.3597
Hypertension	0.2027	0.6667	0.4833	0.5604	0.8463	0.8817	0.0995
Diabetes	0.1959	0.8411	0.7759	0.8072	0.9274	0.9586	0.1837
Hypercholesterolemia	0.1216	0.5694	0.5694	0.5694	0.8953	0.9246	0.0740
Heart failure	0.1014	0.8049	0.5500	0.6535	0.9409	0.9692	0.0638
Myocardial infarction	0.0861	0.8571	0.8235	0.8400	0.9730	0.9857	0.0995
Coronary arteriosclerosis	0.0372	0.3846	0.2273	0.2857	0.9578	0.8307	0.0051
Chronic obstr. lung disease	0.0372	0.7391	0.7727	0.7556	0.9814	0.9665	0.0281
Dyspnea	0.0304	0.5000	0.0556	0.1000	0.9696	0.9068	0.0077
Depression	0.0304	0.6471	0.6111	0.6286	0.9780	0.9555	0.0230
Asthma	0.0287	0.8462	0.6471	0.7333	0.9865	0.9951	0.0230
Cardiomyopathy	0.0236	0.7143	0.7143	0.7143	0.9865	0.9779	0.0128
Heart disease	0.0236	0.0000	0.0000	0.0000	0.9764	0.7058	0.0026
Arthritis	0.0220	0.3636	0.3077	0.3333	0.9730	0.9843	0.0128
Sleep apnea	0.0186	0.6667	0.5455	0.6000	0.9865	0.9937	0.0051

Table 5 Performance of our best RoS abnormality prediction model (UMLS+F2K-RN+CBERT) at predicting abnormalities in each system. CP@1: contribution to precision-at-1, the fraction of times an RoS abnormality was a correct top prediction

System	Prevalence rate	Precision	Recall	F1	Accuracy	AUC	CP@1
Cardiovascular	0.3041	0.5867	0.7333	0.6519	0.7618	0.8475	0.5079
Musculoskeletal	0.2010	0.5893	0.5546	0.5714	0.8328	0.8579	0.2402
Respiratory	0.1571	0.5231	0.3656	0.4304	0.8480	0.8639	0.1063
Gastrointestinal	0.0845	0.5217	0.4800	0.5000	0.9189	0.8636	0.0669
Head	0.0828	0.4412	0.3061	0.3614	0.9105	0.9252	0.0591
Neurologic	0.0574	0.0000	0.0000	0.0000	0.9426	0.7864	0.0000
Skin	0.0389	0.6667	0.1739	0.2759	0.9645	0.8719	0.0197

5.2.1 Performance on Binary Prediction Tasks

Besides calculating the aggregate performance of our models, we also compute the performance of our best model for each task at the binary prediction of each diagnosis/RoS abnormality (Tables 4 and 5). We see that generally diagnoses that are more common are detected better. One exception is hypertension which has a low recall and precision despite affecting around 20% of the patients. The instances of hypertension that are not identified by our model show that it is rarely mentioned explicitly during conversation. Instead, it needs to be inferred by values of blood pressure readings and phrases like "that blood pressure seems to creep up a little bit". This indirect way in which hypertension is mentioned possibly makes it harder to detect accurately. In contrast, atrial fibrillation is usually mentioned explicitly during conversation, which is why even the medical-entity-matching baseline achieves a high recall of 0.83 at predicting atrial fibrillation. The model has the worst performance for predicting heart disease. We think it is due to a combination of low frequency and the generic nature of the class. We found that the heart disease tag is used in miscellaneous situations like genetic defect, weakness in heart's function, or pain related to stent placement.

We also calculate the contribution to precision-at-1 for each class for both tasks. This gives us a sense of how often a diagnosis/RoS abnormality becomes the model's top prediction. We do not want a situation where only the most frequent diagnoses/RoS abnormalities are predicted with the highest probability and the rarer classes do not get any representation in the top prediction. We define the contribution to precision-at-1 for a class as the number of times it was a correct top prediction divided by the total number of correct top predictions made by the model. We see that for both tasks, contribution to precision-at-1 is roughly in proportion to the prevalence rate of each diagnosis (Tables 4 and 5). This suggests that the model predicts even the rarer diagnoses with enough confidence for them to show up as top predictions.

5.3 Experimental Details

The hyperparameters of each learning based model are determined by tuning over the validation set. All models except the neural network based ones are sourced from `scikit-learn` [15]. The UMLS based tagging system is taken from [22]. The BERT-based models are trained in AllenNLP [7]. The vanilla BERT model is the bert-base-uncased model released by Google and the clinical BERT model is taken from [1].

The BERT models have a learning rate of 0.00002. We tuned the probability threshold for predicting noteworthy sentences. The optimal threshold was 0.4 for predicting all noteworthy sentences, 0.1 for predicting diagnosis-related noteworthy sentences and 0.02 for predicting RoS-related noteworthy sentences. Among the FillUptoK predictors for diagnosis prediction, the one using AllNoteworthy sentences had $K = 50$ and the one using diagnosis-noteworthy sentences has $K = 15$. For the FillUptoK predictors used for RoS abnormality prediction, the one using all-noteworthy sentences had $K = 50$ and the predictor using RoS-noteworthy sentences had $K = 20$.

The noteworthy sentence extractors are logistic regression models trained, validated and tested on the same splits of the dataset as the other models. All models are L2-regularized with the regularization constant equal to 1. The AUC scores for the classifiers extracting all, diagnosis-related, and RoS-related noteworthy sentences are 0.6959, 0.6689 and 0.7789 respectively.

6 Conclusion and Future Work

This work is a preliminary investigation into the utility of medical conversations for drafting SOAP notes. Although we have only tried predicting diagnoses and review of systems, there are more tasks that can be attacked using the annotations in the dataset we used - for example, medications, future appointments, lab tests etc. Our work shows that extracting noteworthy sentences out of the conversation improves the performance significantly. However, the performance of noteworthy sentence extractors is poor at the moment and improving it can be a good direction for future work. Currently we are only predicting the organ system that has a reported symptom and not the exact symptom that was reported by the patient. This was because the frequency of occurrence of each symptom was fairly low. Designing models that can predict these symptoms despite their sparsity is an area that we wish to go after in future.

Acknowledgements We gratefully acknowledge support from the Center for Machine Learning and Health in a joint venture between UPMC and Carnegie Mellon University and Abridge AI, who created the dataset that we used for this research.

References

1. Alsentzer, E., et al.: Publicly available clinical BERT embeddings. In: North American Association for Computational Linguistics - Human Language Technologies (NAACL-HLT) (2019)
2. Aronson, A.R.: Effective mapping of biomedical text to the UMLs metathesaurus: the MetaMap program. In: Proceedings of the AMIA Symposium. American Medical Informatics Association (2001)
3. Cheng, Y., Wang, F., Zhang, P., Hu, J.: Risk prediction with electronic health records: a deep learning approach. In: SIAM International Conference on Data Mining (SDM). SIAM (2016)
4. Datla, V., et al.: Automated clinical diagnosis: the role of content in various sections of a clinical document. In: IEEE International Conference on Bioinformatics and Biomedicine (BIBM). IEEE (2017)
5. Dernoncourt, F., Lee, J.Y., Uzuner, O., Szolovits, P.: De-identification of patient notes with recurrent neural networks. J. Am. Med. Inform. Assoc. **24**, 596–606 (2017)
6. Fries, J.: Brundlefly at SemEval-2016 task 12: recurrent neural networks vs. joint inference for clinical temporal information extraction. In: International Workshop on Semantic Evaluation (SemEval) (2016)
7. Gardner, M., et al.: AllenNLP: a deep semantic natural language processing platform. In: Workshop for NLP Open Source Software (NLP-OSS) (2018)
8. Gardner, R.L., et al.: Physician stress and burnout: the impact of health information technology. J. Am. Med. Inform. Assoc. **26**, 106–114 (2018)
9. Guo, D., Duan, G., Yu, Y., Li, Y., Wu, F.X., Li, M.: A disease inference method based on symptom extraction and bidirectional long short term memory networks. Methods **173**, 75–82 (2019)
10. Jagannatha, A.N., Yu, H.: Bidirectional RNN for medical event detection in electronic health records. In: North American Association for Computational Linguistics - Human Language Technologies (NAACL-HLT) (2016)
11. Li, C., Konomis, D., Neubig, G., Xie, P., Cheng, C., Xing, E.: Convolutional neural networks for medical diagnosis from admission notes. arXiv preprint arXiv:1712.02768 (2017)

12. Lipton, Z.C., Elkan, C., Naryanaswamy, B.: Optimal thresholding of classifiers to maximize F1 measure. In: Joint European Conference on Machine Learning and Knowledge Discovery in Databases. Springer (2014)
13. Lipton, Z.C., Kale, D.C., Elkan, C., Wetzel, R.: Learning to diagnose with LSTM recurrent neural networks. In: International Conference on Learning Representations (ICLR) (2016)
14. Murray, G., Carenini, G.: Summarizing spoken and written conversations. In: Empirical Methods in Natural Language Processing (EMNLP). Association for Computational Linguistics (2008)
15. Pedregosa, F., et al.: Scikit-learn: machine learning in Python. J. Mach. Learn. Res. **12**, 2825–2830 (2011)
16. Rajkomar, A., et al.: Automatically charting symptoms from patient-physician conversations using machine learning. JAMA Internal Med. **179**, 836–838 (2019)
17. Roter, D., Frankel, R.: Quantitative and qualitative approaches to the evaluation of the medical dialogue. Soc. Sci. Med. **34**, 1097–1103 (1992)
18. Roter, D.L.: Patient participation in the patient-provider interaction: the effects of patient question asking on the quality of interaction, satisfaction and compliance. Health Educ. Monographs **5**, 281–315 (1977)
19. Roter, D.L., Hall, J.A.: Physicians' interviewing styles and medical information obtained from patients. J. General Internal Med. **2**, 325–329 (1987)
20. Shickel, B., Tighe, P.J., Bihorac, A., Rashidi, P.: Deep EHR: a survey of recent advances in deep learning techniques for electronic health record (EHR) analysis. IEEE J. Biomed. Health Inform. **22**, 1589–1604 (2017)
21. Sinsky, C., et al.: Allocation of physician time in ambulatory practice: a time and motion study in 4 specialties. Ann. Internal Med. **165**, 753–760 (2016)
22. Soldaini, L., Goharian, N.: QuickUMLS: a fast, unsupervised approach for medical concept extraction. In: MedIR Workshop, sigir (2016)
23. Van Asch, V.: Macro-and micro-averaged evaluation measures. Technical report (2013)
24. Wang, L., Cardie, C.: Summarizing decisions in spoken meetings. In: Workshop on Automatic Summarization for Different Genres, Media, and Languages. Association for Computational Linguistics (2011)
25. Wu, Y., Jiang, M., Lei, J., Xu, H.: Named entity recognition in Chinese clinical text using deep neural network. Stud. Health Technol. Inform. **216**, 624 (2015)
26. Wu, Y., et al.: Google's neural machine translation system: bridging the gap between human and machine translation. arXiv preprint arXiv:1609.08144 (2016)

A Multi-talent Healthcare AI Bot Platform

Martin Horn, Xiang Li, Lin Chen, and Sabin Kafle

Abstract AI bots have emerged in various industries hoping to simplify customer communication. However, there are a certain set of challenges for such a product in the healthcare domain, including confidentiality and domain knowledge. We discuss the implementation of a secure, multi-task healthcare chatbot built to provide easy, fast, on-demand conversational access to a variety of health-related resources.

Keywords Virtual assistant · Chatbot · Healthcare · Microservices · Natural Language Understanding

1 Introduction

Conversational agents (AI bots) have been widely popular in both the academic research community [4, 7, 8] and several industries in recent years. While they carry promise [2] for customer engagement, most industrial AI bots have failed to meet expectations, often due to a shallow understanding of user inputs [1, 5, 6], with

M. Horn (✉) · X. Li · L. Chen · S. Kafle
Cambia Health Solutions, 1800 9th Ave, Seattle, WA 98101, USA
e-mail: Martin.Horn@cambiahealth.com

X. Li
e-mail: Xiang.Li@cambiahealth.com

L. Chen
e-mail: Lin.Chen@cambiahealth.com

S. Kafle
e-mail: Sabin.Kafle@cambiahealth.com

© The Editor(s) (if applicable) and The Author(s), under exclusive license to Springer Nature Switzerland AG 2021

A. Shaban-Nejad et al. (eds.), *Explainable AI in Healthcare and Medicine*, Studies in Computational Intelligence 914,
https://doi.org/10.1007/978-3-030-53352-6_15

up to 70% failure rates. Moreover, the domain-specific necessities for AI bots (e.g., confidentiality) add an additional challenge to the task.

Healthcare is a complicated and confusing system, often placing a large workload on customer support services. AI bots present a very useful application of facilitating and optimizing the interactions between the customers and support staff, with bots answering simple questions while deferring complex individualized questions to the support staffs in a timely fashion.

AI bots are developed either through a task-oriented approach based on the domain requirements or an end-to-end data driven approach, which is primarily used to develop social bots. While reinforcement learning has enabled the formulation of AI bot learning as a decision making process, thus providing a unified framework [3, 9] for both task-oriented and social bots, its applicability is limited for many task-oriented bots, especially in sensitive domains such as healthcare. In healthcare it is more useful to have a modular task-centric design for both flexibility and compliance.

We have built a flexible, compliant AI bot platform for the healthcare domain with the hope to overcome some of these obstacles and drive progress on practical, useful conversational agents. We primarily focus on the modular design due to its fine control of both query understanding and answering, which requires components unique to the healthcare domains such as stringent authentication criteria, knowledge of rules, and understanding of healthcare plans. It has been designed to be able to learn more intents and functionalities over time, connect to more services, and wear different personas for different applications and audiences.

2 System Architecture

The AI bot uses a robust microservice architecture (Fig. 1), allowing it to be pluggable, i.e. new components and services can easily be added or replaced.

2.1 Modular Design

Each component is implemented as a standalone Docker-containerized gRPC service using JSON Web Token (JWT)-based authentication. This architecture allows services to be used and re-used independently or in conjunction to form an entire AI bot platform. Our strict service authorization ensures each service is secure and compliant in a sensitive environment containing Protected Health Information (PHI). We also use the concept of *personas*, which are bot configurations that turn on different skills or personalities for a specific product or brand. A new persona can easily be created and configured in order to create an entirely new bot on the platform.

2.2 *The Pipeline*

The platform is orchestrated by the Gateway, which serves as the entry-point to the bot's core functionality. Textual requests are initiated by the interchangeable User Interface (UI). We have a desktop UI for several in-house bots as well as mobile UI for a consumer facing app.

The Gateway passes the query to the Natural Language Understanding (NLU) module in order to detect the user's intent. Each intent corresponds to one app in the App Platform which performs one or more specific skills. This involves extracting useful information from the utterance via NLU, managing the dialog, and consulting a variety of knowledge base services and data stores (simplified as Knowledge Base in Fig. 1). The app then sends the appropriate response to the Gateway to pass back to the UI.

In addition to facilitating the dialog pipeline, the Gateway connects to the Authorization, Monitoring, and Logging services. Authorization ensures that the services are only accessed by those with the right privileges. For example, users can only get user-specific health benefit information about themselves or dependents for which they have been granted access rights. The Monitoring service captures performance metrics about each service and sends out alerts if certain thresholds are exceeded (such as response time). The Logging service captures service events, dialogs, and feedback. Selections of securely logged conversations are then annotated in order to re-train and improve the NLU models.

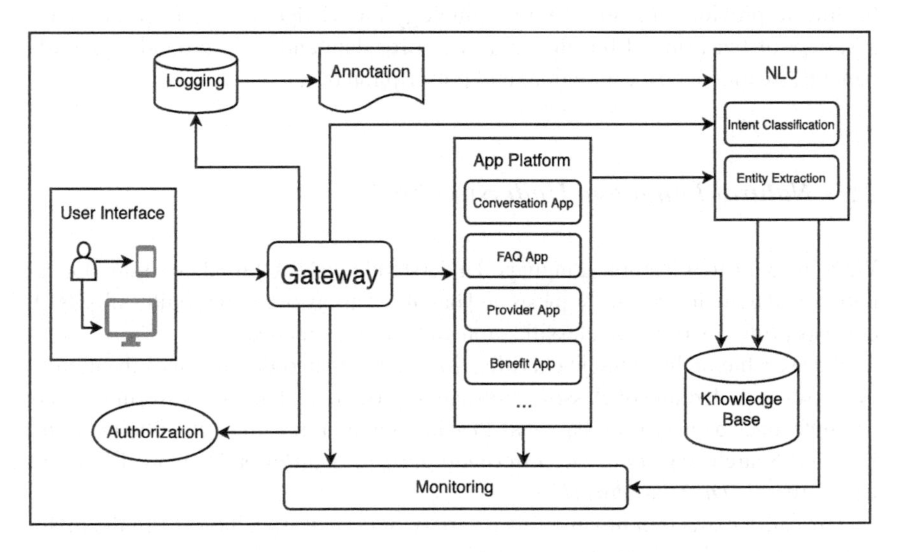

Fig. 1 System Architecture

2.3 App Platform

Our bot platform is multi-talent, meaning it supports multiple different skills and is easily extendable. Each app in the App Platform supplies its own dialog management and business logic so that each skill can be tightly managed and configured for its specific purpose. Once the app has performed its intended task, it generates a custom response depending on the type of service it provides.

Apps like the Conversation App and the FAQ (Frequently Asked Question) App use an information retrieval approach to find fitting responses from a bank of existing question-answer pairs. Others use specific pieces of information from the query to perform a task, such as finding a doctor (Provider App) or looking up insurance benefits (Benefit App). Results from the task are then filtered, sorted, and injected into a response template. Though apps have different dialog logic and response requirements, the addition of new apps is simplified by a standard template and reusable services, such as a dialog state manager and response generator.

2.4 Knowledge Base

The key to creating useful AI bot skills is having a rich knowledge source from which to look up information and draw insights. In a domain such as healthcare, deep and specific business knowledge is crucial. We have codified and operationalized healthcare provider, member, and terminology knowledge bases and connected to a variety of bot-external healthcare services for data and logic around topics like benefits, treatment cost estimation, and prescription drugs.

2.5 Natural Language Understanding

We trained custom Natural Language Understanding (NLU) models so the bot can understand domain-specific requests and handle a variety of custom-built skills. NLU consists primarily of intent classification and entity extraction.

We use a hierarchical intent classification approach in order to reduce the number of classes in each round of classification and improve overall model performance. For example, once a query is classified as Provider intent, it is further classified as either Provider Search (e.g. *Find me a dermatologist in Seattle*) or Network Eligibility Check (e.g. *Is Dr. X in-network?*).

The entity extraction module fills a variety of domain-specific slots in the dialog schema, such as location entities, provider name and specialty entities, and network status for the Provider intents. We treat this as a typical named entity recognition task where tokens are labelled with the IOB (inside-outside-beginning) scheme and then concatenate the relevant adjacent tokens into entities. After entities have been extracted, they are normalized to provide a unified format for downstream processing.

We built and tested a variety of machine learning models for NLU, including a novel joint BiLSTM+CRF intent and entity classification model. For flexibility and performance reasons on new intents and entities with small training data sets, our platform primarily uses a combination of Maximum Entropy (MaxEnt) classifiers for intent classification and Conditional Random Fields (CRF) models for entity extraction.

3 Evaluation

We evaluate our bot platform's performance in a variety of ways. During runtime, we monitor service metrics like job health, response latency, and queries per second. We also collect feedback and calculate custom usage statistics via our logging service.

It is also useful to measure accuracy in order to guarantee the platform's quality. While it is difficult to directly measure an entire bot platform's accuracy, we can isolate and evaluate certain components. Accuracy of information provided by the bot must be tailored to each individual skill. For example, we had domain experts validate in-network provider search performance by comparing results with trusted third-party tools.

However, we were able to perform traditional evaluation on the core NLU components of the system: intent classification and entity extraction. We used a typical train/test split with a hold-out test set of around 1,000 sentences for both tasks.

3.1 Intent Classification

For intent classification, we examined precision, recall, and F1 score for each intent, as well as micro average for each metric and overall accuracy, found in Table 1. The hierarchical nature of the intents is flattened for purposes of evaluation, and only top and bottom four intents (ordered by F1) are displayed due to space constraints.

Results are shown for our MaxEnt models. It is clear that performance varies considerably from intent to intent. For the very low F1 of 0.57 for `faq` and `general`, this is likely due to high variability in the intents.

3.2 Entity Extraction

Separate sequence labeling models were constructed for different domains of data: provider, benefit, and glossary. Entity extraction was evaluated on a strict-match

Table 1 Intent Classification Scores

Intent	Precision	Recall	F1 score	Support
auth	0.93	1.00	0.96	25
copay	0.89	0.96	0.92	25
schedule	0.89	0.96	0.92	25
coinsure	0.85	0.88	0.86	25
cost	0.92	0.62	0.74	71
glossary	0.74	0.71	0.72	41
faq	0.44	0.83	0.57	35
general	0.65	0.50	0.57	26
Avg/Total	0.83	0.81	0.81	747
Overall Accuracy	0.81			

Table 2 Entity Extraction Scores

Domain	Entity Type	Precision	Recall	F1 score
Provider	city	0.96	0.93	0.95
	member_id	1.00	1.00	1.00
	network_status	1.00	0.97	0.99
	facility	0.80	1.00	0.89
	practitioner	0.95	0.92	0.93
	specialty	0.96	0.90	0.92
	state	0.98	0.98	0.98
	zip_code	1.00	1.00	1.00
Benefit	benefit_category	0.95	0.85	0.90
	member_id	1.00	0.98	0.99
Glossary	concept	0.94	0.89	0.92

per-entity basis. Precision, recall, and F1 score for each entity type are found in Table 2.

Results are shown for our CRF models. The relatively low performance of `facility` may be attributed to a lack of support in train and test data and an overlap in distribution to the `practitioner` entity type. `member_id` and `zip_code` may have received perfect F1 scores due to their low variability (they consist exclusively of numerical digits).

4 Conclusion

We introduced an AI bot architecture consisting of upgradable modules which form the foundation for building a stable AI bot with value to customers. Our system consists of a flexible language understanding module, an app platform facilitating dialogs and performing tasks, robust knowledge base services, and a gateway connecting each microservice together. The modularity of the system aids its flexibility, extendability, and compliance in the sensitive healthcare domain. In future work we would like to integrate a more expansive knowledge graph to further empower the AI bot in handling complex, personalized, domain-specific queries.

References

1. Brandtzaeg, P.B., Følstad, A.: Chatbots: changing user needs and motivations. Interactions **25**(5), 38–43 (2018)
2. Dale, R.: The return of the chatbots. Nat. Lang. Eng. **22**(5), 811–817 (2016)
3. Fang, H., Cheng, H., Clark, E., Holtzman, A., Sap, M., Ostendorf, M., Choi, Y., Smith, N.A.: Sounding board–university of washington's alexa prize submission. In: Alexa Prize Proceedings (2017)
4. Gao, J., Galley, M., Li, L.: Neural approaches to conversational AI. In: The 41st International ACM SIGIR Conference on Research & Development in Information Retrieval, pp. 1371–1374. ACM (2018)
5. Jain, M., Kumar, P., Kota, R., Patel, S.N.: Evaluating and informing the design of chatbots. In: Proceedings of the 2018 on Designing Interactive Systems Conference 2018, pp. 895–906. ACM (2018)
6. Piccolo, L., Roberts, S., Iosif, A., Alani, H.: Designing chatbots for crises: a case study contrasting potential and reality (2018)
7. Serban, I.V., Lowe, R., Henderson, P., Charlin, L., Pineau, J.: A survey of available corpora for building data-driven dialogue systems. arXiv preprint arXiv:1512.05742 (2015)
8. Serban, I.V., Sordoni, A., Bengio, Y., Courville, A.C., Pineau, J.: Building end-to-end dialogue systems using generative hierarchical neural network models. AAAI **16**, 3776–3784 (2016)
9. Zhou, L., Gao, J., Li, D., Shum, H.Y.: The design and implementation of xiaoice, an empathetic social chatbot. arXiv preprint arXiv:1812.08989 (2018)

Natural vs. Artificially Sweet Tweets: Characterizing Discussions of Non-nutritive Sweeteners on Twitter

Hande Batan, Dianna Radpour, Ariane Kehlbacher, Judith Klein-Seetharaman, and Michael J. Paul

Abstract This ongoing project aims to use social media data to study consumer behaviors regarding natural and artificial sweeteners Following the recent shifts to natural sweeteners such as Stevia versus artificial, and traditionally-used ones like aspartame in recent years, there has been discussion around potential negative side effects, including memory loss and other chronic illnesses. These issues are discussed on Twitter, and we hypothesize that Twitter may provide insights into how people make nutritional decisions about the safety of sweeteners given the inconclusive science surrounding the topic, how factors such as risk and consumer attitude are interrelated, and how information and misinformation about food safety is shared on social media. As an initial step, we describe a new dataset containing 308,738 de-duplicated English-language tweets spanning multiple years. We conduct a topic model analysis and characterize tweet volumes over time, showing a diversity of sweetener-related content and discussion. Our findings suggest a variety of research questions that these data may support.

Keywords Social media · Public health · Nutrition

H. Batan · D. Radpour · M. J. Paul (✉)
University of Colorado Boulder, Boulder, CO, USA
e-mail: michael.paul@colorado.edu

H. Batan
e-mail: hande.batan@colorado.edu

D. Radpour
e-mail: dianna.radpour@colorado.edu

A. Kehlbacher
University of Reading, Reading, Berkshire, UK

J. Klein-Seetharaman
Colorado School of Mines, Golden, CO, USA

© The Editor(s) (if applicable) and The Author(s), under exclusive license to Springer Nature Switzerland AG 2021

A. Shaban-Nejad et al. (eds.), *Explainable AI in Healthcare and Medicine*, Studies in Computational Intelligence 914,
https://doi.org/10.1007/978-3-030-53352-6_16

1 Introduction

As the general public becomes increasingly aware of the possible associations between artificial sweeteners and various health concerns, natural alternatives have seen a surge in popularity and general usage. However, with the science around both artificial and natural sweeteners being largely inconclusive, data from social media platforms such as Twitter presents itself as a valuable resource for identifying the factors and information that influence the decisions consumers make in choosing what sweeteners to use or avoid.

1.1 Background: New Sweeteners

The law mandates that every new substance that gets introduced into the market for direct human consumption must be approved as safe for entering the human body. Though the consequences for an incorrectly assigned designation could lead to serious health repercussions, the designations are rarely questioned or altered once they are granted, even if that designation is founded in science and research studies from decades ago. This is precisely the phenomenon we are observing today with artificial and low-calorie sweeteners, such as aspartame, sucralose, and saccharin. Though many recent studies suggest that these artificial sweeteners might be linked to a number of negative side effects, including memory loss, cancer, and other chronic illnesses, their regulatory statuses remain unchanged, even with approvals dating as far back as 1981 (aspartame) and 1998 (sucralose). Despite their unchanged regulatory statuses, we have started to observe a shift in widespread consumption of artificial sweeteners to natural alternatives.

1.2 Stevia as a Natural Alternative

Stevia rebaudiana, commonly known as Stevia has, in recent years, become more widely used and drastically grown in market popularity as an everyday sweetener [11]. Stevia is a South American plant native to Paraguay that traditionally has been used as a sugar substitute in tea and other beverages and its safety has been approved by some medical, scientific and regulatory authorities, as well as some countries worldwide, including the World Health Organization, the European Food Safety Authority and the Joint Expert Committee on Food Additives.

While stevia has been approved in over 60 countries and the Food and Drug Administration (FDA) has not questioned the Generally Recognized as Safe (GRAS) status of some specific high-purity steviol glycosides for use in food, stevia leaf and the crude stevia extracts are still not considered GRAS and do not have approval for use in food. While studies, including human studies on safety, metabolism and intake, support their safety, they still await approval from the FDA. In the context of the open marketplace, new substances can be seamlessly integrated into recommended product search results, even when the advertised substances growing in popularity are questionable, with the science around their safety still largely inconclusive. These substances often take the form of supplements, alternative medicines, and questionably derived teas that consumers are taking in large and frequent quantities. While some of these products simply leverage the power of the placebo effect more than any real health effects, the fact that they often lack the basic seal of safety and approval can result in often serious health repercussions.

1.3 The Role of Social Media

With Twitter being one of the world's most prominent social media and microblogging platforms, attracting an approximate 126 million daily users, its potential as a resource in the realm of public health and surveillance continues to prove valuable. However, with its vast influence on millions of users comes the inevitable drawbacks, one being the ability for unreliable health information to spread and lead to the mass consumption of substances that people are exposed to only on Twitter. When faced with conflicting information, how do people make choices about what to consume? Our goal is to use Twitter data to help answer this question in the domain of sweeteners, which are commonly used but also controversial.

Related research has examined social media platforms like Twitter to understand a variety of issues related to health, most commonly for disease forecasting [10]. In the areas of diet and nutrition, multiple studies have looked at mentions of food consumption in Twitter [1, 5, 8], Instagram [7, 12], and search query logs [2, 6, 13]. [4] analyzed food content on Instagram to study how food consumption is related to the availability of food in different locations. Related work has studied weight loss advice posted on social media [9]. To our knowledge, prior work has not specifically examined online discussion of sweeteners.

2 Data

From June 2017 to September 2018, approximately 851k tweets were collected from the Twitter search API matching four relevant terms: stevia, sucralose, aspartame, sweetener. We removed duplicates and non-English tweets. After this filtering step, approximately 309k sweetener-related tweets remained.

We were additionally interested in specifically analyzing tweets related to certain diseases: cancer, diabetes, and Lyme disease. These diseases have been discussed in the context of natural and artificial sweeteners, with cancer and diabetes being linked to the use of artificial sweeteners and claims being made for stevia as a potential cure for Lyme disease (Table 1).

Table 1 Dataset statistics. The bottom three rows are subsampled from the 'filtered without duplicates' set, keyword-filtered for the corresponding disease

Data	# Tweets	# Users
Raw with Duplicates	851319	368503
Filtered without duplicates	308738	154585
Cancer	607	516
Diabetes	1443	833
Lyme	96	69

3 Topic Analysis

Latent Dirichlet Allocation (LDA) [3], a probabilistic topic model, was used to infer topics relevant to the main issues surrounding sweeteners. The output of the model contained 100 "topics," which are clusters of words, with some words in other languages. While the data was filtered for English only, Twitter's language identifier is not always accurate and some tweets contained a mix of multiple languages which is the probable cause behind the multilingual topics.

We used this approach as a way to automatically extract rough representations (lists of related words) of the major themes in the text, to characterize the topics of discussion on Twitter. Domain experts examined the output and identified salient topics. While many topics were hard to understand, some interesting topics were identified.

Examples of relevant topics are provided in Table 2. Topic 7 is that of general health efficacy around natural and plant-based sweeteners. Topic 12 revolves around specific artificial sweeteners' taste in the context of sodas. Topic 40 is similar to Topic 12, but focuses more on the general thoughts around artificial sweeteners. Topic 82 is about sweeteners, artificial as well as sugar, as they relate to health issues of the gut, diabetes, and weight. Topic 32 embodies the discussions around a recent study showing that Stevia is more effective in treating Lyme disease than antibiotics. Topic 90 is about artificial and natural sweeteners in food and drinks, including honey. Topic 96 is about artificial sweeteners, sucralose and aspartame, with reference to their common usage in gums and their effects on insulin levels.

Table 3 shows examples of topics that are arguably irrelevant to discussion of sweeteners. Topic 37 is related to the Ariana Grande album, "Sweetener." Many of the LDA topics were related to this album. Topic 48 contains conversational words related to money, but it is not clear if this is connected to sweeteners. Topic 64 is an example of a topic with non-English words, which we tried to filter out for this analysis. Topic 92 seems to describe advertisements rather than organic discussion of sweeteners.

When manually reviewing samples of tweets, we have observed a very broad array of content, including people describing their usage or abstinence of sweeteners, sharing information/research, and expressing concerns or other opinions.

Table 2 The top words of the relevant topics that were produced by the LDA model

ID	Relevant topics
7	benefits, sugar, health, #stevia, natural, safe, effects, plant, stevia,
12	diet, coke, taste, aspartame, life, sucralose, make, pepsi, salt,
40	artificial, sweeteners, sugar, sucralose, aspartame, drinks, diet, people, think,
82	sugar, artificial, daddy, sweeteners, weight, sucralose, health, blood, diabetes,
32	lyme, disease, study, antibiotics, better, kills, pathogen, confirms,
90	sugar, natural, sweeteners, artificial, helaty, free, drink, honey, drinks,
96	artificial, sweeteners, sucralose, sugar, gum, insulin, aspartame, study, popular

Table 3 The top words of the *irrelevant* topics that were produced by the LDA model

ID	Irrelevant Topics
37	@arianagrande, love, # sweetener, sweetener, album, wait, excited,
48	deal, trade, think, throm, going need, maybe, money, sugar, thats, tax,
64	leche, en, la el, 1 avena, que, caf, una, 2, por, sin, es, le, para, canela, lo, mi
92	#ad, try, @intheraw, giveaway, cake, #stevia, pumpkin, #sweepstake, pecan

4 Temporal Patterns

The volume of tweets was plotted over the time span of data collection to allow for peaks in the chatter to be observed. Figure 1 shows the volumes broken down by different diseases that are mentioned, while Fig. 2 shows the volumes broken down by different sweeteners. We see that there is high temporal variability, and upon inspection we find that spikes in volume are usually aligned with something happening in the news.

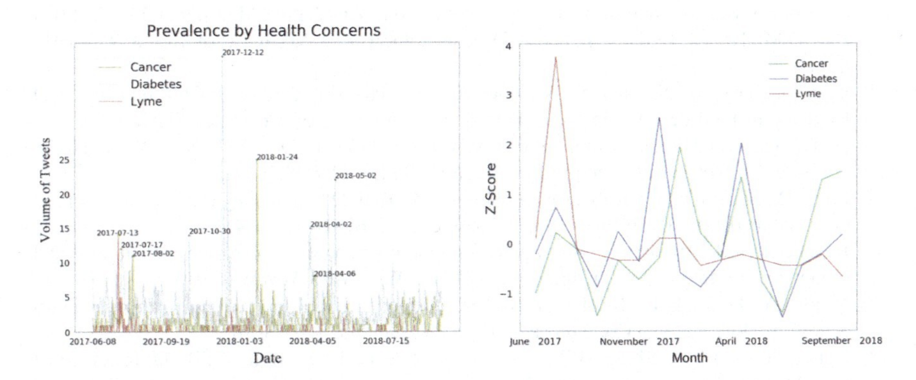

Fig. 1 Volume of tweets (raw volume, left; standardized volume, right) mentioning each of the three diseases we considered

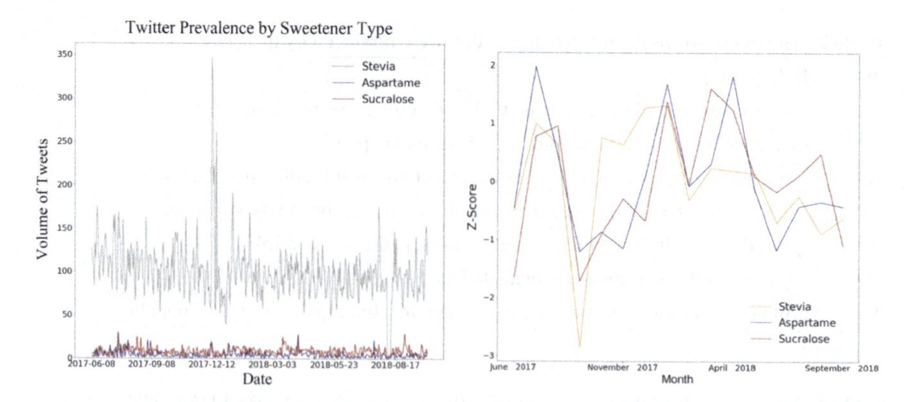

Fig. 2 Volume of tweets (raw, left; standardized, right) mentioning each of the three sweeteners we considered

A thorough search was done for the dates for each of the peaks occurred for the individual health concerns on the news that was being shared. The highest peak for cancer, on January 24th, coincided with a news article that was shared on how artificial sweeteners could someday cause cancer. The peak in July 2017 for Lyme disease coincided with a news article titled, "Stevia the best treatment for Lyme disease, researchers say." Tweets promoting this article were heavily circulated on that day, with people recommending Stevia as a potential cure.

References

1. Abbar, S., Mejova, Y., Weber, I.: You tweet what you eat: studying food consumption through Twitter. In: Conference on Human Factors in Computing Systems (CHI) (2015)
2. Ayers, J.W., Althouse, B.M., Johnson, M., Dredze, M., Cohen, J.E.: What's the healthiest day? circaseptan (weekly) rhythms in healthy considerations. Am. J. Prev. Med. **47**(1), 73–76 (2014)
3. Blei, D.M., Ng, A., Jordan, M.: Latent Dirichlet allocation. J. Mach. Learn. Res. **3**, 993–1022 (2003)
4. De Choudhury, M., Sharma, S., Kiciman, E.: Characterizing dietary choices, nutrition, and language in food deserts via social media. In: Conference on Computer Supported Cooperative Work and Social Computing (CSCW). pp. 1157–1170. ACM, New York (2016). 10.1145/2818048.2819956, http://doi.acm.org/10.1145/2818048.2819956
5. Fried, D., Surdeanu, M., Kobourov, S., Hingle, M., Bell, D.: Analyzing the language of food on social media. In: IEEE International Conference on Big Data (2014)
6. Kusmierczyk, T., Trattner, C., Norvag, K.: Temporality in online food recipe consumption and production. In: International Conference on World Wide Web (WWW) (2015)
7. Mejova, Y., Haddadi, H., Noulas, A., Weber, I.: #foodporn: obesity patterns in culinary interactions. In: International Conference on Digital Health, pp. 51–58 (2015)
8. Nguyen, C.Q., Li, D., Meng, H.W., Kath, S., Nsoesie, E., Li, F., Wen, M.: Building a national neighborhood dataset from geotagged Twitter data for indicators of happiness, diet, and physical activity. JMIR Public Health Surveill. **2**(2), e158 (2016). https://doi.org/10.2196/publichealth.5869, http://www.ncbi.nlm.nih.gov/pubmed/27751984

9. Pagoto, S., Schneider, K.L., Evans, M., Waring, M.E., Appelhans, B., Busch, A.M., Whited, M.C., Thind, H., Ziedonis, M.: Tweeting it off: characteristics of adults who tweet about a weight loss attempt. J. Am. Med. Inform. Assoc. **21**(6), 1032–1037 (2014)
10. Paul, M.J., Dredze, M.: Social monitoring for public health. In: Synthesis Lectures on Information Concepts, Retrieval, and Services, pp. 1–185. Morgan & Claypool (2017)
11. Pawar, R., Krynitsky, A., Rader, J.: Sweeteners from plants-with emphasis on Stevia rebaudiana (bertoni) and Siraitia grosvenorii (swingle). Anal. Bioanal. Chem. **405**, 4397–4407 (2013)
12. Sharma, S., De Choudhury, M.: Detecting and characterizing nutritional information of food and ingestion content in Instagram. In: International Conference on World Wide Web (WWW) (2015)
13. West, R., White, R.W., Horvitz, E.: From cookies to cooks: Insights on dietary patterns via analysis of web usage logs. In: International Conference on World Wide Web (WWW). Republic and Canton of Geneva, Switzerland (2013)

On-line (TweetNet) and Off-line (EpiNet): The Distinctive Structures of the Infectious

Byunghwee Lee, Hawoong Jeong, and Eun Kyong Shin

Abstract The field of epidemic modeling has rapidly grown in recent years. However, the studies to date suffer from empirical limitations and lack the multi-layered relational investigations. The case under scrutiny is the Middle East Respiratory Syndrome (MERS) epidemic episode in South Korea 2015. Linking the confirmed MERS patients data with social media data mentioning MERS, we examine the relationship between the epidemic networks (EpiNet) and the corresponding discourse networks on Twitter (TweetNet). Using network analyses and simulations, we unpack the epidemic diffusion process and the epidemic-related social media discourse diffusion. The on-line discourse structure of the infectious is larger, dense and complex than the off-line epidemic diffusion network and it has its own unique grammar. When we differentiated tweets by their user types, we observed that they display distinct temporal and structural patterns. They showed the divergent sensitivities in the spike timing and retweet patterns compared to simulated RandomNet. High self-clustering patterns by governmental and public tweets can hinder efficient communication/information spreading. Epidemic related social media surveillance should pay customized attentions accordingly to different types of users. This study should generate discussion about the feasibility and future of liability of social media based epidemic surveillance.

This work was supported by the Ministry of Education of the Republic of the Korea and National Research Foundation of Korea (NRF-2018S1A5B6075594).

B. Lee · H. Jeong
Department of Physics, Korea Advanced Institute of Science and Technology, Daejeon, South Korea

E. K. Shin (✉)
Department of Sociology, Korea University, Seoul, South Korea
e-mail: eunshin@korea.ac.kr

© The Editor(s) (if applicable) and The Author(s), under exclusive license to Springer Nature Switzerland AG 2021

A. Shaban-Nejad et al. (eds.), *Explainable AI in Healthcare and Medicine*, Studies in Computational Intelligence 914,
https://doi.org/10.1007/978-3-030-53352-6_17

Keywords Epidemic networks · Health intelligence · Epidemic surveillance · Middle East Respiratory Syndrome (MERS)

1 Introduction

Increased available data sets from the Internet bring many researchers' attention to the heuristic usages of social media data to fully understand epidemic dynamics and to design better surveillance platforms [1–9]. It has been suggested that social media data can be reliable sources for early epidemic detection (for example, Twitter data for H1N1 in 2009 [7, 10]). The contents of social media data have been a popular subject in epidemic surveillance studies [1, 4, 5, 10–13]. Yet, a relational understanding of the epidemic-related social media data is lacking.

A relational understanding of epidemic related social media data is the urgent matter of importance in that not all social media feeds and online activities carry an equal weights of influence. Therefore, identification of central users/feeds are critical in detecting the most critical mass out of exponentially accumulating data. Furthermore, up to date, we have assumed the basic relevance between epidemic diffusion and the social media discourses. However, germaneness between off-line pandemic and related online discussions has not been scientifically tested. To implement actionable surveillance systems using social media data, the aforementioned assumptions require systematic examination, especially in high lethality cases.

In this paper, we examine both epidemic related social media data and epidemic diffusion data to test if preexisting assumptions are valid and to extract the meaningful core clusters from a myriad of social media data, well reflecting off-line epidemic diffusion the case under scrutiny is the MERS epidemic episode in South Korea. Whereas Influenza A virus subtype (H1N1) in 2009 infected 750,000 individuals and 263 deaths in South Korea [13], MERS in 2015 infected 186 citizens and caused 38 deaths in South Korea. Despite the relatively small number of deaths within a relatively short period, unprecedented social fear was widespread [14, 15].

2 Data

In this paper, we examine the relationship between epidemic networks and epidemic related tweet networks based on actual medical data of the confirmed cases of MERS patents (N = 186) and MERS mentioning tweet data (N = 1,840,550). For the medical sets, we have access to the information who affected whom and in what order, which allowed us to build time-stamped epidemic diffusion networks. We have 186 confirmed MERS patients during this episode including 38 mortality. The keyword-based search led to the total of 1,840,550 tweets composed of 334,237 (18.2%) original tweets and 1,506,313 (81.8%) retweets. Among these data, we limited the time scope of analysis from 05/20/2015 to 12/23/2015. For each tweet entry, we have information related to the total number of followers and following, date of entry of the tweet, and retweet, which enable us to reconstruct the network of the MERS discourse on Twitter.

3 Methods

We used network methods to understand and compare the epidemic diffusion network and the epidemic-related social media discourse network. Then we additionally ran simulations to test link preferences in the observed network. First, for the epidemic diffusion network construction, we constructed an epidemic diffusion network using the data from the confirmed patients. This process followed the straightforward one mode network method [16, 17]. Nodes (represented as circles in the network mapping) are patients and links (represented as lines) are the contagion channels. We only look at the connected component (EpiNet) given our interest in the diffusion structures. Therefore we excluded the isolated patients whose infection routes were not clear. Numbers on the nodes are the infection sequences.

Then, we built an online discourse network using tweets that mentioned MERS. Among the total of 334,237 tweets, 246,174 (73.6%) were never retweeted. Due to the large amount of un-retweeted feeds, we used two-mode network method [18, 19] to induce the MERS tweet network using retweet pattern. Nodes are tweet entries and links are retweet relationships. Compared to the follower–following network, our approach can unveil the activated network structure facilitated the MERS discourse on social Media. A tweet is connected to another tweet if a user retweets those tweet entries. Given the large amount of tweet volume, we first sampled our data, filtering tweets that have retweet volume greater than or equal to 50, resulting in 3,536 tweets and 955 mediating users. The 3,536 tweets had been retweeted 776,649 times that was 51.6% of the entire retweet volume (1,506,313). From the bipartite network, we extracted a one-mode network of tweets by tracing out the 'user' nodes. We projected the bipartite network into a network of tweets where the links are weighted by the number of users who co-retweeted two tweets. In the one-mode tweet network, the more link weight between two nodes the more people retweet two tweets together, implying an innate similarity in that the two tweets together resonate with lots of same individuals. The resulted tweet network consists of 3,536 nodes and 4,124,853 links (with the mean degree $\langle k \rangle$ of 2,333) which is a highly dense network where, most of the nodes are connected to all other nodes. If all nodes are fully connected, the network would have $3,536 \times 3,535/2 = 6,249,880$ links.

However, not all links are equally important. Using the significance filter, next we extracted the core network. The distribution of link weight showed the heavy-tailed distribution implying high heterogeneity in weight. Given the heterogeneity in the link weight distribution of the tweet network and high link density, we applied a filtering algorithm to obtain significantly relevant edges for our further detailed investigation. We found locally significant links for each node by employing a local significance filtering algorithm using disparity filter [20]. By imposing a significance level alpha, we could filter out the links that carry weights that is not compatible with a randomly distributed weight with a certain significance. If a link is significant for both end nodes, the link is considered statistically relevant. To decide an optimum alpha, we chose an α that gave a maximum score S-N_b/N-L_b/L, where N and N_b are the number of nodes in the original network before filtering and in the filtered network, and L and L_b are number of links in the original network and filtered network respectively. For the one mode tweet network, we found that alpha = 0.008 produced the maximum score. We took the largest connected cluster of the filtered network as our final system (TweetNet). The TweetNet was composed of 3,491 nodes and 83,168 links. Compared to the original degree distribution (normal distribution shape), the TweetNet exhibited a heavy tailed distribution with hubs.

Finally, we further tested the differential clustering patterns among different tweet users. We coded Twitter users into three types: (1) Governmental agency (such as city governments or Korea Centers for Disease Control and Prevention), (2) News Media and (3) Public. We compared the TweetNet to 100 null model networks (RandomNet), generated by shuffling random nodes holding network characteristics at constant to identify unique clustering patterns in the TweetNet.

4 Results

The basic characteristics are compared in Table 1. Figure 1 shows the temporal changes in the number of infections and overall tweet volume. After an early increase in infections and at the first death, the tweet volume began to increase rapidly. Figure 1A shows the comparison to total tweets and Fig. 1B shows the comparison of our TweetNet. The introduction of new nodes (tweets) in the TweetNet (filtered network) behaves similarly with the total tweet data. The total tweets and infection rates are correlated at 0.589 (P-value at 0.0001) and the TweetNet and infection rates are correlated at 0.554 (P-value at 0.0001). We can conclude that our filtered TweetNet preserved well the original tweet pattern. Our TweetNet is more sensitive to the

Table 1 Comparison of network characteristics

	EpiNet	TweetNet
Number of nodes	16	3,491
Number of edges	181	83,168
Average degree	1.117	47.6471
Network diameter	3	8
Network density	0.007	0.0136
Modularity	0.650	0.247
Number of communities	12	9
Clustering coefficient	0.033	0.779

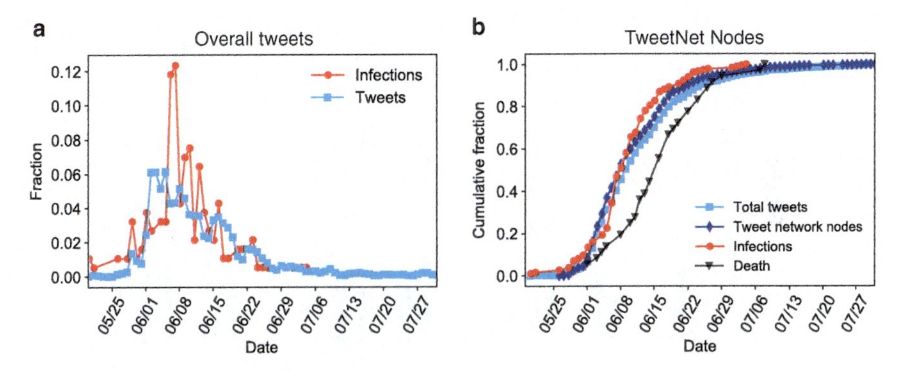

Fig. 1 Temporal changes of infections and tweet activity

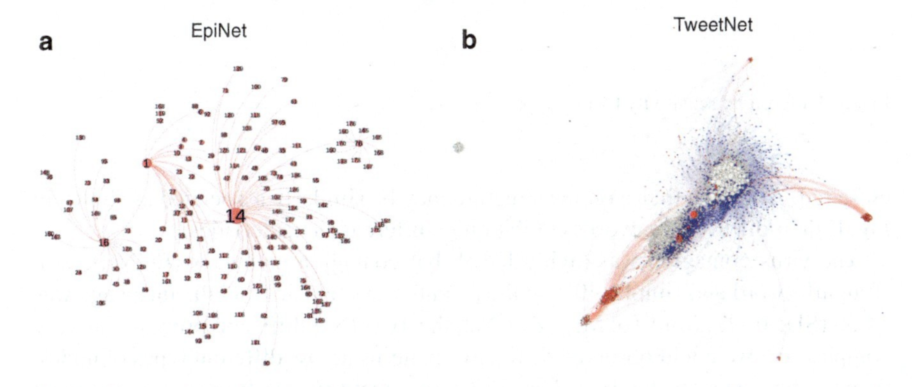

Fig. 2 Epidemic network and tweet network

actual epidemic curve: increases and saturation were slightly earlier than the total tweet data set. The TweetNet, consisted of 1% original tweets which allowed us to closer examination on the user types and temporalities. Please note that both original tweets and the TweetNet were not significantly correlated with the death rate because the mortality cases were sporadic and small in number.

The visualization of the EpiNet and the TweetNet are presented in Fig. 2. In the EpiNet (Fig. 2A), the size of nodes is proportional to the degree centrality, which means the number of people the node had infected. The label shows the order of confirmed date of the patients. In the TweetNet (Fig. 2B), the node size is proportional to the retweet volume of the node and colored by the node type: Government (red), Media (blue), and Public (gray). Compared the EpiNet, the TweetNet is larger and wider (network diameter is 3 for the EpiNet and 8 for the TweetNet). Especially when we compare the average degree, despite of few mega spreaders, EpiNet shows 1.11 ties to the other patients on average. But MERS-mentioning tweets have 47.64 ties to other MERS tweets. The TweetNet is much densely connected and highly

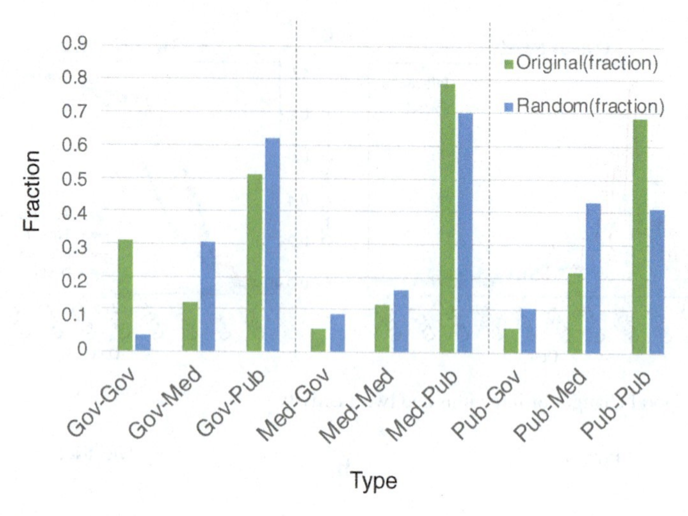

Fig. 3 Differential sensitivity by user types

clustered. Although their volume structure may be similarly associated as shown in Fig. 1, their diffusion networks exhibit quite different characteristics.

The virus contagion was highly lethal, but contagion pathways (EpiNet) were straightforward and simple. 40% of the patents were females and the mean age was 54.58 (SD: 16.27; Min: 16; Max: 87). Yet, the TweetNet diffusion process was very complex and we could observed distinctive patterns across different types of nodes. Among the nodes in the TweetNet, 312 nodes were tweets from the government agencies, 1,095 from the news media, and 2,084 from the public. Figure 3 presents the cumulative fractions by the node type. We can observe differential sensitivity in its spike timing. The public tweets sparked first and then news media and government. Also, the Public and Media tweets were rapidly increased right after the first death case, however the government tweets responded slowly.

To further test the differential linking patterns in the TweetNet, we investigated the link types: Gov-Gov, Gov-Media, Gov-Public, Media-Media, Media-Public, Public-Public. We compared the TweetNet results against 100 randomly shuffled networks. The results are shown in Fig. 4. Despite its small number of proportion in the overall nodes, 3% of government tweets were self-loop, co-retweeted with other government agents' tweets. Compared to the random networks, governmental nodes were notably under-associated with the media feeds. Media nodes were less inclined to be tweeted with other media feeds. Media feeds were more likely to get along with public feeds. Public nodes exhibited more co-retweeted (69.1%). In the RandomNet, we observed more Pub-Med links but in the observed MERS TweetNet, we did not observe a high prevalence of Pub-Med retweeting.

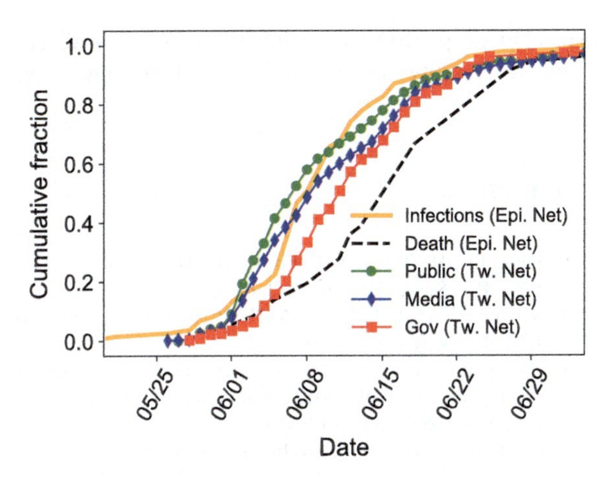

Fig. 4 Linking prevalence comparison TweetNet and RandomNet

5 Findings and Discussions

The growth of the field of epidemic modeling with social media data is an encouraging development and the results have the potential of influencing the directions of critical interventions and surveillance programs. In existing studies, both epidemic network structures and corresponding social media discourses are relatively incomplete. Multiple layers of epidemic data can help us to unpack the associated dynamics between epidemic diffusion and responding discourse diffusion. The strength of the our study was mainly driven by its ability to address the main problems that have plagued past research on social media analysis related to epidemic diffusion: linking social media data and epidemic data directly.

We tested the validity of the assumption about the similarity between on- and off-line diffusion. To date, most research of this kind has suffered from lack of whole network information on epidemic diffusion and corresponding discourse diffusion networks. Tweet activities are known to be correlated with the infection rate. However, we found that the response patterns of the epidemic episode on Twitter were different by the user types (Gov, Media, Public). When we differentiated nodes by their user type into three categories, they displayed distinct temporal and structural behavior. They showed the different sensitivities in spike timing and clustering patterns compared to the RandomNet. A high self-clustering pattern by governmental tweets and the public could possibly hinder efficient communication/information spreading. Although two networks were in accordance in term of the growing pattern, they exhibited the distinguished network characteristics. The TweetNet was denser, highly clustered, and less modular. As noticed from the size comparison of 186 infections versus 1,840,550 tweet mentions, the online MERS network is larger than the actual epidemic network to a great extent. The online discourse network was approximately 10,000 times larger than the confirmed number of MERS cases. Epidemic related social media surveillance should pay customized attentions based on the types of users. Although we have looked at only one epidemic case and one social media platform, this study should generate discussion about the feasibility and future of liability of social media based epidemic surveillance.

References

1. Fung, I.C.-H., Tse, Z.T.H., Fu, K.-W.: The use of social media in public health surveillance. West. Pac. Surveill. Response J. **6**(2), 3–6 (2015)
2. Salathé, M., et al.: Influenza A (H7N9) and the importance of digital epidemiology. New Engl. J. Med. **369**(5), 401 (2013)
3. Zhang, E.X., et al.: Leveraging social networking sites for disease surveillance and public sensing: the case of the 2013 avian influenza A (H7N9) outbreak in China. West. Pac. Surveill. Response J. **6**(2), 66–72 (2015)
4. Gu, H., et al.: Importance of Internet surveillance in public health emergency control and prevention: evidence from a digital epidemiologic study during avian influenza A H7N9 outbreaks. J. Med. Internet Res. **16**(1), e20 (2014)
5. Mollema, L., et al.: Disease detection or public opinion reflection? Content analysis of tweets, other social media, and online newspapers during the measles outbreak in The Netherlands in 2013. J. Med. Internet Res. **17**(5), e128 (2015)
6. Shin, E.K., Shaban-Nejad, A.: Public health intelligence and the Internet: current state of the art, in public health intelligence and the Internet. In: Shaban-Nejad, A., Brownstein, J.S., Buckeridge, D.L. (eds.), pp. 1–17. Springer, Cham (2017)
7. Culotta, A.: Towards detecting influenza epidemics by analyzing Twitter messages. In: Proceedings of the First Workshop on Social Media Analytics. ACM (2010)
8. Broniatowski, D.A., Paul, M.J., Dredze, M.: National and local influenza surveillance through Twitter: an analysis of the 2012–2013 influenza epidemic. PloS One **8**(12), e83672 (2013)
9. Brownstein, J., Freifeld, C.: HealthMap: the development of automated real-time Internet surveillance for epidemic intelligence. Euro Surveill **12**(11), E071129 (2007)
10. Fung, I.C.-H., Wong, K.: Efficient use of social media during the avian influenza A (H7N9) emergency response. West. Pac. Surveill. Response J. **4**(4), 1 (2013)
11. Corley, C.D., et al.: Text and structural data mining of influenza mentions in web and social media. Int. J. Environ. Res. Public Health **7**(2), 596–615 (2010)
12. Chunara, R., Andrews, J.R., Brownstein, J.S.: Social and news media enable estimation of epidemiological patterns early in the 2010 Haitian cholera outbreak. Am. J. Trop. Med. Hyg. **86**(1), 39–45 (2012)
13. Kim, J.Y.: The 2009 H1N1 pandemic influenza in Korea. Tuberc. Respir. Dis. **79**(2), 70–73 (2016)
14. Ro, J.-S., et al.: Worry experienced during the 2015 middle east respiratory syndrome (MERS) pandemic in Korea. PloS One **12**(3), e0173234 (2017)
15. Kim, D.-H.: Structural factors of the middle east respiratory syndrome coronavirus outbreak as a public health crisis in Korea and future response strategies. J. Prev. Med. Public Health **48**(6), 265 (2015)
16. Wasserman, S., Faust, K.: Social Network Analysis: Methods and Applications, vol. 8. Cambridge University Press, Cambridge (1994)
17. Kadushin, C.: Understanding Social Networks: Theories, Concepts, and Findings. OUP, USA (2012)
18. Borgatti, S.P., Everett, M.G.: Network analysis of 2-mode data. Soc. Netw. **19**(3), 243–269 (1997)
19. Breiger, R.L.: The duality of persons and groups. Soc. Forces **53**(2), 181–190 (1974)
20. Serrano, M.Á., Boguná, M., Vespignani, A.: Extracting the multiscale backbone of complex weighted networks. Proc. Natl. Acad. Sci. **106**(16), 6483–6488 (2009)

Medication Regimen Extraction from Medical Conversations

Sai P. Selvaraj and Sandeep Konam

Abstract Extracting relevant information from medical conversations and providing it to doctors and patients might help in addressing doctor burnout and patient forgetfulness. In this paper, we focus on extracting the Medication Regimen (dosage and frequency for medications) discussed in a medical conversation. We frame the problem as a Question Answering (QA) task and perform comparative analysis over: a QA approach, a new combined QA and Information Extraction approach, and other baselines. We use a small corpus of 6,692 annotated doctor-patient conversations for the task. Clinical conversation corpora are costly to create, difficult to handle (because of data privacy concerns), and thus scarce. We address this data scarcity challenge through data augmentation methods, using publicly available embeddings and pre-train part of the network on a related task (summarization) to improve the model's performance. Compared to the baseline, our best-performing models improve the dosage and frequency extractions' ROUGE-1 F1 scores from 54.28 and 37.13 to 89.57 and 45.94, respectively. Using our best-performing model, we present the first fully automated system that can extract Medication Regimen tags from spontaneous doctor-patient conversations with about ~71% accuracy.

1 Introduction

Physician burnout is a growing concern, estimated to be experienced by at least 35% of physicians in the developing world and 50% in the United States [11]. It is found

S. P. Selvaraj (✉) · S. Konam
Abridge AI Inc., Pittsburgh, USA
e-mail: prabhakarsai@abridge.com

S. Konam
e-mail: san@abridge.com

© The Editor(s) (if applicable) and The Author(s), under exclusive license to Springer Nature Switzerland AG 2021
A. Shaban-Nejad et al. (eds.), *Explainable AI in Healthcare and Medicine*, Studies in Computational Intelligence 914,
https://doi.org/10.1007/978-3-030-53352-6_18

that for every hour physicians provide direct clinical facetime to patients, nearly two additional hours are spent on EHR (Electronic Health Records) and administrative or desk work. As per the study conducted by Massachusetts General Physicians Organization (MPGO) [2] and as reported by [12], the average time spent on administrative tasks increased from 23.7% in 2014 to 27.9% in 2017. Both surveys found that time spent on administrative tasks was positively associated with higher likelihood of burnout. Top reasons under administrative burden include working on the ambulatory EHR, handling medication reconciliation (sometimes done by aides), medication renewals, and medical billing and coding. The majority of these reasons revolve around the documentation of information exchanged between doctors and patients during clinical encounters. Automatically extracting such medical information [4] could both alleviate the documentation burden on the physician, and also allow them to dedicate more time directly with patients.

Among all the medical information extraction tasks, Medication Regimen (medication name, dosage, and frequency) extraction is particularly interesting due to its ability to help doctors with medication orders cum renewals, medication reconciliation, verifying of reconciliations for errors, and other medication-centered EHR documentation tasks. In addition, the same information when provided to patients can help them better recall doctor's instructions which might aid in compliance with the care plan. This is particularly important given that patients forget or wrongly recollect 40–80% [8] of what is discussed in the clinic, and accessing EHR data has its own challenges [5].

Spontaneous medical conversations happening between a doctor and a patient have several distinguishing characteristics from a normal monologue or prepared speech: it involves multiple speakers with overlapping dialogues, covers a variety of speech patterns, and the vocabulary can range from colloquial to complex domain-specific language. With recent advancements in Conversational Speech Recognition [9] rendering the systems less prone to errors, the subsequent challenge of understanding and extracting relevant information from the conversations is receiving increased research focus [4, 6, 7, 14].

In this paper, we focus on *local* information extraction in transcribed medical conversations. Specifically, we extract dosage (e.g. 5 mg) and frequency (e.g. once a day) for the medications (e.g. aspirin) from these transcripts, collectively referred to as *Medication Regimen (MR) extraction*. The information extraction is *local* as we extract the information from a segment of the transcript and not the entire transcript since doing the latter is difficult owing to the long meandering nature of the conversations often with multiple medication regimens and care plans being discussed.

The challenges associated with the Medication Regimen (MR) extraction task include understanding the spontaneous dialog with medical vocabulary and understanding the relationship between different entities as the discussion can contain multiple medications and dosages (e.g. doctor revising a dosage or reviewing all the medications).

We frame this problem as a Question Answering (QA) task by generating questions using templates. We base the QA model on pointer-generator networks [17] augmented with Co-Attentions [19]. In addition, we develop models combining QA and Information Extraction frameworks using multi-decoder (one each for dosage and frequency) architecture.

Lack of availability of a large volume of data is a typical challenge in healthcare. A conversation corpus by itself is a rare commodity in the healthcare data space because of the cost and difficulty in handing (because of data privacy concerns). Moreover, transcribing and labeling the conversations is a costly process as it requires domain-specific medical annotation expertise. To address data shortage and improve the model performance, we investigate different high-performance contextual embeddings (ELMo [16], BERT [3] and ClinicalBERT [1]), and pretrain models on a medical summarization task. We further investigate the effects of training data size on our models.

On the MR extraction task, ELMo with encoder multi-decoder architecture and BERT with encoder-decoder with encoders pretrained on the summarization task perform the best. The best-performing models improve our baseline's dosage and frequency extractions ROUGE-1 F1 scores from 54.28 and 37.13 to 89.57 and 45.94, respectively.

Using our models, we present the first fully automated system to extract MR tags from spontaneous doctor-patient conversations. We evaluate the system (using our best performing models) on the transcripts generated from Automatic Speech Recognition (ASR) APIs offered by Google and IBM. In Google ASR's transcripts, our best model obtained ROUGE-1 F1 of 71.75 for Dosage extraction (which in this specific case is equal to the percentage of times dosage is correct, refer to Sect. 4.2 for more details) and 40.13 for Frequency extraction tasks. On qualitative evaluation, we find that for 73.58% of the medications the model can find the correct frequency. These results demonstrate that the research on NLP can be used effectively in a real clinical setting to benefit both doctors and patients.

2 Data

Our dataset consists of a total of 6,693 real doctor-patient conversations recorded in a clinical setting using distant microphones of varying quality. The recordings have an average duration of 9 min 28 s and have a verbatim transcript of 1,500 words on average (written by the experts). Both the audio and the transcript are de-identified (by removing the identifying information) with digital zeros and `[de-identified]` tags, respectively. The sentences in the transcript are grounded to the audio with the timestamps of their first and last words.

The transcript of the conversations are annotated with summaries and Medication Regimen tags (MR tags), both grounded using the timestamps of the sentences from the transcript deemed relevant by the expert annotators, refer to Table 1. The transcript for a typical conversation can be quite long, and not easy for many of the high performing deep learning models to act on. Moreover, the medical information about a concept/condition/entity can change during the conversation after a significant time gap. For example, the dosage of a medication can be different when discussing current

Table 1 Example of our annotations grounded to the transcript segment

(Timestamp) transcript	Summary	Medication name	Dosage	Frequency
(1028.3 s) So, let's, I'm going to have them increase the mg, uh, Coumadin level, so that, uh, like I said, the pulmonary embolism doesn't get worse here (1044.9 s) Yeah (1045.2 s) Yeah (1045.4 s) Increase it to three point five in the morning and before bed	Increase Coumadin to 3.5 mg to prevent pulmonary embolism from getting bigger	Coumadin	3.5 mg	Twice a day

medication the patient is on, versus when they are prescribed a different dosage. For this reason, we have annotations, that are grounded to a short segment of the transcript.

The summaries (#words-μ = 9.7; σ = 10.1) are medically relevant and local. The MR tags are also local and are of the form {`Medication Name`, `Dosage`, `Frequency`}. If dosage ($\mu = 2.0; \sigma = 0$) or frequency ($\mu = 2.1; \sigma = 1.07$) information for a medication is not present in a grounded sentence, the corresponding field will be 'none'.

In the MR tags, `Medication Name` and `Dosage` (usually a quantity followed by its units) can be *extracted* with relative ease from the transcript except for the units of the dosage, which is sometimes inferred. In contrast, due to high degree of linguistic variation with which `Frequency` is often expressed, extracting it requires an additional *inference* step. For example, 'take one in the morning and at noon' from the transcript is tagged as 'twice a day' in the frequency tag, likewise 'take it before sleeping' is tagged as 'at night time'.

Out of 6,693 files, we set aside a random sample of 423 files (denoted as $\mathcal{D}_{test}$) for final evaluation. The remaining 6,270 files are considered for training with 80% train (5016), 10% validation (627), and 10% test (627) split. Overall, the 6,270 files contain 156,186 summaries and 32,000 MR tags, out of which 8,654 MR tags contain values for at least one of the `Dosage` or `Frequency`, which we used for training to avoid overfitting (the remaining MR tags have both `Dosage` and `Frequency` as 'none'). Note that we have two test datasets: '10% test'-used to evaluate all the models, and $\mathcal{D}_{test}$-used to measure the performance of best performing models on ASR transcripts.

3 Approach

We frame the Medication Regimen extraction problem as a Question Answering (QA) task, which forms the basis for our first approach. It can also be considered as a specific inference or relation extraction task, since we extract specific information about an entity (`Medication Name`). For this reason, our second approach is at the intersection of Question Answering (QA) and Information Extraction (IE) domains. Both approaches involve using a contiguous segment of the transcript and the `Medication Name` as input to find or infer the medication's `Dosage` and `Frequency`. When testing the approaches mimicking real-world conditions, we extract `Medication Name` from the transcript separately using medication ontology (refer to Sect. 5.4).

In the first approach, we frame the MR task as a QA task and generate questions using the template: "What is the <dosage/frequency> for `<Medication Name>`?". Here, we use an abstractive QA model based on pointer-generator networks [17] augmented with coattention encoder [19] (QA-PGNet).

In the second approach, we frame the problem as a conditioned IE task, where the information extracted depends on an entity (`Medication Name`). Here, we use a multi-decoder pointer-generator network augmented with coattention encoder (Multi-decoder QA-PGNet). Instead of using templates to generate questions and single decoder to extract different types of information as in the QA approach (which might lead to performance degradation), here we consider separate decoders for extracting specific types of information conditioned on an entity E (`Medication Name`).

3.1 *Pointer-Generator Network (PGNet)*

The network is a sequence-to-sequence attention model that can both copy a word from the input I containing P word tokens or generate a word from its vocabulary $vocab$, to produce the output sequence.

First, embeddings of the input tokens are fed one-by-one to the encoder, a single bi-LSTM layer which encodes the tokens into hidden states-$H = encoder(I)$, where $H = [h_1...h_P]$. For each decoder time step t, in a loop, we compute, (1) attention a_t (using the last decoder state s_{t-1}), over the input, and (2) the decoder state s_t using a_t. Then, at each time step, using both a_t and s_t we can find the probability $P_t(w)$, of producing a word w (from both $vocab$ and I). For convenience, we denote the attention and the decoder as $decoder_{pg}(H) = P(w)$, where $P(w) = [P_1(w)...P_T(w)]$. The output can then be decoded from $P(w)$, which is decoded until it produces an 'end of output token' or the number of steps reach the maximum allowed limit.

3.2 QA PGNet

We first separately encode both the question-$H_Q = encoder(Q)$ and the input-$H_I = encoder(I)$ using encoders (with shared weights). Then, to condition I on Q (and vice versa), we use the coattention encoder [19] which attends to both the I and Q simultaneously to generate the coattention context-$C_D = coatt(H_I, H_Q)$. Finally, using the pointer-generator decoder we find the probability distribution of the output sequence-$P(w) = decoder_{pg}([H_I; C_D])$, which is then decoded to generate the answer sequence.

3.3 Multi-decoder (MD) QA PGNet

For extracting K types of information about an entity E, we first encode the inputs into H_I and $H_E = encoder(E)$. Then in an IE fashion, we use multi-decoder (MD) setup to obtain K probability distributions $P^k(w)$, which can then be decoded to get the corresponding output sequences.

$$C_D^k = coatt^k(H_I, H_E)\ \forall k = 1...K$$

$$P^k(w) = decoder_{pg}^k([H_I; C_D^k])\ \forall k = 1...K$$

All the networks discussed above are trained using a negative log-likelihood loss.

4 Experiments

We initialized MR extraction models' vocabulary from the training dataset after removing words with a frequency lower than 30 in the dataset, resulting in 456 word tokens. Our vocabulary is small because of the size of the dataset, hence we rely on the model's ability to copy words to produce the output effectively. In all our model variations, the embedding and the network's hidden dimension are set to be equal. The networks were trained with a learning rate of 0.0015, dropout of 0.5 on the embedding layer, normal gradient clipping set at 2, batch size of 8, and optimized with Adagrad and the training was stopped using the 10% validation dataset.

4.1 Data Processing

We did the following basic preprocessing to our data: (1) added 'none' to the beginning of the input utterance so that the network could point to it when there was no relevant information in the input, (2) filtered outliers with a large number of grounded transcript sentences (>150 words), and (3) converted all text to lower case.

To improve performance, we (1) standardized all numbers (both digits and words) to words concatenated with a hyphen[1] (e.g. 110 -> one-hundred-ten), in both input and output, (2) removed units from `Dosage` as sometimes the units are not explicitly mentioned in the transcript segment but were written by the annotators using domain knowledge, (3) prepended all medication mentions with 'rx-' tag, as this helps the model's performance when multiple (different) medications are discussed in a segment, and (4) created new data points by randomly shuffling medications and dosages in both input and output (when we have more than one in a transcript segment) to increase the number of training data points. Randomly shuffling the entities increases the number of training MR tags from 8,654 to 11,521. Based on the data statistics after data processing, we fixed the maximum encoder steps to 100, dosage decoder steps to 1, and frequency decoder steps to 3 (for both the QA and Multi-decoder QA models).

4.2 Metrics

For the MR extraction task, we measure the ROUGE-1 scores for both the Dosage and Frequency extraction tasks. It should be noted that since `Dosage` is a single word token (after processing), both the reference and hypothesis are a single token, making its ROUGE-1 F1, Precision and Recall scores equal, which are in turn equal to the percentage of times we find the correct dosage for the medications.

In our annotations, Frequency has conflicting tags (e.g. {'Once a day', 'twice a day'} and 'daily'), hence metrics like Exact Match will be erroneous. To address this issue, we use the ROUGE scores to compare different models on the 10% test dataset and we use qualitative evaluation to measure the top-performing models on $\mathcal{D}_{test}$.

4.3 Model Variations

We consider QA PGNet and Multi-decoder QA PGNet with lookup table embedding as baseline models and improve over the baselines with other variations described below. Apart from learning-based baselines, we also create two naive baselines. For Dosage extraction, the baseline we consider is 'Nearest Number', where we take the number nearest to the `Medication Name` as the prediction, and 'none' if none exist or if the `Medication Name` is not detected in the input.[2] For Frequency

[1]This prevents overfitting and repetition when converting all the numbers to words.

[2]This can happen when a different form of a medication (e.g. abbreviation, generic or brand name) is used in the conversation compared to the annotation.

extraction, the baseline we consider is 'Random Top-3' where we predict a random `Frequency` tag, from top-3 most frequent ones from our dataset-{'none', 'daily', 'twice a day'}.

4.3.1 Embedding

We developed different variations of our models with a simple lookup table embeddings learned from scratch and using high-performance contextual embeddings, which are ELMo [16], BERT [18] and ClinicalBERT [1] (trained and provided by the authors). Refer to Table 2 for the performance comparisons.

We derive embeddings from ELMo by learning a linear combination of its last three layer's hidden states (task-specific fine-tuning [16]). Similarly, for BERT-based embeddings, we take a linear combination of the hidden states from its last four layers, as this combination performs best without increasing the size of the embeddings [18]. Since BERT and ClinicalBERT use word-piece vocabulary and compute sub-word embeddings, we compute word-level embedding by averaging the corresponding sub-word tokens. ELMo and BERT embeddings both have 1024 dimensions, ClinicalBERT have 768 as it is based on BERT base model, and the lookup table have 128–higher dimension models leads to overfitting.

4.3.2 Pertaining Encoder

We trained the PGNet as a summarization task using the medical summaries and used the trained model to initialize the encoders (and the embeddings) of the corresponding QA models. We use a vocab size of 4073 words, derived from the training dataset with a frequency threshold of 30 for the task. We trained the models using Adagrad optimizer with a learning rate of 0.015, normal gradient clipping set at 2 and trained for around 150000 iterations (stopped using validation dataset). On the summarization task PGNet obtained ROUGE-1 F1 scores of 41.42 with ELMo and 39.15 with BERT embeddings. We compare the effects of pretraining the model in Table 2, models with 'pretrained encoder' had their encoders and embeddings pretrained with the summarization task.

Table 2 ROUGE-1 scores of baselines and models for the MR extraction task on the 10% test dataset. PT: using pretrained encoder; B: baseline; MD: Multi-decoder

Models	Dosage	Frequency		
	F1	F1	Recall	Precision
Nearest number (B)	67.13	-	-	-
Random top-3 (B)	-	29.32	29.20	29.49
Lookup table + QA PGNet (B)	54.28	37.13	59.83	29.82
ELMo + QA PGNet	88.67	35.04	59.94	26.91
BERT + QA PGNet	86.34	43.52	70.46	34.81
ClinicalBERT + QA PGNet	84.20	43.92	73.67	34.27
Lookup table + MD QA PGNet (B)	51.97	36.56	71.37	25.00
ELMo + MD QA PGNet	88.64	37.41	58.66	30.56
BERT + MD QA PGNet	85.59	42.57	71.05	33.33
ClinicalBERT + MD QA PGNet	84.82	44.04	71.50	33.05
ELMo + QA PGNet(PT)	89.17	42.78	70.61	33.54
BERT + QA PGNet(PT)	88.62	**45.94**	74.54	36.70
ELMo + MD QA PGNet(PT)	**89.57**	44.82	75.71	34.62
BERT + MD QA PGNet(PT)	86.98	44.47	74.43	34.75

5 Results and Discussion

5.1 Difference in Networks and Approaches

5.1.1 Embeddings

On Dosage extraction, in general, ELMo obtains better performance than BERT, refer to Table 2. This could be because we concatenated the numbers with a hyphen, and because ELMo uses character-level tokens it can learn the tagging better than BERT. Similar observations are found in recent literature. On the other hand, on Frequency extraction, without pretraining, ELMo's performance lags by a big margin of ~8.5 ROUGE-1 F1 compared to BERT-based embeddings.

Although ClinicalBERT performed the best in the Frequency extraction task (by a small margin) in cases without encoder pretraining, in general it does not perform as well as BERT. This could also be a reflection of the fact that the language and style of writing used in clinical notes is very different from the way doctors converse with patients and the embedding dimension difference. Lookup table embedding performed decently in the frequency extraction task, but lags behind in the Dosage extraction task.

5.1.2 Other Variations

Considering various models' performance (without pretraining) and the resource constraint, we choose ELMo and BERT embeddings to analyze the effects of pretraining the encoder. When the network's encoder (and embedding) is pretrained with the summarization task, we (1) see a small decrease in the average number of iterations required for training, (2) improvement in individual performances of all models for both the sub-tasks, and (3) get best performance metrics across all variations, refer to Table 2. Both in terms of performance and the training speed, there is no clear winner between shared and multi-decoder approaches. Medication tagging and data augmentation increase the best-performing model's ROUGE-1 F1 score by $\sim$1.5 for the Dosage extraction task.

We also measure the performance of Multitask Question Answering Network (MQAN) [15] a QA model trained by the authors on the Decathlon multitask challenge. Since MQAN was not trained to produce our output sequence, it would not be fair to compute ROUGE scores, hence we haven't included them in the tables. Instead, we randomly sample the MQAN's predictions from the 10% test dataset and qualitatively evaluate it. From the evaluations, we find that MQAN can not distinguish between frequency and dosage, and mixed the answers. MQAN correctly predicted the dosage for 29.73% and frequency for 24.24% percent of the medications compared to 84.12% and 76.34% for the encoder pretrained BERT QA PGNet model trained on our dataset. This could be because of differences in the training dataset, domain, and the tasks in the Decathlon challenge compared to ours.

Almost all our models perform better than the naive baselines and the ones using lookup table embeddings, and our best performing models outperform them significantly. Among all the variations, the best performing models are ELMo with Multi-decoder (Dosage extraction) and BERT with shared-decoder QA PGNet architecture (Frequency extraction) with pretrained encoder. We chose these two models for our subsequent analysis.

5.2 Breakdown of Performance

We categorize the 10% test dataset into different categories based on the complexity and type of the data and analyze the breakdown of the system's performance in Table 3. We breakdown the Frequency extraction into two categories: (1) None: ground truth Frequency tag is 'none', and (2) NN (Not None): ground truth Frequency tag is not 'none'. Similarly, the Dosage extraction into 4 categories: (1) None: ground truth dosage tag is 'none', (2) MM (Multiple Medicine): input segment has more than one Medication mentioned, (3) MN (Multiple Numbers): input segment has more than one number present, and (4) NBM (Number between correct Dosage and Medicine): between the `Medication Name` and the correct Dosage in the input segment there are other numbers present. Note that the categories of the Dosage extraction task are not exhaustive, and one tag can belong to multiple categories.

Table 3 Performance (ROUGE-1 F1) breakdown of the best performing models measured on the 10% test dataset for the MR extraction task, refer Results section for more details. PT: using pretrained encoder; MD: Multi-decoder

Models	Dosage				Frequency	
	None	MM	MN	NBM	None	NN
ELMo + MD QA PGNet	**93.46**	81.73	69.87	**66.68**	42.98	34.58
BERT + QA PGNet	93.12	78.38	65.51	53.32	45.10	42.13
ELMo + MD QA PGNet(PT)	92.29	**84.17**	**74.01**	60.02	**45.77**	45.02
BERT + QA PGNet(PT)	**93.46**	79.95	68.77	60.02	43.31	**46.11**

From the performance breakdown of Dosage extraction task, we see that (1) the models are able to better identify when a medication's dosage is absent ('none') than other categories, (2) there is a performance dip in hard cases (MM, MN, and NBM), (3) the models are able to figure out the correct dosage (decently) for a medication even when there are multiple numbers/dosage present, and (4) the model struggles the most in the NBM category. The models' low performance in NBM could be because we have a comparatively lower number of examples to train in this category. The Frequency extraction task performs equally well when the tag is 'none' or not. In most categories, we see an increase in performance when using pretrained encoders.

5.3 *Training Dataset Size*

We vary the number of MR tags used to train the model and analyze the model's performance when training the networks using publicly available contextual embeddings, compared to using pretrained embeddings and encoder (pretrained on the summarization task). Out of the 5,016 files in the 80% train dataset, only 2,476 have at least one MR tag. Therefore, out of the 2476 files, we randomly choose 100, 500, and 1000 files and trained the best performing model variations to observe the performance differences (refer to Fig. 1). For all these experiments we used the same vocabulary size (456), the same hyper/training parameters, and the same 10% test split of 627 files.

As expected, we see that the encoder pretrained models have higher performance across all the training data sizes, when compared to their non-pretrained counterparts (refer to Fig. 1). The difference, as expected, shrinks as the training data size increases.

5.4 *Evaluating on ASR Transcripts*

To test the performance of our models in real-world conditions, we use commercially available ASR services from Google and IBM to transcribe the $\mathcal{D}_{test}$ files and measure

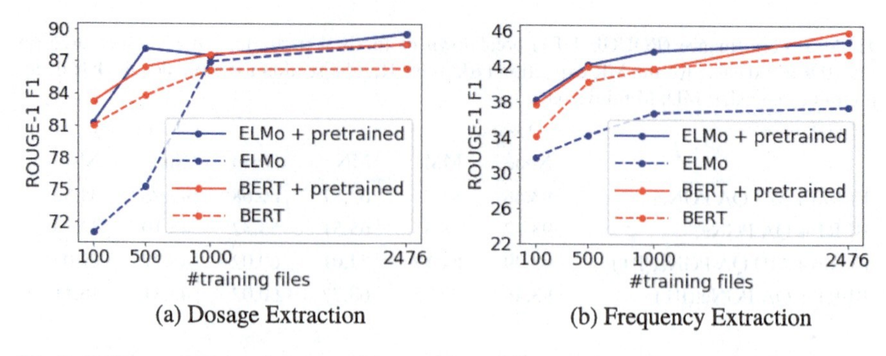

(a) Dosage Extraction (b) Frequency Extraction

Fig. 1 Difference in the performance of the models on different training data sizes (100, 500, 1,000, and 2,476) on the 10% test dataset for the MR extraction task

the performance of our models without assuming any annotations (except when calculating the metrics). It should be noted that this is not the case in our previous evaluations using '10% test' dataset where we use the segmentation information. For ground truth annotations on ASR transcripts, we aligned the MR tags from human written transcripts to the ASR transcript using their grounded timing information. Additionally, since ASR is prone to errors, if a medication from an MR tag is not recognized correctly in the ASR transcript, during the alignment we remove the corresponding MR tag.

In our evaluations, we use Google Cloud Speech-to-Text[3] (G-STT) and IBM Watson Speech to Text[4] (IBM-STT) as these were among the top-performing ASR APIs on medical speech [10] and were readily available to us. We used G-STT, with the 'video model' with punctuation settings. Unlike our human written transcripts, the transcript provided by G-STT is not verbatim and does not have disfluencies. IBM-STT, on the other hand, does not give punctuation so we used the speaker changes to add end-of-sentence punctuation.

In our $\mathcal{D}_{test}$ dataset, on initial study we see a Word Error Rate of ~50% for the ASR APIs. This number is not accurate because of, (1) de-identification, (2) disfluency (verbatim) differences between the human written and ASR transcript, and (3) minor alignment differences between the audio and the ground truth transcript.

During this evaluation, we followed the same preprocessing methods we used during training. Then, we auto segment the transcript into small contiguous segments similar to the grounded sentences in the annotations for tags extraction. To segment the transcript, we follow a simple procedure. First, we detect all the medications in a transcript using RxNorm [13] via string matching.[5] For all the detected medications, we select $2 \leq x \leq 5$ nearby sentences as the input to our model. We

[3]https://cloud.google.com/speech-to-text/.

[4]https://www.ibm.com/cloud/watson-speech-to-text.

[5]Since we had high quality human written transcripts and our ASR transcripts did not contain spelling mistakes (as long as the word was correctly recognized), string matching worked well during testing.

Table 4 ROUGE-1 scores of the best performing models on the ASR (Google: G-STT and IBM: IBM-STT) and human written transcripts (HW) on the $\mathcal{D}_{test}$ dataset for the MR extraction task. PT: using pretrained encoder; HS: Human segmentation; AS: Auto segmentation; MD: Multi-decoder; P: Precision; R: Recall

Transcripts	Models	Dosage	Frequency		
		F1	F1	R	P
HW	BERT + QA PGNet(PT)	85.69	49.25	72.25	41.20
with HS	ELMo + MD QA PGNet(PT)	85.79	43.99	69.27	35.63
HW	BERT + QA PGNet(PT)	75.43	41.62	62.29	35.01
with AS	ELMo + MD QA PGNet(PT)	75.71	37.83	61.41	29.84
G-STT	BERT + QA PGNet(PT)	70.51	39.90	61.32	33.07
with AS	ELMo + MD QA PGNet(PT)	71.75	40.13	67.22	31.14
IBM-STT	BERT + QA PGNet(PT)	73.93	30.90	52.53	24.10
with AS	ELMo + MD QA PGNet(PT)	78.93	36.58	67.03	26.52

increased x iteratively until we encounter a quantity entity–detected using spaCy's entity recognizer,[6] and we set x as 2 if we did not detect any entities in the range.

We show the model's performance on ASR transcripts and human written transcripts with automatic segmentation, and human written transcripts with human (defined) segmentation, in Table 4. Since the number of recognized medications in IBM-STT is only 95 compared to 725 (human written), we mainly consider the models' performance on G-STT's transcripts (343).

On the Medications that were recognized correctly, the models can perform decently on ASR transcripts in comparison to human transcripts (within 5 points ROUGE-1 F1 for both tasks, refer to Table 4). This shows that the models are robust to ASR variations discussed above. The lower performance compared to human transcripts is mainly due to incorrect recognition of `Dosage` and other medications in the same segments (changing the meaning of the text). By comparing the performance of the model on the human written transcripts with human (defined) segmentation and the same with auto segmentation, we see a 10 point drop in Dosage and 6 point drop in Frequency extraction tasks. This points out the need for more sophisticated segmentation algorithms.

With G-STT, our best model obtained ROUGE-1 F1 of 71.75 (which equals to percentage of times dosage is correct in this case) for Dosage extraction and 40.13 for Frequency extraction tasks. To measure the percentage of times the correct frequency was extracted by the model, we qualitatively compared the extracted and predicted frequency. We find that the model can find the correct frequency from the transcripts for 73.58% of the medications.

[6]https://spacy.io/api/entityrecognizer.

6 Conclusion

In this paper, we explore the Medication Regimen (MR) extraction task of extracting dosage and frequency for the medications mentioned in a doctor-patient conversation transcript. We explore different variations of abstractive QA models and a new architecture at the intersection of QA and IE frameworks and provide a comparative performance analysis of the methods along with other techniques like pre-training to improve the performance. Finally, we demonstrate the performance of our best-performing models by automatically extracting MR tags from spontaneous doctor-patient conversations (using commercially available ASR). Our best model can correctly extract the dosage for 71.75% (interpretation of ROUGE-1 score) and frequency for 73.58% (on qualitative evaluation) of the medications discussed in the transcripts generated using Google Speech-To-Text. In summary, we demonstrate that our research can be translated into real clinical settings to realize its benefits for both doctors and patients.

Using ASR transcripts in training to improve the performance and extracting other important medical information can be interesting lines of future work.

Acknowledgements We thank: University of Pittsburgh Medical Center (UPMC) and Abridge AI Inc. for providing access to the de-identified data corpus; Dr. Shivdev Rao, CEO, Abridge AI Inc. and a practicing cardiologist in UPMC's Heart and Vascular Institute, and Prlof. Florian Metze, Associate Research Professor, Carnegie Mellon University for helpful discussions; Ben Schloss, Steven Coleman, and Deborah Osakue for data business development and annotation management.

References

1. Alsentzer, E., Murphy, J., Boag, W., Weng, W.H., Jindi, D., Naumann, T., McDermott, M.: Publicly available clinical BERT embeddings. In: Proceedings of the 2nd Clinical Natural Language Processing Workshop, Association for Computational Linguistics, Minneapolis, Minnesota, USA, pp. 72–78 (2019)
2. del Carmen, M.G., Herman, J., Rao, S., Hidrue, M.K., Ting, D., Lehrhoff, S.R., Lenz, S., Heffernan, J., Ferris, T.G.: Trends and factors associated with physician burnout at a multi-specialty academic faculty practice organization. JAMA Netw. Open **2**(3), e190554–e190554 (2019)
3. Devlin, J., Chang, M.W., Lee, K., Toutanova, K.: BERT: pre-training of deep bidirectional transformers for language understanding. In: Proceedings of the 2019 Conference of the North American Chapter of the Association for Computational Linguistics: Human Language Technologies, vol. 1 (Long and Short Papers), pp. 4171–4186 (2019)
4. Finley, G., Edwards, E., Robinson, A., Brenndoerfer, M., Sadoughi, N., Fone, J., Axtmann, N., Miller, M., Suendermann-Oeft, D.: An automated medical scribe for documenting clinical encounters. In: Proceedings of the 2018 Conference of the North American Chapter of the Association for Computational Linguistics: Demonstrations, pp. 11–15 (2018)
5. GAO: Medical records: fees and challenges associated with patients' access. United States Government Accountability Office, Report to Congressional Committees GAO-18-386 (2018)

6. Jeblee, S., Khattak, F.K., Crampton, N., Mamdani, M., Rudzicz, F.: Extracting relevant information from physician-patient dialogues for automated clinical note taking. In: Proceedings of the Tenth International Workshop on Health Text Mining and Information Analysis (LOUHI 2019), pp. 65–74 (2019)
7. Kannan, A., Chen, K., Jaunzeikare, D., Rajkomar, A.: Semi-supervised learning for information extraction from dialogue. In: Interspeech, pp. 2077–2081 (2018)
8. Kessels, R.P.: Patients' memory for medical information. J. R. Soc. Med. **96**(5), 219–222 (2003)
9. Kim, S., Dalmia, S., Metze, F.: Cross-attention end-to-end ASR for two-party conversations. arXiv preprint arXiv:190710726 (2019)
10. Kodish-Wachs, J., Agassi, E., Kenny III, J.P.: A systematic comparison of contemporary automatic speech recognition engines for conversational clinical speech. In: AMIA Annual Symposium Proceedings, American Medical Informatics Association, vol. 2018, p. 683 (2018)
11. Kumar, S.: Burnout and doctors: prevalence, prevention and intervention. Healthcare **4**(3), 37 (2016). Multidisciplinary Digital Publishing Institute
12. Leventhal, R.: Physician burnout addressed: how one medical group is (virtually) progressing. Healthcare Innovation (2018)
13. Liu, S., Ma, W., Moore, R., Ganesan, V., Nelson, S.: RxNorm: prescription for electronic drug information exchange. IT Prof. **7**(5), 17–23 (2005)
14. Liu, Z., Lim, H., Suhaimi, N.F.A., Tong, S.C., Ong, S., Ng, A., Lee, S., Macdonald, M.R., Ramasamy, S., Krishnaswamy, P., et al.: Fast prototyping a dialogue comprehension system for nurse-patient conversations on symptom monitoring. In: Proceedings of the 2019 Conference of the North American Chapter of the Association for Computational Linguistics: Human Language Technologies, vol. 2 (Industry Papers), pp. 24–31 (2019)
15. McCann, B., Keskar, N.S., Xiong, C., Socher, R.: The natural language decathlon: multitask learning as question answering. arXiv preprint arXiv:180608730 (2018)
16. Peters, M.E., Neumann, M., Iyyer, M., Gardner, M., Clark, C., Lee, K., Zettlemoyer, L.: Deep contextualized word representations. In: Proceedings of NAACL-HLT, pp. 2227–2237 (2018)
17. See, A., Liu, P.J., Manning, C.D.: Get to the point: summarization with pointer-generator networks. In: Proceedings of the 55th Annual Meeting of the Association for Computational Linguistics (Long Papers), vol. 1, pp. 1073–1083 (2017)
18. Vaswani, A., Shazeer, N., Parmar, N., Uszkoreit, J., Jones, L., Gomez, A.N., Kaiser, Ł., Polosukhin, I.: Attention is all you need. In: Advances in Neural Information Processing Systems, pp. 5998–6008 (2017)
19. Xiong, C., Zhong, V., Socher, R.: Dynamic coattention networks for question answering. arXiv preprint arXiv:161101604 (2016)

Quantitative Evaluation of Emergency Medicine Resident's Non-technical Skills Based on Trajectory and Conversation Analysis

Kei Sato, Masaki Onishi, Ikushi Yoda, Kotaro Uchida, Satomi Kuroshima, and Michie Kawashima

Abstract In this paper, we propose a quantitative method for evaluating non-technical skills (e.g., leadership skills, communication skills, and decision-making skills) of Emergency Medicine Residents (EMRs) who are participating in a simulation-based training. This method creates a workflow event database based on the trajectories of and conversations among the medical personnel and scores an EMR's non-technical skills based on that database. We installed a data acquisition system in the emergency room of Tokyo Medical University Hospital to obtain trajectories and conversations. Our experimental results show that the method can create a workflow event database for cardiac arrest. In addition, we evaluated EMRs who are beginners, intermediates, and experts to show that our method can correctly represent the differences in their skill levels.

K. Sato (✉)
University of Tsukuba, Ibaraki, Japan
e-mail: s1820455@s.tsukuba.ac.jp

M. Onishi · I. Yoda
National Institute of Advanced Industrial Science and Technology, Ibaraki, Japan
e-mail: onishi-masaki@aist.go.jp

I. Yoda
e-mail: i-yoda@aist.go.jp

K. Uchida
Tokyo Medical University, Tokyo, Japan
e-mail: ktr21277@gmail.com

S. Kuroshima
Tamagawa University, Tokyo, Japan
e-mail: skuroshi@lab.tamagawa.ac.jp

M. Kawashima
Kyoto Sangyo University, Kyoto, Japan
e-mail: kawashima411@gmail.com

© The Editor(s) (if applicable) and The Author(s), under exclusive license to Springer Nature Switzerland AG 2021

A. Shaban-Nejad et al. (eds.), *Explainable AI in Healthcare and Medicine*, Studies in Computational Intelligence 914,
https://doi.org/10.1007/978-3-030-53352-6_19

Keywords Non-technical skills evaluation · Emergency medicine · Resident education · Trajectory analysis · Conversation analysis

1 Introduction

Non-technical skills, such as leadership skills, communication skills, decision-making skills, and time-management skills, are important for emergency physicians. This is because most emergency treatments are performed by a team that is led by the physician. Lately, simulation-based training for Emergency Medicine Residents (EMRs) has attracted significant attention. Several studies [1] have reported that treating a dummy patient for an actual treatment is one of the most effective ways of educating EMRs. However, few studies have reported on a quantitative evaluation of the non-technical skills of the EMRs participating in the simulation-based training.

Therefore, we propose a quantitative method that evaluates the non-technical skills of an EMR participating as medical team leaders in simulation-based training. This method scores the treatment workflow of the evaluation target as the quantitative evaluation of the non-technical skills of the evaluation target. The treatment workflow is a series of events performed during emergency treatment (e.g., airway management, epinephrine administration, and defibrillation). We assume that the treatment workflow in emergency medicine is determined by the leader's non-technical skills, non-technical skills of team members other than the leader, all team members' manipulation skills, and the patient's condition. In the simulation-based training, we can conduct trainings under the same conditions, except the leader's non-technical skills, as described in Sect. 3.1. Therefore, the score of the treatment workflow performed by the evaluation target is equal to the score of the non-technical skills of the leader, i.e., evaluation target. This method scores the treatment workflow of the evaluation target using the workflow event database and the average number of training days. The workflow event database is an organized collection of events in the treatment workflow performed by EMRs who have various skill levels participating in the simulation-based training as a leader. Each event in the database has the average number of training days. The average number of trained days of each event is the average number of days spent in training in the emergency department of each EMR who participated in the event. Although it is difficult to judge the non-technical skills of an EMR by the EMR's number of training days alone, the average number of training days has the potential to be a criterion of skillfulness because the number of training days usually represents the skillfulness of EMRs. For example, EMRs who spend several days in the emergency department are typically more skillful. The method refers to the workflow event database and finds out the most similar event in the database to each event in the evaluation target's workflow. Subsequently, each event in the evaluation target's workflow is scored based on the average number of training days of the most similar event. The overall score of the evaluation target's workflow, that is, the score of the evaluation target's non-technical skills is the average score of all events in the workflow of the evaluation target.

In this study, we extract the events in the treatment workflow using trajectories of and conversations among medical team members for the database creation. Nara et al. [4] presented a workflow event extraction method for neurosurgical operation by trajectory analysis. Although trajectory analysis is also effective for emergency medicine because multiple medical personnel move frequently in an emergency room during the emergency treatment, trajectory analysis alone is not enough to extract the events of treatment workflow. The reason is that the medical personnel proceed with the treatment simultaneously and the treatment strategy is different depending on the situation. Therefore, we also use conversations among medical team members for the workflow event extraction. Furthermore, our proposed method represents the events in the workflow with the data I related to instruction from the EMR with the leader's role because as these instructions are frequently used during a treatment, we can assume that the instructions from the leader strongly affect the treatment workflow. I is a quantitative data comprising the instructee's trajectory, instruction utterance's text, and instruction utterance timing. We define each event in the evaluation target's workflow as an I uttered by the evaluation target. In addition, each event in the workflow event database is defined as a cluster of I because there is a possibility that similar Is could be uttered by different EMRs. The workflow event database has clusters as events in the treatment workflow.

First, our system acquires trajectories of and conversations among medical team members to obtain Is. Then, the clustering method creates the workflow event database. Finally, the non-technical skills of the leader EMR are quantitatively evaluated based on the database and the average number of training days.

2 Non-technical Skills Evaluation Method for Emergency Medicine Residents

2.1 Data Acquisition Process

In this section, we describe the data acquisition process to obtain Is from trajectories and conversations for the workflow event database creation.

Sensors of the data acquisition system installed in the emergency room and trajectories captured by the system are shown in Fig. 1. The system tracks medical team members using four stereo cameras and the two-stage clustering method [5]. Furthermore, a microphone of the system records their conversations. The recorded conversations are manually transcribed. The transcripts are segmented into utterances and labeled according to their context (e.g., instruction, informing, and response) by sociologists. We use only instruction utterances from the leader EMR. Table 1 shows examples of the instructions.

The quantitative data I comprises of a trajectory $\boldsymbol{t}$ after I, the vector $\boldsymbol{u}$ of the I's text, and the I's timing τ. The trajectory $\boldsymbol{t}$ is the trajectory (i.e., a series of x–y coordinates) of an instructee until he/she stops subsequent to moving in response to

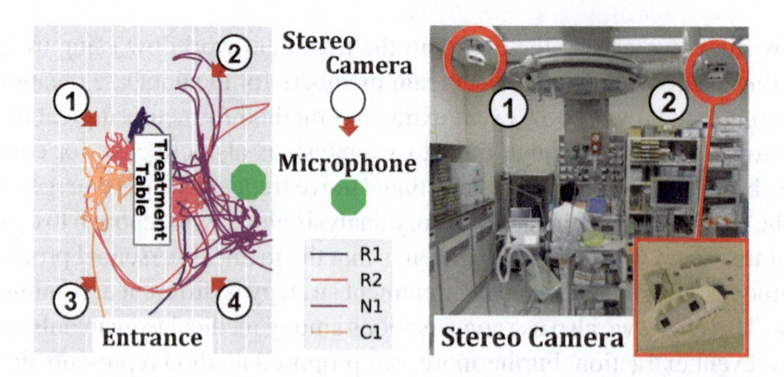

Fig. 1 Four stereo cameras and a microphone installed in emergency room are used as a data acquisition system. The lines in the left panel are real trajectories of the medical team members participating in a simulation-based training, where R1 is an EMR with a leader's role, R2 is an assistant EMR, N1 is a nurse, and C1 is an ambulance crew member

Table 1 Examples of instructions from an EMR with a leader's role

Timing τ [s]	Utterance of instructions
16.7	*"Pass me airway management tools."*
68.3	*"Let's give a milligram of epinephrine."*
182.5	*"Stand clear."*
310.2	*"Deliver the shock."*

an I. $\boldsymbol{u}$ is the vector of I's text. The I's text is split into words, which are tagged with parts-of-speech by the text segmentation library MeCab [2]. We only use nouns for vectorization because the text, which is a conversation transcript, contains too much noise. Then, the I's text is vectorized by the bag-of-words model [7] using the extracted nouns. The bag-of-words model represents a sentence by the frequency of each word in it. The I's timing τ is recorded as the time elapsed from the start of treatment, such as shown in Table 1. The start time is the time when the patient was transferred to the treatment table from the stretcher.

Instructions from different EMRs participating in the training are collected using this process. An instruction I_n is given by $I_n = \{\boldsymbol{t}_n, \boldsymbol{u}_n, \tau_n\}$, where n denotes the index of the instruction, $\boldsymbol{t}_n$ is $\boldsymbol{t}$ of I_n, $\boldsymbol{u}_n$ is $\boldsymbol{u}$ of I_n, and τ_n is τ of I_n. Then, all instructions acquired by the system is given by $\boldsymbol{I} = \{I_n | 1 \leq n \leq N, n \in \mathbb{N}\}$, where N is the total number of instructions. Clustering method with weighted distance partitions $\boldsymbol{I}$ into each cluster, i.e., each event in the database.

2.2 Clustering Method for Workflow Event Database Creation

We use the k-means algorithm [3] to partitioned $\boldsymbol{I}$ into each event in the workflow. The k-means algorithm clusters data according to the metric value of $\boldsymbol{I}$, which is defined as the distance between an instruction I_α and an instruction I_β, given by

$$d_I(I_\alpha, I_\beta) = \boldsymbol{D} \cdot \boldsymbol{W}^T,$$

where $\boldsymbol{D} = [d_t, d_u, d_\tau]$ is a distance matrix of $\boldsymbol{t}$, $\boldsymbol{u}$, and τ, $\boldsymbol{W} = [w_t, w_u, w_\tau]$ is a coefficient matrix of d_t,d_u, and d_τ, and d_I is the metric of $\boldsymbol{I}$. d_t is a distance between $\boldsymbol{t}_\alpha$ and $\boldsymbol{t}_\beta$. d_t is calculated by Dynamic Time Warping [6]. d_u is the distance between $\boldsymbol{u}_\alpha$ and $\boldsymbol{u}_\beta$, and d_τ is the distance between τ_α and τ_β. d_u and d_τ are calculated by the Euclidean metric. Each coefficient indicates the importance of each distance in the clustering.

By k-means clustering with the metric d_I, every instruction I_n is assigned to any one of the clusters (i.e., an event in the database) $\boldsymbol{I}^c = \{I_n^c | n \in L^c\}$, where c denotes the index of the cluster, and L^c is the set of indexes of instructions assigned to cluster c. Finally, we could obtain the workflow event database $\boldsymbol{F} = \{\boldsymbol{I}^c | 1 \leq c \leq K, c \in \mathbb{N}\}$, where K is the total number of clusters. The evaluation target is scored based on the F and the average number of training days.

2.3 Scoring Method Based on Workflow Event Database

Our proposed method scores the target's non-technical skills as quantitative evaluation. An event in the database $\boldsymbol{I}^c \in \boldsymbol{F}$ and an event in the workflow of an evaluation target (i.e., an I of the evaluation target) are comparable quantitatively, because metric d_I is defined mathematically. However, we cannot measure the evaluation target's non-technical skills by the comparison, because the difference between the two events is not a criterion of skillfulness. Therefore, this evaluation method uses the number of training days in the emergency department T. Although it is difficult to judge the non-technical skills of evaluation target only by T, we use the average number of training days $\overline{T^c}$ as criterion. $\overline{T^c}$ is the average T of the EMRs who instructed $I_n^c \in \boldsymbol{I}^c$.

Instructions from an evaluation target (i.e., events in the treatment workflow of an evaluation target) are given as $\hat{\boldsymbol{I}} = \{\hat{I}_m | 1 \leq m \leq M, m \in \mathbb{N}\}$, where $\hat{I}_m$ is an instruction of the evaluation target, m denotes the index of each instruction, and M is the total number of instructions from the evaluation target during simulation-based training. Then, the nearest cluster number γ_m to $\hat{I}_m$ is given by

$$\gamma_m = \text{arg min}_c d_I(\overline{\boldsymbol{I}^c}, \hat{I}_m),$$

where $\overline{T^{c}}$ is a centroid of cluster I^{c}. A score of $\hat{I_m}$ is given by

$$\hat{S}_m = \frac{\overline{T^{\gamma_m}} - \overline{T^{c_{min}}}}{\overline{T^{c_{max}}} - \overline{T^{c_{min}}}},$$

where $\overline{T^{\gamma_m}}$ is the average number of training days of the instructors in cluster I^{γ_m}; $\overline{T^{c_{min}}}$ is the average number of training days of the instructors in cluster $I^{c_{min}}$, which has the minimum average number of training days in all clusters; $\overline{T^{c_{max}}}$ is the average number of training days of the instructors in cluster $I^{c_{max}}$, which has the maximum average number of training days in all clusters; and $\hat{S}_m$ is the score of $\hat{I_m}$. Finally, a score of the evaluation target $\hat{S}$ is given by

$$\hat{S} = \frac{\sum_{m=1}^{M} \hat{S}_m}{M}.$$

When $\hat{S}$ is high, $\hat{I}$ is similar to I^{c}, which is a cluster of instructions of EMRs who have been in the emergency department for more days. Likewise, when $\hat{S}$ is low, $\hat{I}$ is similar to I^{c}, which is a cluster of instructions of EMRs who have been in the emergency department for fewer days. Thus, the non-technical skills of the evaluation target can be evaluated by $\hat{S}$ because $\hat{S}$ represents how skillful EMRs similar to the evaluation target are by [0.0, 1.0].

3 Experiment and Discussion

3.1 Experimental Design

We conduct an experiment using real data to confirm the validity of our proposed method by correctly clarifying the difference in non-technical skills of EMRs who have clearly different lengths of experience.

We acquired data from a simulation-based training conducted in the emergency room of Tokyo Medical University Hospital. All subjects treated a dummy patient with cardiac arrest in simulation-based training. All training was performed in a team. The team is usually composed of an EMR or physician with the leader's role, an assistant EMR, a nurse, and an ambulance crew member. The acquired data were comprised of 32 simulations, 31 lead by an EMR and one by an emergency physician. All data are partitioned into a data for database creation and evaluation target shown in Table 2. All evaluation targets are scored using the proposed method based on the workflow event database created by data for database. The evaluation target groups were divided into three according to the length of their experience (i.e., their number of training days). Therefore, we assume that there is a difference in non-technical skill levels between the three groups. This experiment aims to represent the differences in non-technical skills among the groups.

Table 2 Explanation of data. The data for database were used for the database creation and each evaluation target that belonged to every group was scored using the scoring method. Group I contains data of first-month EMRs as beginners, Group II contains data of second-month EMRs as intermediates, and Group III contains data of EMRs with experience of more than three months in the emergency department or emergency physician as experts

Data name	No. subjects	No. instructions	No. training days	Experience
Data for Database	22	302	44–61	Intermediate
Evaluation Target Group I	3	39	23, 27, 28	Beginner
Evaluation Target Group II	4	86	44, 50, 54, 59	Intermediate
Evaluation Target Group III	3	40	-	Expert

The patient's condition, manipulation skills, and cooperative skills are normalized as follows to conduct all trainings under the same conditions, except the leader's non-technical skills. All trainings are conducted with the same scenario (e.g., patient's initial condition, changes in the patient's condition, and timing of changes in condition). Medical personnel can perform manipulations such as airway management, injection, and chest compression. The trainees are informed of only the patient's initial condition and that they will not be judged based on whether the manipulations were done correctly or not. All medical personnel except the leader are told to act only according to the instructions given by their leader.

Firstly, we acquired Is when the 32 subjects participated in the simulation-based training with the above setting. Then, the clustering method created the workflow event database using instructions of data for database $\boldsymbol{I}$. The number of clusters $K = 22$ and the coefficient matrix $\boldsymbol{W} = [0.710, 0.997, 0.555]$ are experimentally determined to be a large $\overline{\boldsymbol{T}^{c_{max}}} - \overline{\boldsymbol{T}^{c_{min}}}$ because the difference between $\overline{\boldsymbol{T}^{c_{max}}}$ and $\overline{\boldsymbol{T}^{c_{min}}}$ is the score range. Finally, the scoring method scored the evaluation targets using the targets' instructions $\hat{\boldsymbol{I}}$.

3.2 *Results of Workflow Event Database Creation and Scoring Non-technical Skills*

Proposed clustering method assigned $I_n \in \boldsymbol{I}$ to each cluster $\boldsymbol{I}^c$ for the database creation. Figure 2 shows all clusters' average number of training days and average timing of instructions. In addition, the figure illustrates the details of the instructions of the top four clusters assigned the most I_n. Each cluster that is a set of instructions is regarded as each event in the workflow event database. The color of points on the central panel represents the skillfulness of each event (i.e., skillfulness of the EMRs who issues instructions included in the event) because the clusters are colored based

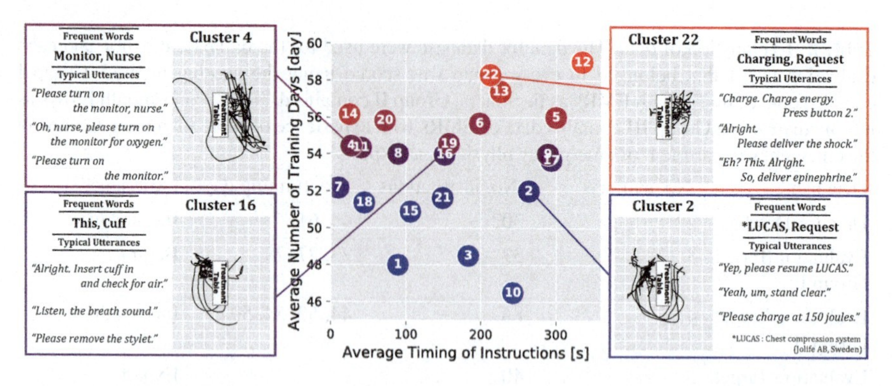

Fig. 2 The points on the central panel correspond to the all cluster's average timing of instructions and the average number of training days. The number of points is the cluster number. Each cluster's point is colored based on the average number of training days. The boxed trajectories are the trajectories of the instructees who move according to instructions included in each cluster. The boxed frequent words and typical utterances are the top two most frequently used words by instructors and the instruction utterances of each cluster

Table 3 Scores of the evaluation targets of each group

Group	I (Beginners)			II (Intermediates)				III (Experts)		
No. Training Days	23	27	28	44	50	54	59	-	-	-
Score $\hat{S}$	**0.550**	**0.539**	**0.456**	**0.601**	**0.594**	**0.503**	**0.616**	**0.605**	**0.560**	**0.590**
Average	0.518			0.579				0.582		
Standard Deviation	0.179			0.115				0.111		

on the cluster's average number of training days. Clusters that have a color close to red have a high average number of training days, whereas those having a color close to blue have a low average number of training days. Workflows of the evaluation targets are scored using the workflow event database. Therefore, the evaluation target who issues instructions that are similar to the red event's instructions scores highly and the evaluation target who issues instructions that are similar to the blue event's instructions scores lowly.

The scoring results of evaluation targets according to the database are shown in Table 3. In addition, Table 4 shows results that differences between average $\hat{S}$ were tested with the Student's t-test. Accordingly, we found a difference in mean values between I and II at 5% level of significance. In the case of I vs III, the null hypothesis is more than likely rejected as well, because the p-value is 0.059. However, there is no difference between the mean values of II vs III.

Table 4 Results of Student's t-test. There is a detectable difference between the average $\hat{S}$ of I and II at the 5% significance level. When we test I and III, a mean difference is detectable at the 10% significance level. However, there is no detectable difference between the means of II vs III

	Degrees of freedom	t-value	p-value
I vs II	123	−2.290	**0.024****
I vs III	77	−1.916	**0.059***
II vs III	124	0.137	0.891

* $p < 0.1$; ** $p < 0.05$

3.3 *Discussion*

We have confirmed that a workflow event database can be created using clustering trajectories, utterances, and timings of instructions. In addition, the actions of EMRs with different skill levels are specifically observed by events in the database and the average number of training days. For example, cluster 22 is a comparatively skillful event (according to average number of training days), and it is based on an instruction about defibrillation (according to utterances) to a person who stands beside the treatment table (according to trajectories) in the middle phase of the treatment (according to utterance timings). As each cluster is explainable and scorable as described above, the database is expected to also be applied in a debriefing of the simulation-based training.

The statistical testing results demonstrate the potential of our proposed method for evaluating EMRs' non-technical skills. However, this method could not represent the difference between intermediates and experts in this experiment. There are two possible reasons for this: 1) a shortage of data for database, and 2) difficulty in training. The reliability of this method depends on the number and variety of data for database because we use the training days of the EMRs included in the data for database as a criterion. Thus, it is possible that our database could not represent the features of the experts' events in the workflow. Another possibility is that the workflow did not change depending on the leader's non-technical skills because the training difficulty was too low.

4 Conclusion

We proposed a method for evaluating EMRs participating as medical team leaders in simulation-based training. As an experiment, we evaluated EMRs and a physician using real data. The statistical testing results of beginners versus intermediates and beginners versus experts indicated that the method could adequately represent the differences between the skill levels. However, the problem of this method is that the number and variety of the data for database are related to the reliability of the evaluation criterion. Acquiring more data for database or efficiently creating a database will be part of future work. Furthermore, the reliability of the scoring requires further examination.

References

1. Bond, W.F., Spillane, L.: The use of simulation for emergency medicine resident assessment. Acad. Emergency Med. **9**(11), 1295–1299 (2002)
2. Kudo, T., Yamamoto, K., Matsumoto, Y.: Applying conditional random fields to japanese morphologiaical analysis. IPSJ SIG Notes **161**, 89–96 (2004), https://ci.nii.ac.jp/naid/110002911717/en/
3. MacQueen, J.: Some methods for classification and analysis of multivariate observations. In: Proceedings of the Fifth Berkeley Symposium on Mathematical Statistics and Probability, vol. 1: Statistics, pp. 281–297. University of California Press, Berkeley (1967). https://projecteuclid.org/euclid.bsmsp/1200512992
4. Nara, A., Izumi, K., Iseki, H., Suzuki, T., Nambu, K., Sakurai, Y.: Surgical workflow monitoring based on trajectory data mining. In: JSAI International Symposium on Artificial Intelligence, pp. 283–291. Springer (2010)
5. Onishi, M.: Analysis and visualization of large-scale pedestrian flow in normal and disaster situations. ITE Trans. Media Technol. Appl. **3**(3), 170–183 (2015)
6. Sakoe, H., Chiba, S.: Dynamic programming algorithm optimization for spoken word recognition. IEEE Trans. Acoustics, Speech, Signal Process. **26**(1), 43–49 (1978)
7. Zhang, Y., Jin, R., Zhou, Z.H.: Understanding bag-of-words model: a statistical framework. Int. J. Mach. Learn. Cybern. **1**, 43–52 (2010). https://doi.org/10.1007/s13042-010-0001-0

Implementation of a Personal Health Library (PHL) to Support Chronic Disease Self-Management

Nariman Ammar, James E. Bailey, Robert L. Davis, and Arash Shaban-Nejad

Abstract This paper shows a work-in-progress on the implementation of an integrated Personal Health Library (PHL) for Chronic disease self-management (CDSM) using the Social Linked Data framework. The proposed method is fully decentralized and follows the linked open data platform specifications and grants patients true ownership over their data while giving them fine-grained access control mechanisms to exchange and share their data. It also supports interoperability and portability by following standard protocol, format, and vocabulary provided by ontologies.

Keywords Personal health library · Interoperability · Self-care · Decision support systems

1 Introduction

Historically, medicine has largely been healthcare provider centered rather than patient-centered [1, 2]. However, patients often express the need to have an active role in managing their data [3, 4], including data stored in their personalized health records. This seems far from trivial in the digital health era, considering the amount and variety

N. Ammar · R. L. Davis · A. Shaban-Nejad (✉)
Department of Pediatrics, College of Medicine, The University of Tennessee Health Science Center - Oak-Ridge National Lab (UTHSC-ORNL), Center for Biomedical Informatics, Memphis, TN, USA
e-mail: ashabann@uthsc.edu

N. Ammar
e-mail: nammar@uthsc.edu

R. L. Davis
e-mail: rdavis88@uthsc.edu

J. E. Bailey
Center for Health System Improvement, College of Medicine, University of Tennessee Health Science Center, Memphis, TN, USA
e-mail: jeb@uthsc.edu

© The Editor(s) (if applicable) and The Author(s), under exclusive license to Springer Nature Switzerland AG 2021

A. Shaban-Nejad et al. (eds.), *Explainable AI in Healthcare and Medicine*, Studies in Computational Intelligence 914,
https://doi.org/10.1007/978-3-030-53352-6_20

of data coming from diverse sources, not to mention legal, privacy, and ethical issues. Patients' data might include their profile, contacts, financial data, medications, test results, appointment, etc. Add to this other data coming from other external sources such as online media and social networks, family histories, activity trackers, videos, and public datasets (e.g. census data). Current EMR systems are mostly centralized, with patients' data distributed across multiple networks in data silos owned and maintained by different providers. While patients wish to share their data, they are concerned with their privacy and ownership over their data [5, 6]. The effective use of digital technologies can improve patients' engagement to improve their health awareness and adoption of healthy behaviors [7]. A Personal Health Library (PHL) [8] can provide a single searchable resource of secure access to patients' personal digital health life. The Social Linked Data (Solid) framework [9], from the other side, provides a solution for decentralized data exchange by separating the patient's data space from applications space. This paper reports on the early implementation of the PHL for diabetes patients using the Solid infrastructure.

2 Method

The proposed solution enables patients to distribute their data among multiple distributed Personal Online Data stores (PODs) and to selectively authenticate applications to access specific resources within those PODs. Applications are just front-end services that render different aspects of the same snapshot of the users' data. The main features supported by the Solid framework, including defining agents and WebIDs, generating Web of Trust and Web of Resources, Hierarchical resource representation, Linkability, Portability, and Interoperability, Live Notifications, Authentication, and Privacy Management have been thoroughly described in our previous work [7].

3 The PHL in Action

Let us consider a scenario that involves three actors, two patients (Alice and Bob) and a physician (Mary) interacting through the Solid-based PHL for chronic disease self-management.

WebIDs and Profile Documents. Alice is a cancer patient and a university student. She links her WebID to her public profile document, which shows that Alice is of type *Person*, and specifies her role as a *Student*. Also, it displays her contact information and a list of trusted apps and people (e.g. her links to Bob and Mary via the FOAF:*knows* vocabulary). Mary is Alice's primary care physician, and Bob is Alice's friend and a diabetes patient. The three actors generate unique WebIDs and set up their PODs, and keep the list of vCard URIs of their contacts in their extended profiles using FOAF:*knows*.

Hierarchical Resource Representation and ACLs. Alice wants to notify her primary care physician, Mary, of blood test results that she received from a clinic. She also wants to integrate information that she got from her oncologist about treatment options with the information from blogs shared by other patients in her network. Thus, she adds Bob and Mary to her contacts list and gives them fine-grained permissions to resources within her extended profile (Fig. 1). Mary needs to read and edit permissions to annotate the test results document and share annotations with other physicians, while Bob needs to add permission to share blogposts with Alice. Alice can also share a notepad with Mary, who grants her access permission to the cancer patients' messages resource within her inbox (Fig. 2). She also wants to integrate

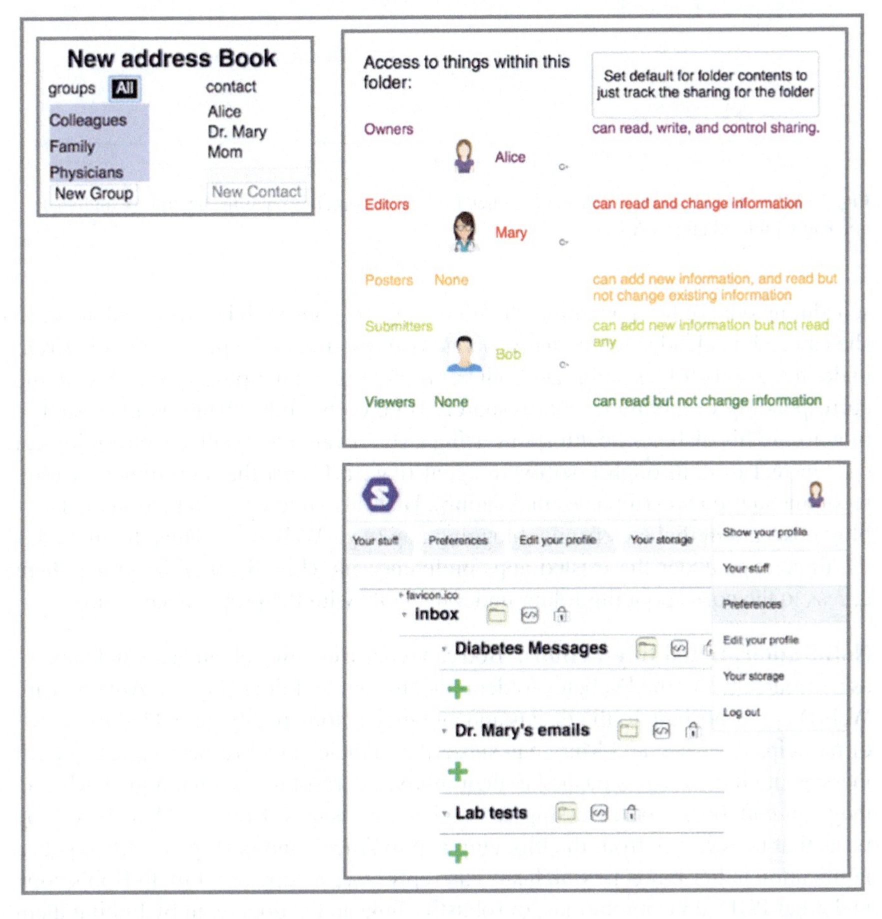

Fig. 1 Alice can add her contacts to an address book and grant them access control permissions to resources within her profile document

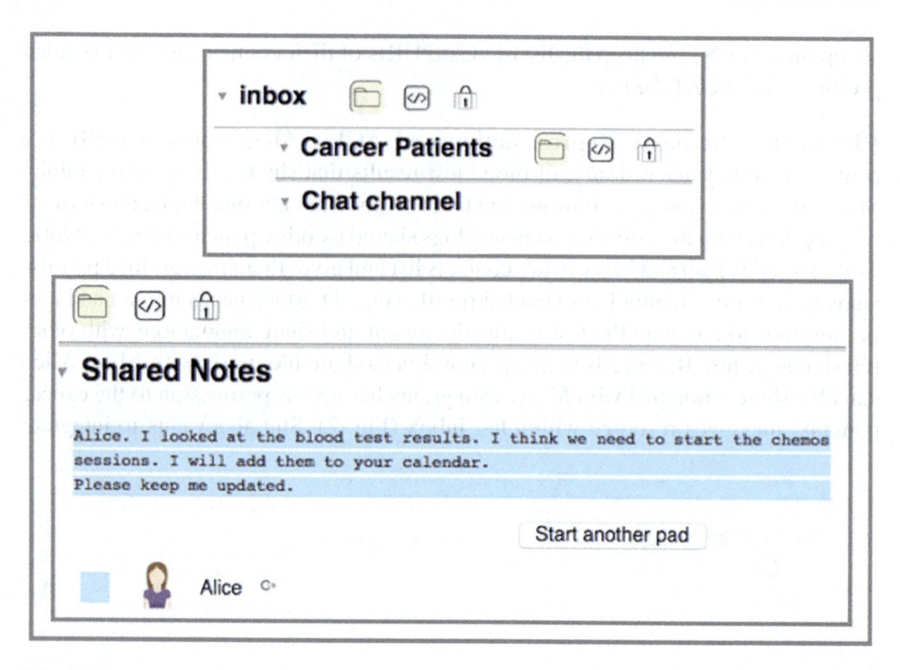

Fig. 2 Mary, the physician, can manage her folders in containers within her inbox and share a notepad with her Patient, Alice

scheduling data of her upcoming chemotherapy sessions with her work calendar, so she can add a calendar folder and generate two resources of type *SchedulableEvent* under her events folder and grant both her professor and her physician access to the corresponding events under that resource. Bob, on the other hand, is interested in monitoring his diabetes health status using a smartwatch that collects physiological data in real-time through a software agent (e.g., a Fitbit) that coordinates with a recommendation app (diabetes messaging). He is also interested in getting the latest blog posts about diabetes from a blogging app (e.g., Webalyst). Thus, he first adds the three apps under the trusted apps preference file (Fig. 3), then he grants them access to the corresponding folder under his POD with the proper access mode.

Notifications-based Interactions. Bob can set up a chatting channel about Diabetes as a resource under the Diabetes folder under his public folder (Fig. 3). Anyone with WebID can contribute to the discussion and they automatically get added to the list of participants. Alice and Mary can subscribe to the channel respectively, and each message notification gets pushed to their inboxes as resources with unique URIs and they can collaborate on assessing each other's messages. Bob can also share blog posts that he receives from the blogging app to Alice's inbox (Fig. 1). Alice gets a notification in her inbox to which she can reply, save a copy or a link to Bob's copy under her POD, add annotations, or robustify links in the document by linking them to other concepts or sources of knowledge.

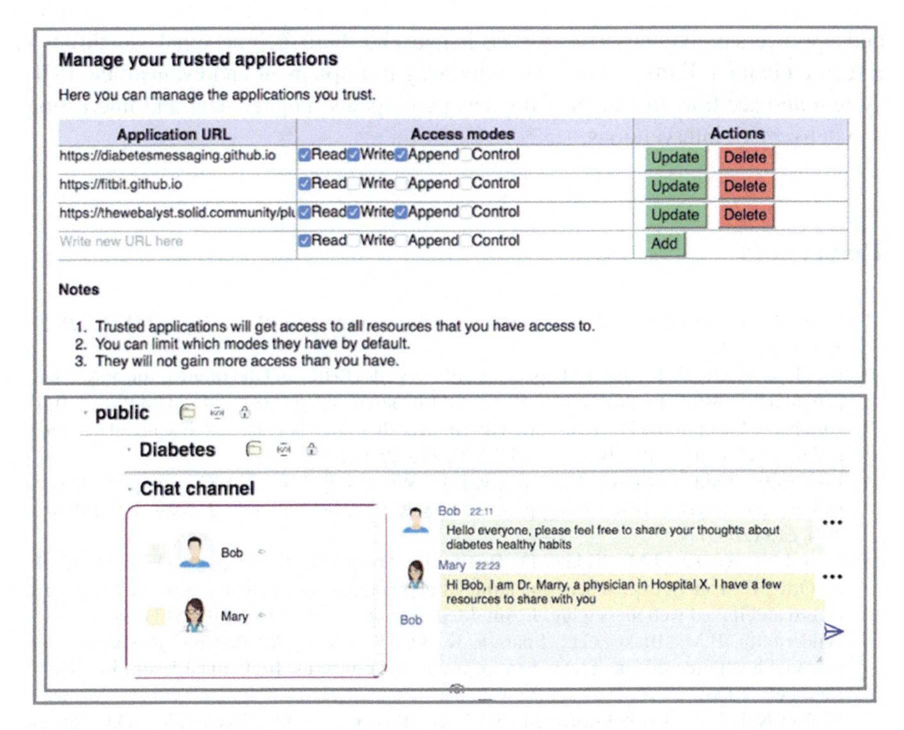

Fig. 3 Bob can manage a list of trusted applications to feed data into certain resources under his profile document. He can also manage a chatting channel about diabetes

Interoperability. Everyone in the Solid ecosystem can store any piece of the data they produce wherever they want. If Alice comments on one of the messages shared through Bob's channel, her comment gets stored in her POD while the message remains in Bob's POD. This has two consequences: one needs to connect Alice's comment to Bob's message, and give Alice's comment a universal meaning that different scanning apps use to read that message. Both features are supported by Solid.

Linkability. Assuming a message stored in Bob's profile is identified by https://bob.solid.community/messages/diabetes/1234, then Alice's comment at https://alice.solid/comments/36756 links back to Bob's message by *hasTarget* link type defined in the Web Annotation Ontology (www.w3.org/ns/oa).

4 Conclusion

The proposed approach facilitates data integration and interoperability between different data sets, consisting of patients both biomarkers and sociomarkers [10], and empowers individuals to access, exchange, and share their personal data through a secure channel. Further works are underway to implement and evaluate the PHL system and see how this method impacts participants' engagement and interaction and delivery of interventions.

References

1. Laine, C., Davidoff, F.: Patient-centered medicine: a professional evolution. JAMA **275**(2), 152–156 (1996)
2. Baird, A., North, F., Raghu, T.: Personal health records (PHR) and the future of the physician-patient relationship. In: Proceedings of the 2011 iConference, pp. 281–288. ACM Press (2011)
3. Auerbach, S.M.: Should patients have control over their own health care? Empirical evidence and research issues. Ann. Behav. Med. **22**(3), 246–259 (2000)
4. Ross, S.E., Todd, J., Moore, L.A., Beaty, B.L., Wittevrongel, L., Lin, C.T.: Expectations of patients and physicians regarding patient-accessible medical records. J. Med. Internet Res. **7**(2), e13 (2005)
5. Hassol, A., Walker, J.M., Kidder, D., Rokita, K., Young, D., Pier-don, S., Deitz, D., Kuck, S., Ortiz, E.: Patient experiences and attitudes about access to a patient electronic health care record and linked web messaging. J. Am. Med. Inform. Assoc. **11**(6), 505–513 (2004)
6. Wildermuth, B.M., Blake, C.L., Spurgin, K., Oh, S., Zhang, Y.: Patients' perspectives on personal health records: an assessment of needs and concerns. In: Critical Issues in eHealth Research (2006)
7. Shaban-Nejad, A., Michalowski, M., Peek, N., Brownstein, J.S., Buckeridge, D.L.: Seven pillars of precision digital health and medicine. Artif. Intell. Med. **103**, 101793 (2020). https://doi.org/10.1016/j.artmed.2020.101793
8. Ammar, N., Bailey, J., Davis, R.L., Shaban-Nejad, A.: The personal health library: a single point of secure access to patient digital health information. In: Studies in Health Technology and Informatics (2020)
9. Sambra, A.V., Mansour, E., Hawke, S., Zareba, M.,Greco, N., Ghanem, A., Zagidulin, D., Aboulnaga, A., Berners-Lee, T.: Solid: a platform for decentralized social applications based on linked data. Technical Report (2017). www.emansour.com/publications/paper/solid_protocols.pdf
10. Shin, E.K., Mahajan, R., Akbilgic, O., Shaban-Nejad, A.: Sociomarkers and biomarkers: predictive modeling in identifying pediatric asthma patients at risk of hospital revisits. NPJ Digit Med. **1**, 50 (2018). https://doi.org/10.1038/s41746-018-0056-y. eCollection 2018

KELSA: A Knowledge-Enriched Local Sequence Alignment Algorithm for Comparing Patient Medical Records

Ming Huang, Nilay D. Shah, and Lixia Yao

Abstract Sequence alignment methods have the promise to reserve important temporal information in electronic health records (EHRs) for comparing patient medical records. Compared to global sequence alignment, local sequence alignment is more useful when comparing patient medical records. One commonly used local sequence alignment algorithm is Smith-Waterman algorithm (SWA), which is widely used for aligning biological sequence. However directly applying this algorithm to align patient medical records will obtain suboptimal performance since it fails to consider complex situations in EHRs such as the temporality of medical events. In this work, we propose a new algorithm called Knowledge-Enriched Local Sequence Alignment algorithm (KELSA), which incorporates meaningful medical knowledge during sequence alignments. We evaluate our algorithm by comparing it to SWA on synthetic EHR data where the reference alignments are known. Our results show that KELSA aligns better than SWA by inserting new daily events and identifying more similarities between patient medical records. Compared to SWA, KELSA is more suitable for locally comparing patient medical records.

1 Introduction

Transforming patient medical records, either structured or unstructured to a computable patient representation is required for explainable artificial intelligence applications in healthcare, such as disease prognosis and medication outcomes prediction, and patient similarity comparison, which identifies patients similar to a target patient (e.g., a rare, hard-to-diagnose patient or a patient with multiple chronic conditions [3, 16]. Building computable patient representation is still at primitive stage and becomes the bottleneck, relative to fast development of new machine learning algorithms and advanced predictive models. Most applications adopt one-hot

M. Huang · N. D. Shah · L. Yao (✉)
Department of Health Sciences Research, Mayo Clinic, Rochester, MN, USA
e-mail: lixia.cn.yao@gmail.com

© The Editor(s) (if applicable) and The Author(s), under exclusive license to Springer Nature Switzerland AG 2021

A. Shaban-Nejad et al. (eds.), *Explainable AI in Healthcare and Medicine*, Studies in Computational Intelligence 914,
https://doi.org/10.1007/978-3-030-53352-6_21

vector representation and totally ignore the temporal information in patient medical records, which is very critical for diagnosis and treatment. For example, cough is a common symptom associated with allergy, cold, flu, pneumonia or bronchitis. However, persistent cough over months can also be the first sign of lung cancer. Physicians use both the present symptoms and temporal information explicitly or implicitly, for legitimate diagnosis, treatment, and prognosis decisions.

Mathematically medical records of a patient can be treated as a temporal sequence of medical events (e.g., diagnosis, procedure, medication, and lab test). Recently researchers start to consider the use of sequence alignment methods when analyzing and comparing patient medical records. Computable patient representation with proper sequence alignment also suggests a novel solution to reserve temporal information in EHRs [5, 6]. Given the fact that medical records of two or more patients are often of different time spans and different frequencies of visits, global sequence alignment methods (e.g., dynamic time warping [12] and Needleman–Wunsch algorithm [14]) that align multiple sequences from beginning to end, is not truly useful in many practical application scenarios. For example, some patient medical records may be incomplete and/or contain noise. Junior doctors and residents often need to figure out the diagnosis or treatment for patients of rare or complex situations. Local sequence alignment methods (e.g., Smith-Waterman Algorithm [17] and BLAST [1]) that align subsections of various lengths for optimal local similarity is more meaningful in these cases (See Fig. 1).

Smith-Waterman algorithm (SWA) is the most commonly used local sequence alignment algorithm based on dynamic programming [17]. SWA has been widely

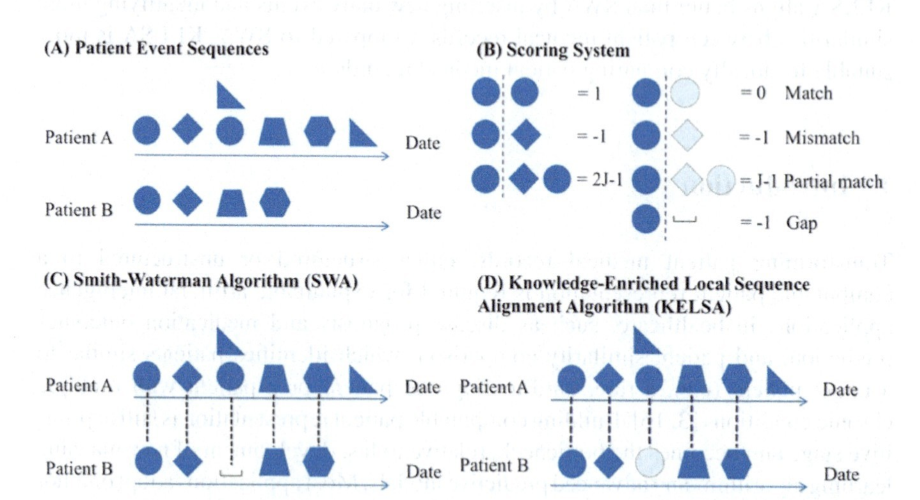

Fig. 1 An illustration demonstrating the significance of sequence alignment: two simplified temporal event sequences (**A**), scoring function to calculate the pairwise patient similarity (**B**), local sequence alignment algorithms, SWA (**C**) and KELSA (**D**). The shapes with light blue and dash border are extra medical events inserted by KELSA during sequence alignment. "_" is a gap spot inserted by SWA or KELSA during sequence alignment. The different shapes (e.g., circle, diamond, and triangle) represent different medical events. Circle shape denotes a chronic disease

used in bioinformatics, in particularly at comparing protein, DNA or RNA sequences to identify regions of similarity that may be a consequence of functional, structural or evolutionary relationships between the sequences [4, 20, 22]. This algorithm is guaranteed to find the optimal local alignment with respect to the predefined scoring system, thus it outperformed FASTA [11] and BLAST [1] on accuracy [2]. However directly applying this algorithm to align patient medical records will obtain suboptimal performance, because time spans of different medical events (e.g., acute and chronic diseases) vary dramatically and chronic conditions cannot be treated as events at a single time point.

In this work, we propose a new algorithm called Knowledge-Enriched Local Sequence Alignment (KELSA). It incorporates meaningful medical knowledge (e.g., acute and chronic diseases) and inserts daily events with inferred hidden medical events when aligning two patient medical records (See Fig. 1). We compare it with the SWA for evaluation using some synthetic patient medical records generated from real world patient medical records. Thus, we know the reference alignment (gold standard). For objective comparison, the same similarity scoring system is used for SWA and KELSA, which maximizes the similarity between sequences.

2 Methods

2.1 *Knowledge-Enriched Local Sequence Alignment Algorithm (KELSA)*

Similar to SWA, KELSA consists of three main steps: (a) Initialization, (b) Scoring, and (c) Traceback as shown in Fig. 2. Specifically, given two temporal sequences of medical events, **X** ([$X_1, X_2, \ldots, X_i, \ldots, X_n$]) and **Y** ([$Y_1, Y_2, \ldots, Y_j, \ldots, Y_m$]) of two patients P_X and P_Y, KELSA initializes and calculates an accumulated score matrix $\boldsymbol{A}_{(n+1)\times(m+1)}$ and then tracks back to identify an optimal alignment path.

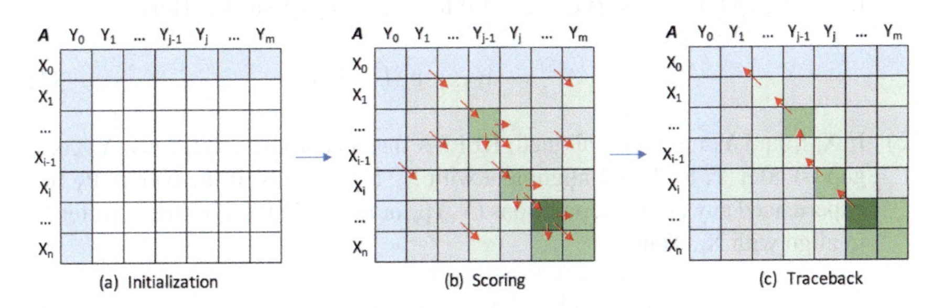

Fig. 2 An illustration of local sequence alignment with an accumulated score matrix

2.2 Initialization of the Accumulated Score Matrix

In this step, a gap (i.e., "-") is inserted into X_0 and Y_0. KELSA sets boundary elements (i.e., the blue colored matrix elements in Fig. 2a) in the accumulated score matrix to zero, namely,

$$A_{i,j} = 0 \quad i = 0 \; or \; j = 0 \tag{1}$$

2.3 Iterative Scoring Process to Fill the Accumulated Score Matrix

Let $s(X_i, Y_j)$ denotes the similarity score between two medical events X_i and Y_j. According to the alignments between previous medical events in **X** and **Y**, there are three possible alignment strategies to align X_i and Y_j for calculating the accumulated score element $A_{i,j}$:

(a) If X_{i-1} and Y_{j-1} are better aligned, KELSA imposes X_i and Y_j to align. Thus,

$$A^{(a)}_{i,j} = A_{i-1,j-1} + s\left(X_i, Y_j\right) \tag{2}$$

(b) If X_i and Y_{j-1} are better aligned, KELSA inserts chronic conditions (X_c) or a gap (-) after X_i for better alignment with Y_j. Specifically, if the patient P_X has experienced any chronic conditions (X_c) prior to X_i, the algorithm inserts X_c after X_i to align with Y_j. Thus,

$$A^{(b)}_{i,j} = A_{i,j-1} + s\left(X_c, Y_j\right) \tag{3}$$

If the patient P_X has experienced no chronic conditions prior to X_i, KELSA inserts a gap (-) after X_i (i.e., X_c is set to "-") to align with Y_j. Thus,

$$A^{(b)}_{i,j} = A_{i,j-1} + s\left(-, Y_j\right) \tag{4}$$

(c) If X_{i-1} and Y_j are better aligned, KELSA inserts chronic conditions (Y_c) or a gap (-) after Y_j for better alignment with X_i. Specifically, if the patient P_Y has experienced any chronic conditions (Y_c) prior to Y_j, KELSA inserts Y_c after Y_j to align with X_i. Thus,

$$A^{(c)}_{i,j} = A_{i-1,j} + s(X_i, Y_c) \tag{5}$$

If the patient P_Y has experienced no chronic conditions prior to Y_j, KELSA inserts a gap (-) after Y_j (i.e., Y_c is set to "-") to align with X_i. Thus,

$$A_{i,j}^{(c)} = A_{i-1,j} + s(X_i, -) \tag{6}$$

After the calculations with the three alignment strategies, KELSA compares them with zero to choose the largest value to fill the accumulated score matrix $A_{i,j}$:

$$A_{i,j} = \max(A_{i,j}^{(a)}, A_{i,j}^{(b)}, A_{i,j}^{(c)}, 0) \tag{7}$$

where 0 is used to mask certain mismatched alignments and render locally matched alignments visible. The iterative scoring process continues until all elements of accumulated score matrix are filled up as illustrated in Fig. 2b.

2.4 Traceback to Identify a Local Alignment

By starting at the element with the highest accumulated score in the $\boldsymbol{A}_{(n+1)\times(m+1)}$, KELSA identifies the local alignment path with the highest similarity by tracking back and choosing the path with maximal accumulated score (See Fig. 2c).

The corresponding pseudo codes for KELSA algorithm are,

Pseudo codes for KELSA

```
Function KELSA(X, Y)
  A[0, 0…m], A[0…n, 0] ← 0

  for i=1 to n, j=1 to m
     Xc ← Chronic_condition(X[1…i])
     Yc ← Chronic_condition(Y[1…j])
     match ← Fa(i,j) where Fa(i,j) = A[i-1,j-1] + s(X[i], Y[j])
     insertX ← Fb(i,j) where Fb(i,j) = A[i,j-1] + s(Xc, Y[j])
     insertY ← Fc(i,j) where Fc(i,j) = A[i-1,j] + s(X[i], Yc)
     A[i,j] ← max(match, insertX, insertY, 0)

  AlignX, AlignY ← ""
  i, j← argmax(A)
  while (i >0 or j > 0) and A[i,j] > 0
     if i > 0 and j > 0 and A[i,j] == Fa(i,j) then
        AlignX ← X[i] + AlignX
        AlignY ← Y[j] + AlignY
        i ← i - 1
        j ← j - 1
     else if i > 0 and A[i,j] == Fc(i,j) then
        AlignX ← Chronic_condition(X[1…i]) + AlignX
        i ← i – 1
     else
        AlignY ← Chronic_condition(Y[1…j]) + AlignY
        j ← j - 1
```

3 Evaluation

3.1 Evaluation Design

In this section, we evaluated our KELSA algorithm by comparing the KELSA algorithm to traditional SWA commonly used in biological sequence alignment. The evaluation is a challenging task for several reasons: (1) Patient medical records are complex. Patient medical records involve thousands of diagnosis codes, whereas nucleic acid (or protein sequences) have only 4 (or 20) types of nucleic acids (or amino acids). In addition, all the diagnosis documented in patient medical records in the same way can have different semantic meaning (e.g., acute and chronic diseases). (2) There is no gold standard patient data available for evaluating sequence alignment algorithms. One possible solution is to ask medical experts to evaluate and rank the results from different sequence alignment algorithms, but it can be very subjective and expensive.

To solve the evaluation issue, we propose to synthesize simulated patient medical records based on representative patients carefully chosen from a large real-world EHR database. This way, we can design and control the differences between patient medical records for objective and comprehensive algorithm evaluation.

The rest of the section is organized as follows: Firstly, we introduced a real-world EHR database for selecting representative seed patients. We then described the strategies to select seed patients and synthesize simulated EHRs. Finally, we described the evaluation metrics for comparing sequence alignment algorithms.

3.2 Real-World EHR Database for Evaluation

In the mid-1960s, Dr. Leonard T. Kurland established the Rochester Epidemiology Project (REP) [15, 18, 19]. Since then, the REP data has been well tested in over 2,700 scientific reports. It includes patient demographic and comprehensive coded information such as medical diagnoses, surgical procedures, prescriptions, and laboratory test results in both outpatient (e.g., office visit, urgent care, and emergency room) and inpatient settings across all medical facilities in Olmsted County, MN, regardless of where the care was delivered or of insurance status. In 2016, REP contained approximately 2 million patient medical records from more than 577,000 individuals. REP enables investigators to conduct long-term, population-based studies of disease incidence, prevalence, risk factors, medical outcomes, cost-effectiveness, and health services utilization.

In this work, we used all patient medical records during 1995–2015 in the REP database to select seed patients for evaluation. Without loss of generality, we only considered diagnosis information in these patient medical records, because all other medical information such as procedures, lab tests, medications, and clinical notes have dependency on diagnoses. Without considering the dependency on diagnoses and the underlying medical rational, we cannot easily synthesize procedures, lab tests, medications, and clinical notes to meaningfully simulate real world situations.

REP uses the International Classification of Diseases, Ninth Revision, Clinical Modification (ICD-9-CM) [13] to code diagnoses and classify diseases and medical conditions. As our purpose in this section is to evaluate various sequence alignment methods for comparing patient medical records, we aggregated the ICD-9-CM codes to the PheCode [21], which represents a level of disease granularity close to what is used in clinical practice [7, 9]. PheCode has a two-level hierarchy structure and we only used the root PheCode to capture broader disease categories. In total, 14,335 diseases and medical conditions coded by ICD-9-CM in the REP database were grouped into 582 diseases and medical conditions defined by PheCode.

3.3 *Synthesis of Patient Medical Records*

3.3.1 Selection Criteria of Patient Medical Records

Medical care is highly specialized, complicated, and heterogenous. In order to sample representative patients from the REP database and synthesize patient medical records that simulate real world situations, we consider the following characteristics and criteria when selecting medical records of seed patients:

(1) Multiple scenarios of a daily event (i.e., patient clinical encounters on a single day). Patient clinical encounters include one or multiple visits per day, and one or multiple diagnoses per visit. When we select the seed patients, we make sure that they have these types of daily encounter scenarios on their medical records.
(2) Nature of diseases. An acute disease on patient medical records can be considered as an event on a specific time point, whereas a chronic disease covers a longer time span. The selected patients must include both acute and chronic diseases on their medical records.
(3) The lengths of medical records of different patients vary significantly. The selected patient medical records should cover a wide range of length, in terms of total event dates.

Totally, we identified 76,699 patients satisfying the criteria (1) and (2) in the REP database. The medical records of these patients have a wide range of length in terms of total event dates. Figure 3 shows the length distribution of these patient medical records. We found that 99.3% patients had the lengths of medical records less than 210 dates. Thus, we randomly selected 100 seed patients with the lengths of their medical records ranging in [10, 12, 14, …, 206, 208] dates.

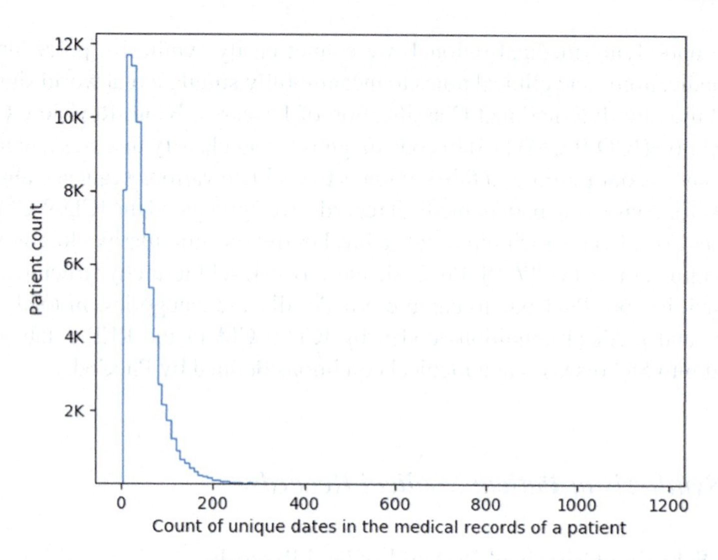

Fig. 3 The distribution of medical record lengths in terms of count of unique dates for patient medical records with multiple types of daily clinical encounters, and acute and chronic diseases in the REP database

3.3.2 Synthesis Methods of Patient Medical Records

We performed three types of operations (i.e., deleting, updating, and switching) on the medical records of the 100 seed patients at the level of single-day event and multiple-day events (event block). In the context of sequence alignment, the operation of inserting in one sequence is equivalent to deleting in another sequence, so we only kept the deleting operation. In the end, we synthesized 20 new patient medical records with various similarity to seed patients by applying one or more deleting, updating and switching operations, for each of the 100 seed patients. The second column of Table 1 specifies the operations we performed. The size of event block is determined by the maximum of (2, N/10), where N is the number of event dates for a seed patient.

More details regarding our criteria to select seed patients and methods to synthesize new patient medical records can be found in our previous work [8, 10].

3.4 *Evaluation Metrics*

The scoring system s(X, Y) widely used in the biological sequence alignment was deployed to measure the similarity between two aligned daily events (X and Y) (See Fig. 1B). The scoring system assigns the score of matching to 1 as a reward. It also sets the same score of −1 to both mismatching and gap insertion as a penalty. For a daily event (X or Y) containing multiple medical codes, we used Jaccard index

Table 1 Count of pairwise local sequence alignments by category

Count		KELSA vs SWA				KELSA vs REF				SWA vs REF			
ID	Operation	$C_k > C_s, S_k > S_s$	$C_k > C_s, S_k = S_s$	$C_k = C_s, S_k > S_s$	$C_k = C_s, S_k = S_s$	$C_k > C_r, S_k > S_r$	$C_k > C_r, S_k = S_r$	$C_k = C_r, S_k > S_r$	$C_k = C_r, S_k = S_r$	$C_s > C_r, S_s > S_r$	$C_s > C_r, S_s = S_r$	$C_s = C_r, S_s > S_r$	$C_s = C_r, S_s = S_r$
1	x	0	0	25	75	90	0	0	10	90	0	0	10
2	x x	1	0	46	53	100	0	0	0	100	0	0	0
3	u	0	0	0	100	97	1	0	2	97	1	0	2
4	u u	0	1	0	99	99	1	0	0	99	1	0	0
5	s	0	0	3	97	92	0	0	8	91	1	0	8
6	s s	0	0	3	97	100	0	0	0	100	0	0	0
7	x u	0	0	30	70	99	0	0	1	99	0	0	1
8	x s	0	0	22	78	100	0	0	0	100	0	0	0
9	u s	1	0	0	99	100	0	0	0	100	0	0	0
10	x u s	0	1	24	75	99	0	0	1	99	0	0	1
11	X	0	0	50	50	79	4	0	17	77	6	0	17
12	X X	2	0	56	42	87	1	0	12	85	3	0	12
13	U	0	1	5	94	77	4	0	19	77	3	0	20
14	U U	0	1	10	89	91	1	0	8	91	0	0	9
15	S	3	0	25	72	94	1	0	5	91	2	0	7
16	S S	3	0	47	50	91	1	0	8	89	1	0	10
17	X U	1	0	40	59	89	0	0	11	89	0	0	11
18	X S	8	1	52	39	94	1	0	5	93	1	0	6
19	U S	1	1	31	67	87	1	0	12	86	2	0	12
20	X U S	11	0	44	45	92	2	0	6	90	2	0	8
All		31	6	513	1450	1857	18	0	125	1843	23	0	134

[a]KELSA, SWA, and REF stand for Knowledge-Enriched Local Sequence Alignment algorithm, Smith-Waterman Algorithm, and baseline REFerence

[b]C_K (or C_S, C_R) denote coverage of the longest aligned subsequences between seed patient and synthetic patient made by KELSA (or SWA, REF). S_K (or S_S, S_R) is similarity of the longest aligned subsequences between seed patient and synthetic patient calculated with KELSA (or SWA, REF)

[c]The lower csase letters “x”, “u”, and “s” denote deleting, updating, and switching a daily event, respectively. The upper case letters “X”, “U”, and “S” stand for deleting, updating, and switching multi daily events (event block)

J(X, Y) to measure their similarity s(X, Y) defined by the following equation,

$$s(X, Y) = 2 * J(X, Y) - 1 \quad (8)$$

$$J(X, Y) = \frac{|X \cap Y|}{|X \cup Y|} \quad (9)$$

We also penalized similarity between an original daily event in a patient sequence and an extra daily event inserted into another patient sequence by setting the score range between −1 (mismatching) and 0 (matching). In other words, the similarity s(X, Y) between them is defined as,

$$s(X, Y) = J(X, Y) - 1 \quad (10)$$

We calculated the coverage (C) and similarity score (S) of the longest aligned subsequences between seed patient and synthetic patient. S is the summation of the similarity scores s(X, Y) of daily events in the aligned subsequences divided by the number of event dates in the seed patient sequence. C is the coverage of the seed patient sequence aligned to the synthetic patient sequence. Specifically, C is the ratio of the number of daily events in the seed patient sequence aligned to a synthetic patient sequence and the total number of daily events in the seed patient sequence.

4 Results and Discussion

After creating 2000 synthetic patients based on 100 seed patients, we used KELSA and SWA to align locally medical records of each seed patient and each synthetic patient and identify the longest aligned patient subsequences. We calculated coverage and similarity score for each pair of the longest aligned patient subsequences. KELSA alignments are compared with SWA and reference alignments (REF) in terms of coverage and similarity scores of the aligned patient subsequences as shown in Table 1.

Our results show that the coverage and similarity scores of both KELSA and SWA alignments were better than or at least as good as those of reference alignments. It suggests the necessity of sequence alignments. More specifically, 92.85% (or 92.15%) of 2000 alignments made by KELSA (or SWA) had larger coverage and higher similarity scores than reference alignments. 0.90% KELSA (or 1.15% SWA) alignments showed equal similarity scores but larger coverage than reference alignments. For the rest 7.15% (or 7.85%) alignments, KELSA (or SWA) gave the same coverage and similarity scores as reference alignments.

KELSA alignments were also better than SWA alignments. We found that 27.50% of 2000 KELSA alignments had either larger coverage or higher similarity scores than SWA alignments while the rest 72.50% KELSA and SWA alignments had identical coverage (and similarity scores). Particularly, 1.55% KELSA alignments had superior coverage and similarity scores than SWA alignments. 25.65% alignments made by KELSA received equal coverage but higher similarity scores, compared to SWA.

We then carefully examined over $3 \times 4 \times 20$ typical cases out of the 2000 raw alignments in terms of 3 alignment algorithms, 4 combinations of coverage and similarity score, and 20 synthetic operations. We found some subtle differences among these raw alignments and designed some illustrative cartoons in Fig. 4 for case-by-case discussion.

Figure 4A, B show the alignments between seed sequence and synthetic sequence created by only deleting operation. In Fig. 4A, the reference alignment and SWA alignment had the first two aligned daiy events due to a deletion of the 3rd daily event in the seed sequence. Their coverage and similarity score were 0.50. In contrast, KELSA was able to insert a chronic (circle) daily event in the right position and created a new sequence identical to the seed sequence for better alignment. Thus, KELSA alignment had the highest coverage (1.00) and similarity score (0.75). Compared to the seed sequence in Fig. 4A, the seed sequence in Fig. 4B contained one more triangle

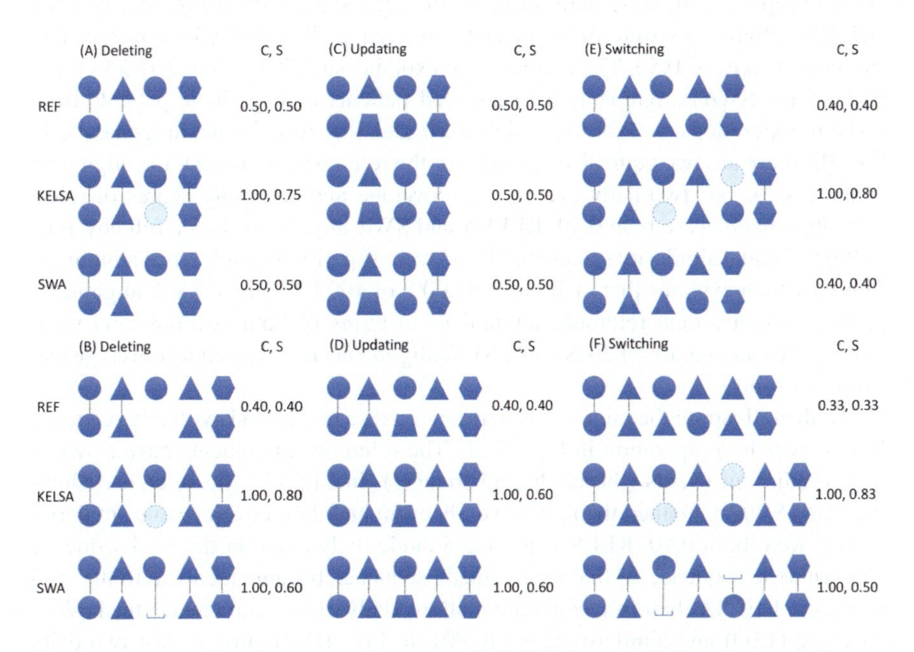

Fig. 4 Scenarios of local sequence alignments. REF, KELSA, and SWA refer to as reference alignment, and alignment with Knowledge-Enriched Local Sequence Alignment algorithm, and alignment with Smith-Waterman Algorithm, respectively. In each pair, seed sequence is listed on the top and aligned synthetic sequence is listed on the bottom. The coverage (C) and similarity scores (S) between seed sequence and synthetic sequence are also listed on the right side of each pair. The different shapes (e.g., circle, diamond, and triangle) represent different medical events. The shapes with light blue and dash border are extra medical events inserted by KELSA during sequence alignment. "_" is a gap spot inserted by SWA or KELSA during alignment. Circle shape denotes a chronic disease

daily event. The reference alignment could be alignment of the first or last two daily events. The coverage and similarity score of both cases were 0.40. However, KELSA or SWA could insert a chronic (circle) daily event or gap spot in the middle position to make a better alignment of seed sequence and synthetic sequence with full coverage and higher similarity score. Due to the inserted chronic daily event, the similarity score of KELSA alignment (0.80) is higher than that of SWA alignment (0.60). As shown in Table 1, among 400 alignments between seed patients and synthetic patients made only by the deleting operation, 89.0% KELSA alignments and 88.0% SWA alignments had larger coverage and higher similarity scores than reference alignments. 44.3% KELSA alignments received higher similarity scores than SWA alignments while they had identical coverage. The larger coverage or higher similarity score can be attributed to event or gap insertion by KELSA or SWA. Especially, KELSA uses a chronic event to fill a gap position in a sequence and gains better performance than SWA.

Alignments between seed sequences and synthetic sequences with a daily event updating are shown in Fig. 4C, D. The synthetic sequence had a trapezoidal daily event to replace a triangle daily event in the seed sequence. In Fig. 4C, KELSA and SWA alignments were the same as the reference alignment with coverage and similarity scores of 0.50. The cartoon case explains such 7.25% (or 7.75%) of 400 KELSA (or SWA) alignments between seed patients and synthetic patients from updating operations in Table 1, which were identical to reference alignments. In Fig. 4D, there are two equivalent options for the reference alignment: the alignment of the first or last two daily events. The coverage and similarity scores of reference alignments were both 0.40. KELSA and SWA alignments had a full coverage (1.00) and equal similarity scores (0.60) and were superior than reference alignment. This case then explains that in Table 1, 91.00% of 400 KELSA or SWA alignments performed better than reference alignments in terms of both coverage and similarity scores and 95.50% KELSA and SWA alignments received equal coverage and similarity scores.

We then illustrate the alignments between seed sequence and synthetic sequence from a switching operation in Fig. 4E, F. The reference alignments have a switch of two adjacent events (the circle and triangle). In Fig. 4E, the reference alignment and SWA contained the first two daily event and their coverage and similarity scores were both 0.40. KELSA inserted a circle daily event in the seed sequence and synthetic sequence so that the synthetic sequence became identical to the seed sequence. KELSA therefore could align all the daily events and receive the highest coverage (1.00) and similarity score (0.80). In Fig. 4F, the first or last two daily events could be aligned as the reference alignment, whose coverage and similarity scores are 0.33. Both KELSA and SWA alignments could receive full coverage of 1.00 and higher similarity score by inserting a circle daily event and gap spot. KELSA even made the alignment with higher similarity score than SWA by filling circle daily events rather than gap spots. In, among 400 alignments between seed patients and synthetic patients created with switching operations, 94.25% KELSA alignments and 92.75% SWA alignments demonstrated better coverage and similarity scores than reference alignments. 21% KELSA alignments achieved higher similarity scores than SWA alignments.

We further discuss some practical issues when applying KELSA to identify locally similar patients:

Since sequence alignment methods only consider the sequential order of events rather than their temporal intervals, sequence alignment methods can be used for prescreening of temporal interval sensitive cases. However, sequence alignment methods could use different time spans of elements (e.g., daily or weekly events) in the sequences to differentiate adjacent events that are temporally close or distant.

Here we use the equal weights for different diagnosis codes and daily events. In practical scenarios, the users can assign various weighting schemes between diagnosis codes and between daily events. The weighting schemes can be easily incorporated into calculations of similarity score between daily events and accumulated similarity matrix.

We demonstrated and validated our KELSA algorithm by identifying the longest subsequences between two given patient event sequences. Practically, the users can select any sub-region with interested events like rare diseases of a patient event sequence to identify locally similar patients.

5 Conclusion

Recently researchers began to consider sequence alignment methods for comparing patient medical records. Local alignment is truly more useful than global alignment for identifying similar patients. In this work, we propose a novel Knowledge-Enriched Local Sequence Alignment (KELSA) algorithm, which allows the incorporation of medical knowledge (e.g., event chronicity) while aligning patient medical records. We then objectively evaluated our KELSA algorithm by comparing it to standard SWA on synthetic patient data where the reference alignments are known. The results show that KELSA can align better by inserting hidden daily events and identify more similarities between patient medical records than SWA. Compared to SWA, KELSA is more suitable for comparing patient medical records.

Acknowledgements Funding for this study was provided by NLM (5K01LM012102) and the Center for Clinical and Translational Science (UL1TR002377) from the NIH/NCATS.

References

1. Altschul, S.F., Gish, W., Miller, W., et al.: Basic local alignment search tool. J. Mol. Biol. **215**, 403–410 (1990)

2. Bello, H.K., Gbolagade, K.A.: Residue number system: an important application in bioinformatics. Int. J. Comput. Applicat. **975**, 8887
3. Brown, S.A.: Patient similarity: emerging concepts in systems and precision medicine. Front. Physiol. **7**, 561 (2016)
4. Buchfink, B., Xie, C., Huson, D.H.: Fast and sensitive protein alignment using DIAMOND. Nat. Methods **12**, 59 (2015)
5. Che, C., Xiao, C., Liang, J., et al.: An RNN architecture with dynamic temporal matching for personalized predictions of Parkinson's disease. In: Proceedings of the 2017 SIAM International Conference on Data Mining, pp. 198–206. SIAM (2017)
6. Giannoula, A., Gutierrez-Sacristán, A., Bravo, Á., et al.: Identifying temporal patterns in patient disease trajectories using dynamic time warping: A population-based study. Sci. Rep. **8**, 4216 (2018)
7. Huang, M., Eltayeby, O., Zolnoori, M., et al.: Public opinions toward diseases: infodemiological study on news media data. J. Med. Internet Res. **20**, e10047 (2018)
8. Huang, M., Shah, N.D., Yao, L.: Evaluating global and local sequence alignment methods for comparing patient medical records. BMC Med. Inform. Decis. Mak. **19**, 263 (2019)
9. Huang, M., Zolnoori, M., Balls-Berry, J.E., et al.: Technological innovations in disease management text mining US patent data from 1995 to 2017. J. Med. Internet Res. **21**, e13316 (2019)
10. Huang, M., Zolnoori, M., Shah, N., et al.: Temporal sequence alignment in electronic health records for computable patient representation. In: 2018 IEEE International Conference on Bioinformatics and Biomedicine (BIBM), pp. 1054–1061. IEEE (2018)
11. Lipman, D.J., Pearson, W.R.: Rapid and sensitive protein similarity searches. Science **227**, 1435–1441 (1985)
12. Müller, M.: Dynamic time warping. In: Information Retrieval for Music and Motion, pp. 69-84 (2007)
13. National Center for Health Statistics International classification of diseases, ninth revision, clinical modification (ICD-9-CM). Centers for Disease Control Prevention, Atlanta, Georgia, USA (2013)
14. Needleman, S.B., Wunsch, C.D.: A general method applicable to the search for similarities in the amino acid sequence of two proteins. J. Mol. Biol. **48**, 443–453 (1970)
15. Rocca, W.A., Grossardt, B.R., Brue, S.M., et al.: Data resource profile: expansion of the Rochester epidemiology project medical records-linkage system (E-REP). Int. J. Epidemiol. **47**, 368–368j (2018)
16. Shickel, B., Tighe, P.J., Bihorac, A., et al.: Deep EHR: a survey of recent advances in deep learning techniques for electronic health record (EHR) analysis. IEEE J. Biomed. Health Inform. **22**, 1589–1604 (2018)
17. Smith, T.F., Waterman, M.S.: Identification of common molecular subsequences. J. Mol. Biol. **147**, 195–197 (1981)
18. St Sauver, J.L., Grossardt, B.R., Yawn, B.P., et al.: Data resource profile: the Rochester Epidemiology Project (REP) medical records-linkage system. Int. J. Epidemiol. **41**, 1614–1624 (2012)
19. St Sauver, J.L., Grossardt, B.R., Yawn, B.P., et al.: Use of a medical records linkage system to enumerate a dynamic population over time: the Rochester epidemiology project. Am. J. Epidemiol. **173**, 1059–1068 (2011)
20. Sun, J., Chen, K., Hao, Z.: Pairwise alignment for very long nucleic acid sequences. Biochem. Biophys. Res. Commun. **502**, 313–317 (2018)
21. Wei, W.Q., Bastarache, L.A., Carroll, R.J., et al.: Evaluating phecodes, clinical classification software, and ICD-9-CM codes for phenome-wide association studies in the electronic health record. PLoS ONE **12**, e0175508 (2017)
22. Zong, N., Kim, H., Ngo, V., et al.: Deep mining heterogeneous networks of biomedical linked data to predict novel drug–target associations. Bioinformatics **33**, 2337–2344 (2017)

Multi-Level Embedding with Topic Modeling on Electronic Health Records for Predicting Depression

Yiwen Meng, William Speier, Michael Ong, and Corey W. Arnold

Abstract Deep learning methods have exhibited impressive performance in many classification and prediction tasks in healthcare using electronic health record (EHR) data. However, effectively handling the heterogeneous nature and sparsity of EHR data remain as challenges. In this work, we present a model that utilizes heterogeneous data and attempts to address sparsity using temporal Multi-Level Embeddings of diagnoses, procedures, and medication codes with demographic information and Topic modeling (MLET). MLET aggregates various categories of EHR data and learns inherent structure based on hospital visits in a patient's medical history. We demonstrate the potential of the approach in the task of predicting depression using different time windows prior to a clinical diagnosis. We found that MLET outperformed all baseline methods with a highest improvement from 0.6122 to 0.6808 in precision recall area under the curve (PRAUC). Our results demonstrate the model's ability to utilize heterogeneous EHR information to predict depression, which may have future implications for screening and early detection.

Y. Meng · W. Speier · C. W. Arnold (✉)
Computational Diagnostics Lab, Department of Radiological Sciences, 924 Westwood Blvd, Suite 420, Los Angeles, CA 90024, USA
e-mail: cwarnold@ucla.edu

Y. Meng
e-mail: lanyexiaosa@ucla.edu

W. Speier
e-mail: speier@ucla.edu

Y. Meng · C. W. Arnold
Department of Bioengineering, 924 Westwood Blvd, Suite 420, Los Angeles, CA 90024, USA

M. Ong
Department of Medicine, 924 Westwood Blvd, Suite 420, Los Angeles, CA 90024, USA
e-mail: mong@mednet.ucla.edu

C. W. Arnold
Department of Pathology, 924 Westwood Blvd, Suite 420, Los Angeles, CA 90024, USA

© The Editor(s) (if applicable) and The Author(s), under exclusive license to Springer Nature Switzerland AG 2021

A. Shaban-Nejad et al. (eds.), *Explainable AI in Healthcare and Medicine*, Studies in Computational Intelligence 914,
https://doi.org/10.1007/978-3-030-53352-6_22

1 Introduction

Electronic Health Records (EHRs) have been broadly adopted for documenting a patient's medical history [1]. They are composed of data from various sources, including diagnoses, procedures, medications, clinical notes, and laboratory results, which contribute to their high dimensionality and heterogeneity. However, only a few studies have attempted to use data from a broad set of categories. Thus, data heterogeneity remains a technical barrier for combining various types of EHR data in one model. Several studies have utilized EHR data to predict future diagnosis of chronic diseases such as heart failure [2] and depression [3]. The goal of this work is to overcome the heterogeneity and sparsity of EHR data and construct a predictive model for depression. To achieve this goal, we propose Multi-Level Embedding with Topic modelling (MLET), which aggregates diagnoses, procedure codes, medications, and demographic information together with topic modelling of clinical notes. MLET follows the method of building a hierarchical structure on different categories of EHR data with multiple embedding levels [2], while preserving its sequential nature. In this way, it learns the inherent interaction between EHR data from various sources within each visit and across multiple visits for an individual patient. Furthermore, MLET extends the hierarchical model by including clinical notes, processed by topic modelling, while studying the contribution of different EHR data modalities on prediction accuracy. As a result, we found that MLET consistently outperformed all baselines models in sequential prediction of depression.

2 Methodology

For this analysis, patients were identified from our EHR in accordance with an institutional review board (IRB) approved protocol. Each patient visit was represented by multiple data modalities in the EHR, including diagnosis codes in ICD-9 (International Classification of Disease, ninth revision) format, procedure codes in CPT (Current Procedural Terminology) format, medication lists, demographic information, and clinical notes. All patient records coded with ICD-9 values for myocardial infarction (MI), breast cancer, or liver cirrhosis from 2006–2013 were included. In total, 10,148 patients were identified, where 3,047 were diagnosed with depression.

EHR data contains a variety of data resources. ICD-9 codes, CPT codes, medication lists, and patient's gender are all categorical variables, while ages are numerical. Therefore, an intuitive approach is to encode these features in a multi-hot binary vector, where each column corresponds to a specific code or data element. Embedding is a technique that has been widely adopted in NLP to project long and sparse feature vectors into a dense lower dimensional space [4]. This approach efficiently reduces the dimensionality of a model's parameters and decreases training time, allowing our model to process categorical data from the EHR.

Previous studies [2, 5, 6] used data from the entire EHR for future disease prediction. This method could add bias for patients with longer medical histories in the EHR. Hence, we narrowed the data window for each patient to six months. As predicting the future risk of a disease in a prospective setting is an ongoing task, the time window of a patient's EHR is highly varied. Therefore, as a similar approach to [3], we defined four decision points in advance of the diagnosis of depression: two weeks, three months, six months, and one year. We compared the prediction accuracy with random forest(RF) and a modified version of MiME. The MiME model demonstrated state-of-the-art performance in predicting heart failure onset [2] and we adapted it for the task of depression prediction. We changed the structure of MiME as there was no direct linked relationship between ICD-9 codes, CPT codes, and medication lists in our EHR data, which is more common in clinical documentation. Two data sets were generated to compare the contribution to prediction of depression for demographics and topic features. There were both applied to baseline models as well as MLET: all data types (ICD-9 codes, CPT codes, medication lists, demographics, and topic features); ICD-9 codes, CPT codes, and medication lists.

Figure 1 illustrates the hierarchical structure of MLET. The ultimate goal of the model is to predict the probability of a chronic disease for patient i given the feature embedding representing a sequence of visits, $\mathbb{P}(y_i|\overrightarrow{\mathrm{h_i}})$, where y_i is the disease prediction for patient i and $\overrightarrow{\mathrm{h_1}}$ stands for the patient level embedding of the patient's EHR. While the model is designed to be generalizable, we focus here on the prediction of depression. Each patient has multiple hospital visits from $\overrightarrow{v_1}$ to $\overrightarrow{v_t}$, which compose the visit level embedding. During one visit $\overrightarrow{v_t}$, the code level embedding $\overrightarrow{e_t}$ is the ensemble of multiple ICD-9 and CPT codes, medications, demographic information, and topic features extracted from associated clinical notes. Since there are five

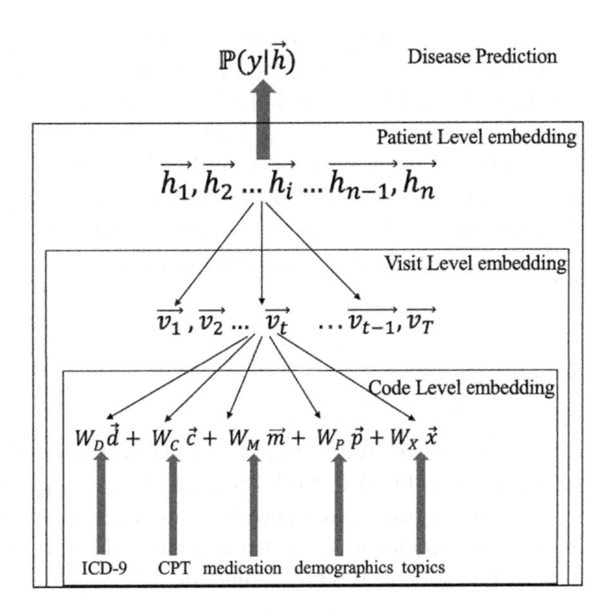

Fig. 1 Illustration of for EHR data. There are three levels of embedding: patient level, visit level and code level

categories of EHR data, we built individual embedding for each first and aggregated them together.

Equation (1), Eq. (2 and Eq. (3) describe mathematical formulation of MLET in the top-down view, denoting the *Patient level*, *Visit level* and *Code level* embeddings, respectively.

$$\vec{h_1} = f(\vec{v_1} \ldots \vec{v_t} \ldots \vec{v_T}) \tag{1}$$

Equation (1) shows the method to process temporal information of various visit level embeddings to compute a patient level embedding, where f stands for the function to input visit information in a sequential order. As mentioned before, RNN, LSTM and GRU have been widely used to fulfill this task. Since RNN often encounters the vanishing gradient problem and better performance has been shown for GRU over LSTM in previous work [2], we used GRU in the current model.

$$\vec{v_t} = \alpha(\boldsymbol{W}_e \vec{e_t}) + \vec{e_t} \tag{2}$$

In Eq. (2), a visit level embedding is generated by first doing a matrix transformation with weight $\boldsymbol{W}_e \, \varepsilon \, \mathbb{R}^{z \times z}$, followed by a non-linear ReLU (rectified linear unit) transformation function α, where z is the embedding size. We omitted the bias term here to formulate a residual connection [7].

$$\vec{e_t} = \beta(F) + F \tag{3}$$

$$F = W_D\vec{d} + W_C\vec{c} + W_M\vec{m} + W_P\vec{p} + W_X\vec{x} \tag{4}$$

Equations (3) and (4) define the code level embedding by summing up individual embeddings from five EHR data resources with a non-linear transformation function β. As in Eq. (2), we use ReLu for β. The $W_D \, \varepsilon \, \mathbb{R}^{z \times z}$, $W_C \, \varepsilon \, \mathbb{R}^{z \times z}$, $W_M \, \varepsilon \, \mathbb{R}^{z \times z}$, $W_P \, \varepsilon \, \mathbb{R}^{z \times z}$ and $W_X \, \varepsilon \, \mathbb{R}^{z \times z}$ represent the weight matrices for transforming the embeddings of ICD-9 codes, CPT codes, medication lists, demographics, and topic features into a latent space with the same dimension. In the same manner as Eq. (2), all of the corresponding bias terms were omitted to denote residual connections. Finally, binary cross entropy was used as the loss function.

3 Results

Table 1 displays the results from four models at four time points in advance of diagnosis in terms of PRAUC. MLET with all EHR data types outperformed other models for every prediction period. The results indicate a consistent decrease in accuracy as the prediction point moves further away from the time of diagnosis. Results from the two MLET models all outperformed the RF, which further confirms the advantage

Table 1 Comparison of prediction performance in PRAUC for different models

Prediction window	RF (codes)	RF (codes+demo+topic)	MLET (codes)	MLET (codes+demo+topic)
Two weeks	0.6506 (0.0215)	0.6679 (0.0261)	0.6700 (0.0180)	**0.7304*** **(0.0158)**
Three months	0.6078 (0.0141)	0.6244 (0.0275)	0.6463 (0.0211)	**0.7075*** **(0.0172)**
Six months	0.5782 (0.0137)	0.5908 (0.0204)	0.6122 (0.0140)	**0.6808*** **(0.0217)**
One year	0.5756 (0.029)	0.5839 (0.0316)	0.6092 (0.0113)	**0.6575*** **(0.0153)**

MLET(codes) was a modified version from MiME. Codes denote data from ICD-9, CPT, and medication lists, while demo stands for demographic information. Values in parenthesis refer to standard deviations across randomizations. Bold values denotes the highest in each column * denotes the statistical significance ($p < 0.05$) between PRAUC values from MLET(codes) and MLET(codes+demo+topic)

of using a temporal model over non-temporal methods. Adding demographic information and topic features also improved the performance for both RF and MLET, which demonstrates their significant contribution in predicting depression as well as emphasizes the advantage of building a model being able to aggregate EHR data from multiple sources. MLET with all types of EHR data achieved the highest ROCAUC and PRAUC at each prediction time as shown in bold in Table 1. In particular, it generated the highest mean PRAUC of 0.7304 when predicting two weeks prior to the diagnosis, and the value dropped to 0.6575 when predicting one year in advance. Moreover, the model with all data types achieved the highest gain of 0.07 in PRAUC when predicting six months prior to diagnosis when compared to MLET (code).

A common issue in deep learning models like MLET that it is often unclear which features were used for classification. This ambiguity limits the adoption of such methods in a clinical setting where physicians need to understand how models reach diagnostic conclusions from the data. One approach for making deep models more intuitive is incorporating an attention mechanism, which has been used to provide the importance of features in on predicting heart failure [6] and sentence generation [8]. Future studies can utilize attention weights on the EHR categories or visit embeddings to show the predictive value of the various data sources and individual patient visits.

4 Conclusion

We have developed a temporal deep learning model MLET that was able to integrate five types of EHR data during multiple visits for depression prediction. MLET consistently outperformed the baseline models tested, achieving an increase in PRAUC of

0.07 over the best baseline model. The results demonstrate the ability of MLET as an approach to deal with data heterogeneity and sparsity in modelling the EHR. In future work, MLET could possibly be used as the basis for constructing a screening tool by utilizing the models' predictions to intervene with individuals who have a higher risk of developing depression.

References

1. Press, D.: Benefits and drawbacks of electronic health record systems, pp. 47–55 (2011)
2. Choi, E., Xiao, C., Stewart, W., Sun, J.: MiME : Multilevel Medical Embedding of Electronic Health Records for Predictive Healthcare. Adv. Neural Inf. Process Syst. **31** (NIPS 2018) (2018)
3. Huang, S.H., LePendu, P., Iyer, S.V., Tai-Seale, M., Carrell, D., Shah, N.H.: Toward personalizing treatment for depression: predicting diagnosis and severity. J. Am. Med. Inform. Assoc. **21**(6), 1069–1075. https://academic.oup.com/jamia/article-lookup/doi/10.1136/amiajnl-2014-002733. Accessed 8 June 2019
4. Mikolov, T., Chen, K., Corrado, G., Dean, J.: Distributed representations of words and phrases and their compositionality. Neural Inf. Process Syst. (2013)
5. Lipton, Z.C., Kale, D.C., Elkan, C., Wetzel, R.: Learning to diagnose with LSTM recurrent neural networks. 2015, 1–18. http://arxiv.org/abs/1511.03677
6. Choi, E, Bahadori, M.T., Kulas, J.A., Schuetz, A., Stewart, W.F., Sun, J.: RETAIN: an interpretable predictive model for healthcare using reverse time attention mechanism. Adv. Neural Inf. Process Syst. **29** (NIPS 2016), 2865 (2016)
7. He, K., Sun, J.: Deep Residual Learning for Image Recognition, pp. 1–9 (2015)
8. Vaswani, A., Brain, G., Shazeer, N., Parmar, N., Uszkoreit, J., Jones, L., et al.: Attention is all you need. Adv. Neural Inf. Process Syst. [Internet]. (Nips), 5998–6008 (2017). http://papers.nips.cc/paper/7181-attention-is-all-you-need.pdf

Faster Clinical Time Series Classification with Filter Based Feature Engineering Tree Boosting Methods

Yanke Hu, Wangpeng An, Raj Subramanian, Na Zhao, Yang Gu, and Weili Wu

Abstract Electronic Health Record (EHR) has been widely adopted in the US in the past decade, which provides an increasing interest for predictive analysis with clinical time series data. Recently, deep learning approaches such as Recurrent Neural Network (RNN) and attention based networks have achieved the state-of-the-art accuracy on several clinical time series tasks, but one shortcoming of the RNN is the slow processing due to its sequential nature. In this paper, we propose a Filter based Feature Engineering method and a two-phase auto hyperparameter optimization method, which fit very well to clinical time series scenario. Combined with two widely used tree boosting methods: XGBoost and LightGBM, we demonstrated that our approach achieved the state-of-the-art results with more than 100X speed

The first two authors contribute equally to this work

Y. Hu (✉) · R. Subramanian
Humana, Irving, TX 75063, USA
e-mail: yhu@humana.com

R. Subramanian
e-mail: RSubramanian5@humana.com

W. An
Tsinghua University, Beijing 100084, China
e-mail: anwangpeng@gmail.com

N. Zhao
Peking University School and Hospital of Stomatology, Beijing 100081, China
e-mail: nanamozhao88@gmail.com

Y. Gu
Suning USA, Palo Alto, CA 94304, USA
e-mail: yang.gu@ussuning.com

W. Wu
Department of Computer Science, The University of Texas at Dallas, Richardson, TX 75080, USA
e-mail: weiliwu@utdallas.edu

© The Editor(s) (if applicable) and The Author(s), under exclusive license to Springer Nature Switzerland AG 2021

A. Shaban-Nejad et al. (eds.), *Explainable AI in Healthcare and Medicine*,
Studies in Computational Intelligence 914,
https://doi.org/10.1007/978-3-030-53352-6_23

acceleration compared with RNN methods on two MIMIC-III benchmark tasks: In Hospital Mortality Prediction and 25-Phenotype Classification. Due to its superior accuracy and faster speed advantages, our approach has broad clinical application prospect, especially assisting doctors to make right diagnosis and treatment prognosis in shorter invaluable time.

Keywords Time series · Deep learning · Tree boosting · Hyperparameter optimization

1 Introduction

Electronic Health Record (EHR) adoption in the US has increased from 13% to more than 90% in the past decade since 2009's Health Information Technology for Economic and Clinical Health (HITECH) Act [1]. EHR contains rich text, visual and time series information such as a patient's medical and diagnose history, radiology images, etc which is the major source for managing and predicting a patient's health status. Meanwhile, "`Deep Learning`" has been a buzz word since its big success in ImageNet 2012 competition [2], which has greatly pushed forward the research frontier of computer vision, speech recognition and natural language processing since then. There is an increasing interest in applying the state-of-the-art deep learning techniques to healthcare industry, especially EHR, from the combined effort of industry and academia. In 2012, IBM started to apply their Watson DeepQA technology to the diagnostic support from a patient's EHR and claim data, after this technology beat the highest ranked human players in the open-domain question answering show - Jeopardy! [3]. In 2018, Amazon unleashed their HIPAA-compliant language processing service called Comprehend Medical, which can help make clinical decisions, identify a patient's symptoms and reduce cost from unstructured medical data [4]. In 2019, the United States Patent Office published a patent application from Google [5], which unveiled its intention to build predictive medicare service with EHR data.

A big portion of these EHR applications belong to the category of multivariate time series classification, such as Intensive Care Unit (ICU) mortality prediction, physiologic decompensation prediction, ICU length of stay prediction, phenotype classification and so on. Traditional approaches for clinical time series classification tasks involve feature engineering on timestamp attributes and then applying task-specific classification or regression models [6]. Later, Recurrent Neural Network (RNN) based approaches such as Long Short-Term Memory (LSTM) [7, 15] demonstrated to be powerful even when being trained on raw time series data without feature engineering [8]. Most recently, as RNN architectures being criticized as less effective in parallel computing and time consuming, certain attention based modeling architectures were proposed and evaluated to achieve the state-of-the-art performance [9].

Despite more sophisticated model architectures being applied, the accuracy gain of these deep learning approaches is very slight. Moreover, the speed performance analysis of these deep learning approaches is missing. Practically, if a model costs too much time in training and inference but only gives a little gain in the accuracy improvement, it may not be considered as a good option for the machine learning system. In this paper, we propose a Filter based Feature Engineering method and a two-phase auto hyperparameter optimization method, which fit very well for clinical time series scenario. Combined with two widely used tree boosting methods: XGBoost [10] and LightGBM [11], we demonstrated that our approach achieved the state-of-the-art results with more than 100X speed acceleration compared with RNNs such as LSTM and Gated Recurrent Unit (GRU) [12, 16] on two MIMIC-III benchmark tasks: In Hospital Mortality Prediction and 25-Phenotype Classification [13]. The major contributions of this work are summarized as the following:

- We proposed an efficient Filter based Feature Engineering method and a two-phase auto hyperparameter optimization method, which fit very well to clinical time series scenario.
- Combined with two widely used tree boosting methods: XGBoost and LightGBM, We demonstrated that our approach achieved the state-of-the-art results on two MIMIC-III benchmark tasks: In Hospital Mortality Prediction and 25-Phenotype Classification.
- We conducted detailed speed performance analysis with LSTM, GRU and our proposed approach on the two MIMIC-III benchmark tasks, and demonstrated that our approach achieves more than 100X speed acceleration in training and inference, which means doctors could make right diagnosis and treatment prognosis in shorter invaluable time.

2 Related Work

Clinical time series classification is challenging due to irregular distribution of the sampling, missing values and wrong timestamp measurements. Recent year research shows that deep learning methods nearly always outperform traditional machine learning methods such as logistic regression, MultiLayer Perceptron (MLP), etc. The seminal work of applying LSTM to clinical time series data was by Lipton et al. [8], which demonstrated that a simple LSTM network with additional training strategies can outperform several strong baselines. With the growing interest and need of reproducibility of published methods to EHR data, Medical Information Mart for Intensive Care (MIMIC-III) database [14] became the widely accepted public dataset for evaluating competing methods, due to its large size of de-identified clinical data of patients admitted to a single-center Intensive Care Unit (ICU) such as vital signs, medications, laboratory measurements, imaging reports, and more. Based on MIMIC-III dataset, [13] proposed four clinical time series analysis task including in hospital mortality prediction, physiological decompensation prediction, length of stay prediction and 25-phenotype classification. They benchmarked and demonstrated the performance advantages of LSTM on these four tasks toward the traditional methods such as logistic regression, and joint training LSTM on the four tasks will further improve the performance. Most recently, [9] proposed the first solely attention based sequence modeling architecture for multivariate time series classification, and demonstrated its performance advantages on the four MIMIC-III benchmark tasks.

3 MIMIC-III Benchmark Task

In this study, we used the MIMIC-III v1.4, which was released on September 2016. The database contains a cohort of 46520 unique patients with a total of 58976 admissions. We followed [13] to transform the data from original format to time series format. The sample input data for these two tasks are quite similar, which is 17 vital signs in time sequence order. Figure 1(a) shows a positive sample of in hospital mortality, with obvious vital signs out of normal range. Figure 1(b) shows a negative sample of in hospital mortality, with normal vital signs. Figure 1(c) shows that

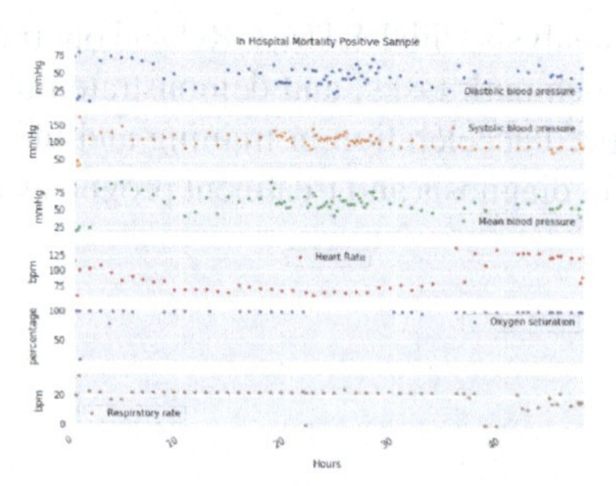

(a) In Hospital Mortality Positive Sample

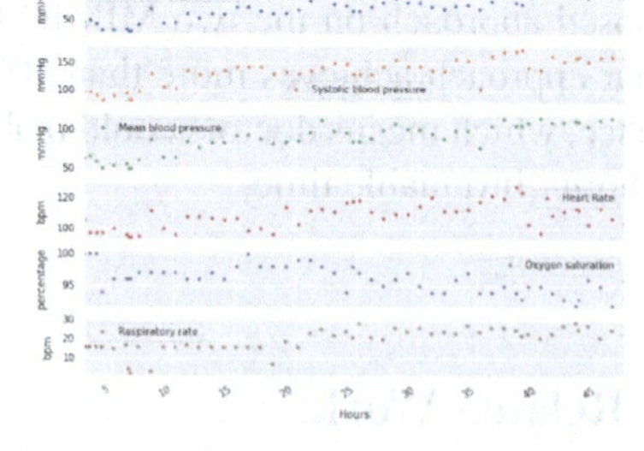

(b) In Hospital Mortality Negative Sample

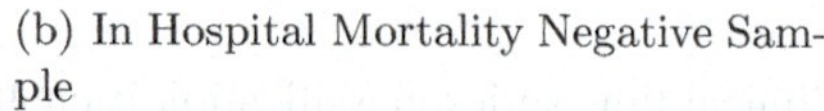

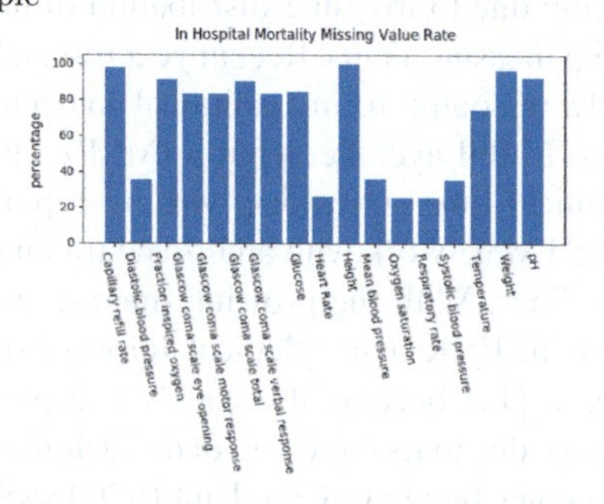

(c) In Hospital Mortality Missing Value Rate

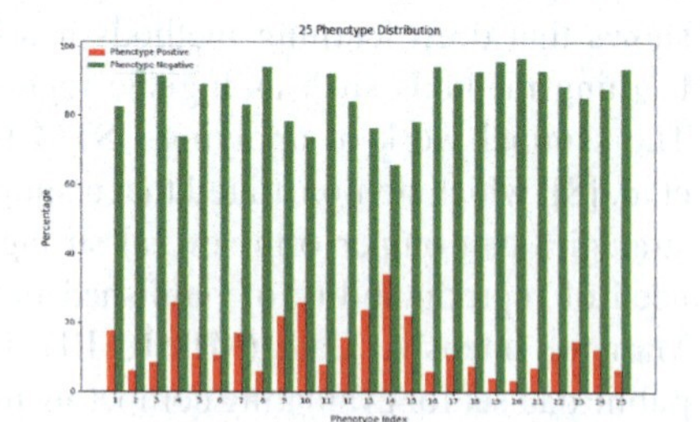

(d) 25 Phenotype Distribution

Fig. 1 MIMIC III Data

11 vital signs have huge percentage of missing values for the in hospital mortality prediction task, which is a similar case for 25-phenotype classification task.

3.1 In Hospital Mortality Prediction

This benchmark task is to predict the in hospital mortality from clinical time series variables recorded in the first 48 h of the ICU admission. ICU mortality rates are the highest among hospital units (10% to 29%). This is a binary classification task, with the ground truth label indicate whether the patient died before the hospital discharge. *In-hospital-mortality* dataset is generated with the root cohort with further excluding all ICU stays whose *Length-of-Stay* is unknown or less than 48 h. Our training dataset contains 17939 samples, validation dataset contains 3222 samples and test dataset contains 3236 samples. The ground truth label is determined by checking if the patient's date of death is between the ICU admission and discharge time. The overall mortality rate in the dataset is 11.60% (2830 of 24397 ICU stays). Since it's a very imbalanced labeled dataset, we use 3 metrics for the evaluation: (i) Area Under Receiver Operator Curve ($AUROC$), (ii) Area Under Precision-Recall Curve ($AUPRC$), and (iii) Minimum of Precision and Sensitivity ($Min(Se, P+)$).

3.2 25 Acute Care Phenotype Classification

This benchmark task is to detect if the patient has any of 25 conditions that are common in adult ICU from clinical time series variables recorded in a single ICU stay episode. These 25 conditions contain 12 critical conditions, such as respiratory failure, 8 chronic conditions, such as diabetes, and 5 mixed conditions since they are chronic with periodic acute episode. Because more than 99% of patients in the benchmark dataset have more than one diagnosis, this task is formulated as a multi-label binary classification. The phenotype labels are the codes in MIMIC-III ICD-9 diagnosis table, and we only consider the 25 categories matching their HCUP CCS categories. Since MIMIC-III ICD-9 codes are associated with hospital visits, not ICU stays, this benchmark task excludes the hospital admissions with multiple ICU stays for reducing the ambiguity samples: We only consider the samples that have only one ICU stay per hospital admission. Our training dataset contains 36020 samples, validation dataset contains 6371 samples and test dataset contains 6281 samples. As Fig. 1(d) shows, this is also a very imbalanced labeled dataset, so we use 3 metrics for the evaluation: (i) micro-averaged Area Under the ROC Curve($AUROC$), which computes single $AUROC$ irrespective of the categories, (ii) macro-averaged $AUROC$, which averages $AUROC$ per label, (iii) weighted $AUROC$, which considers each disease prevalence.

4 Methods

Figure 2 describes the method architecture in this work. Each sample input data is 17 clinical variables in the time sequence order. These 17 clinical variables will be further processed into a one dimension vector of length 76, after the categorical variables are encoded using a one-hot vector, and the numeric variables are standardized by subtracting the mean and dividing by the standard deviation. The final raw data sample input now can be treated as a two dimension vector. Deep learning methods are very good at processing the raw data input and producing the features with their internal structure. Tree boosting methods can also take raw data sample as the input with necessary data reshaping, but to get better accuracy, appropriate feature engineering strategies have to be employed. In hospital mortality prediction has the binary classification ground truth for each sample input, while 25-phenotype classification has the multiple label binary classification ground truth for each sample input.

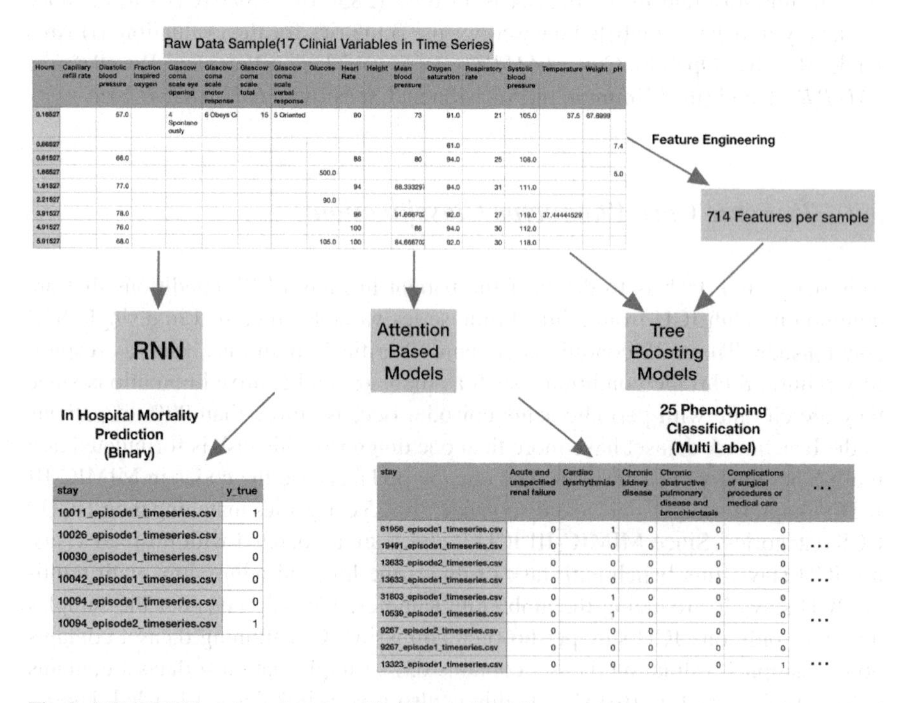

Fig. 2 MIMIC-III benchmark tasks with different Models

4.1 RNN on Raw Data

Given a series of observations $\{x_t\}_{t\geq 1}^{T}$, an LSTM model learns to generate the prediction $\hat{y}$ of the ground truth y. Here t stands for a timestamp, and T stands for the length of the series.

$$\begin{cases} u_t = \sigma(x_t W^{(xu)} + h_{t-1} W^{(hu)}), \\ f_t = \sigma(x_t W^{(xf)} + h_{t-1} W^{(hf)}), \\ o_t = \sigma(x_t W^{(xo)} + h_{t-1} W^{(ho)} + b^{(o)}), \\ c_t = f_t \odot c_{t-1} + \\ \quad u_t \odot tanh(x_t W^{(xc)} + h_{t-1} W^{(hc)} + b^{(c)}), \\ h_t = o_t \odot \sigma(c_t), \end{cases} \tag{1}$$

Here $h_0 = 0$, and the σ (sigmoid) and $tanh$ are element-wise functions. As the formula (1) shows, a LSTM unit has three gates: update gate, forget gate and output gate. Based on the hidden states $\{h_t\}_{t\geq 1}^{T}$, we add a task specific layer.

$$\hat{y} = \sigma(w^{(y)} h_t + b^{(y)}) \tag{2}$$

GRU model is using the following formula for its forward pass procedure:

$$\begin{cases} u_t = \sigma(x_t W^{(xu)} + c_{t-1} W^{(hu)}), \\ r_t = \sigma(x_t W^{(xr)} + c_{t-1} W^{(hr)}), \\ c_t = (1 - u_t) \odot c_{t-1} + \\ \quad u_t \odot tanh(x_t W^{(xc)} + r_t \odot c_{t-1} W^{(hc)} + b^{(c)}), \\ h_t = o_t \odot \sigma(c_t), \end{cases} \tag{3}$$

As we can see, GRU is not using separate memory cells, and it also use less gates (only update gate and reset gate). In practice, GRU can achieve the LSTM comparable accuracy on many use cases with faster speed (Chung et al., 2014).

4.2 Filter Based Feature Engineering (FBFE)

RNN is good at capturing the long range dependencies of time series data, but it also brings one drawback: data has to be processed time stamp by time stamp both in training phase and inference phase, which negatively affect the speed performance. In clinical time series scenario, missing observations are very common, so the input sample matrix is very sparse. The past study [13] utilized the subsequence and sub-timeframe based feature engineering method with the logistic regression. For any given time series input sample, they computed six different sample statistic features (minimum, maximum, mean, standard deviation, skew and number of measurements) on seven different subsequences (full time series, first 10% of time, first 25% of time, first 50% of time, last 50% of time, last 25% of time, and last 10% of time). Thus each time series input sample will generate $17 \times 7 \times 6$ features. This method mainly captures the statistic attribute of the data, but not the sequence dependency, so its accuracy performance is always outperformed by deep learning approaches, moreover, its running time is non-trivial. Here we propose a Filter based Feature Engineering method from the inspiration of the filters in Convolutional Neural Networks(CNN). We take the time series input sample after the one-hot encoding and standardization processing, which is essentially a $T \times 76$ matrix, then we apply convolution operation with a $M \times N$ filter matrix to the input sample, and reshape the output matrix into a one dimension feature vector. Experiments show that our approach can generate more fine-grained features with faster speed.

4.3 Tree Boosting Methods with FBFE

Tree Boosting algorithms work on generated features from each sample. After the feature engineering of one time series sample $\{x_t\}_{t\geq 1}^{T}$, it turns into $\{x_i'\}(x_i' \in R^m)$, where m stands for the number of the features. A tree boosting model uses K additive functions to predict the output.

$$\hat{y}_i = \phi(x_i') = \sum_{k=1}^{K} f_k(x_i'),\ f_k \in F \tag{4}$$

where F is the space of regression tree. For any given sample $\{x_i'\}(x_i' \in R^m)$, the decision rules in ϕ classify it into the leaves and summing up all the scores in the corresponding leaves into a final prediction. The model training is to minimize the following regularized objective.

$$L(\phi) = \sum_i l(\hat{y}_i, y_i) + \sum_k \Omega(f_k) \tag{5}$$

Here l is the loss function that measures that difference between the prediction $\hat{y}_i$ and the ground truth y_i. The function Ω penalizes the complexity of the whole model, which helps smooth the learnt weights to prevent over-fitting.

Two tree boosting implementations we used in this study are XGBoost and LightGBM. The major difference is their ways of splitting tree nodes. XGboost's original implementation was based on the pre-sorted algorithm, which is to find the best split node based on its pre-sorted feature values. LightGBM adopted a histogram-based algorithm, which is to bucket continuous feature values into discrete bins and use them to construct feature histograms during training.

4.4 Two-Phase Auto Hyperparameter Optimization

The performance of most machine learning algorithms hinges on their hyperparameters, which are used to control the training process and set by data scientist before training. For example, RNN methods are sensitive to hyperparameters like learning rate, number of depths, dropout rates, batch size, etc. The performance of tree boosting methods mainly depends on hyperparameters like learning rate, maximum depth, number of leaves in one tree, maximum number of trees, etc.

Given a machine learning algorithm A, which has hyperparameters λ_1, ..., λ_n with respective domains Λ_1, ..., Λ_n. We define its hyperparameter space as $\Lambda = \Lambda_1 \times \cdots \times \Lambda_n$. We use A_λ to denote the learning algorithm A using the hyperparamter setting $\lambda \in \Lambda$, and $l(\lambda) = \mathcal{L}(A_\lambda, \mathcal{D}_{train}, \mathcal{D}_{valid}))$ to denote the validation loss that A_λ achieves on data $\mathcal{D}_{valid}$ after trained on data $\mathcal{D}_{train}$. Our goal is to find the $\lambda \in \Lambda$ that can minimize $l(\lambda)$.

Most commonly used auto hyperparameter optimization methods include grid search, random search [17] and bayesian optimization [18]. In this study, we divide the tree boosting auto hyperparameter optimization process into two phases as described in Algorithm 1: In the first phase, we use random search to evaluate the best range of each hyperparameter; In the second phase, we use grid search on a fine-grained scale.

Algorithm 1 Two-Phase Auto Hyperparameter Optimization for Tree Boosting Methods

Input: Target function l; limit T_1, T_2; hyperparameter space Λ;
Result: Best byperparameter configuration in this process λ^*

1: **for** $i = 1$ to T_1 **do**
2: $\quad Random\ Search\ \lambda_i$
3: $\quad \hat{\lambda_i} \leftarrow Evaluate\ l(\lambda_i)$
4: **end for**
5: **for** $j = 1$ to T_2 **do**
6: $\quad Grid\ Search\ \lambda_j\ from\ \hat{\Lambda}$
7: $\quad \hat{\lambda_j} \leftarrow Evaluate\ l(\lambda_j)$
8: **end for**
9: **return** $\lambda^* \in arg\,min_{\lambda_j \in \{\lambda_1, \ldots, \lambda_{T2}\}}$

5 Experiments

In this section, we compare the accuracy and speed performance between deep learning methods and our proposed approach on two MIMIC-III benchmark tasks: In Hospital Mortality Prediction and 25-Phenotype Classification. Our experiments were conducted on a server with hardware configuration: CPU Intel® Core™ i7-8700K CPU @ 3.70 GHz 12, memory: 32 GB, GPU: GeForce GTX 1080 Ti/PCIe/SSE2.

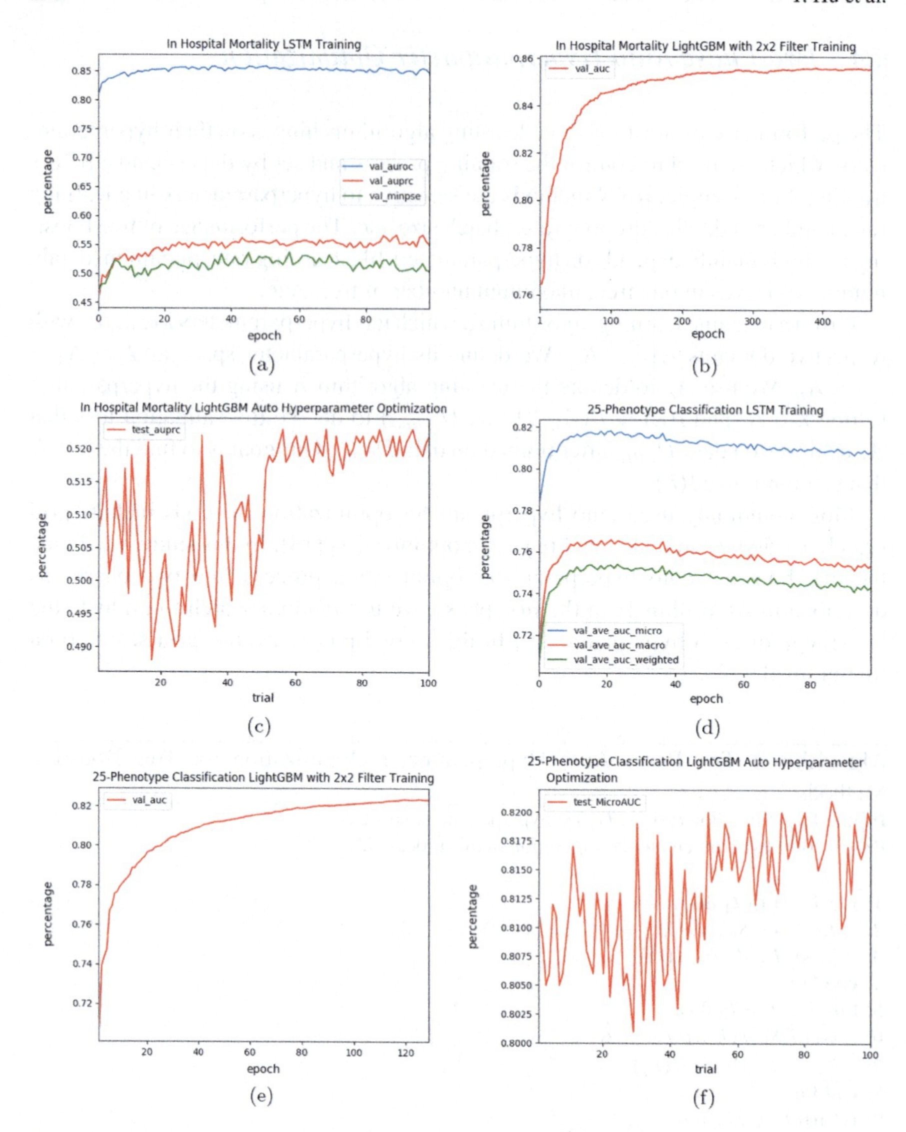

Fig. 3 Experiments

5.1 In Hospital Mortality Prediction

Figure 3(a) shows training a one layer LSTM model for 100 epochs, with input dimension of 16, dropout rate of 0.3, batch size of 8. The model converges at the 25th epoch. For this trial, it takes around 2125 s to find the best model. Figure 3(b) shows training a LightGBM model with a 2×2 dimension filter, with the early stopping round of 80, number of leaves of 11, learning rate of 0.07, number of estimators of

10000. The model converges at the 390th epoch. For this trial, it takes around 61 s to find the best model. Figure 3(c) shows the auto hyperparameter optimization process for LightGBM. In the first 50 trials, it does random search according to the $AUPRC$ of the test set and estimate the trial range of early stopping round as [50, 100], the trial range of number of leavs as [5, 15], the trial range of learning rate as $[0.01, 0.1]$, and the trial range of number of estimators as [1000, 40000]. In the second 50 trials, it does the grid search and find several best hyperparameter configurations that can make the model achieve $AUPRC$ of **0.523** on the test set, beating the previous state-of-the-art $AUPRC$ of **0.518**. One of these configurations with the early stopping round as 80, number of leaves as 11, learning rate as 0.07, number of estimators as 10000 can make the model achieve $min(Se, P+)$ of **0.508** on the test set, beating the previous state-of-the-art $min(Se, P+)$ of **0.500**.

Table 1 shows the accuracy performance of our methods and previously published works. Table 2 shows averaged training epoch time and averaged test inference time for the single layer LSTM model, the single layer GRU model, XGBoost model and LightGBM model on this use case. As we can see, for the MIMIC-III in-hospital-mortality prediction benchmark task, tree boosting methods are more than 100X faster than RNN methods on both training speed and inference speed. Though tree boosting methods may spend more epochs to reach the best model, but the overall training time of that is still around $\frac{1}{30}$ of RNN methods.

5.2 25 Phenotype Classification

Figure 3(d) shows training a one layer LSTM model for 100 epochs, with input dimension of 256, dropout rate of 0.3, batch size of 8. The model reached the best

Table 1 Accuracy Comparison for MIMIC-III In Hospital Mortality Prediction

Metrics	AUROC	AUPRC	min(Se, P+)
LSTM [13]	0.854	0.516	0.491
SAndD [9]	**0.857**	0.518	0.5
FBFE_XGBoost(Ours)	0.856	0.517	0.487
FBFE_LightGBM(Ours)	0.850	**0.523**	**0.508**

Table 2 Speed Comparison for MIMIC-III In Hospital Mortality Prediction

Metrics	Training Epoch (s)	Inference on Test (s)
LSTM_1	85	4.292
GRU_1	65	3.443
FBFE_XGBoost	0.422	0.048
FBFE_LightGBM	**0.156**	**0.006**

accuracy at the 15th epoch, and then started to overfit on the training set. For this trial, it takes around 25200 s to find the best model. Figure 3(e) shows training a LightGBM model with a 2×2 dimension filter, with the early stopping round of 10, number of leaves of 12, learning rate of 0.08, number of estimators of 10000. The model converges at the 119th epoch. For this trial, it takes around 134 s to find the best model. Figure 3(f) shows the auto hyperparameter optimization process for LightGBM. In the first 50 trials, it does random search according to the $MicroAUC$ of the test set and estimate the trial range of early stopping round as [5, 30], the trial range of number of leaves as [10, 20], the trial range of learning rate as [0.05, 0.1], and the trial range of number of estimators as [1000, 30000]. In the second 50 trials, it does the grid search and find one hyperparameter configuration that can train the model to tie the previous state-of-the-art on Micro-Averaged $AUROC$ of **0.821** and Macro-Averaged $AUROC$ of **0.770** on the test set, and beat the previous state-of-the-art on Weighted $AUROC$ of **0.757** by a little gap with Weighted $AUROC$ of **0.760**

Table 3 shows the accuracy performance of our methods and previously published works. Table 4 shows averaged training epoch time and averaged test inference time for the single layer LSTM model, the single layer GRU model, XGBoost model and LightGBM model on this use case. As we can see, for the MIMIC-III 25-phenotype classification benchmark task, tree boosting methods are more than 100X faster than RNN methods on both training speed and inference speed. Though tree boosting methods may spend more epochs to reach the best model, but the overall training time of that is still around $\frac{1}{180}$ of RNN methods.

Table 3 Accuracy Comparison for MIMIC-III 25-Phenotype Classification

Metrics	Micro AUC	Macro AUC	Weighted AUC
LSTM [13]	**0.821**	**0.770**	0.757
SAndD [9]	0.816	0.766	0.754
FBFE_XGBoost(Ours)	0.819	0.768	0.758
FBFE_LightGBM(Ours)	**0.821**	**0.770**	**0.760**

Table 4 Speed Comparison for MIMIC-III 25-Phenotype Classification

Metrics	Training Epoch (s)	Inference on Test (s)
LSTM_1	1680	17.466
GRU_1	1330	14.139
FBFE_XGBoost	6.160	0.919
FBFE_LightGBM	**1.124**	**0.154**

6 Conclusion

Although the recent studies have shown that deep learning approaches have achieved the state-of-the-art results on several clinical time series tasks in the aspect of accuracy performance, yet their speed performance analysis is missing. Practically, if a model costs too much time in training and inference but only gives a little gain in the accuracy improvement, it may not be considered as a good option for the machine learning system. In this work, we developed an efficient Filter based Feature Engineering method and a two-phase auto hyperparameter optimization method, which fit very well to clinical time series scenario. Combined with two widely used tree boosting methods: XGBoost and LightGBM, we demonstrated that our approach achieved the state-of-the-art results with more than 100X faster speed compared with RNN methods on two MIMIC-III benchmark tasks: In Hospital Mortality Prediction and 25-Phenotype Classification. Due to its superior accuracy and faster speed advantages, our approach has broad clinical application prospect, especially assisting doctors to make right diagnosis and treatment prognosis in shorter invaluable time. In the future, we will continue improving this approach and apply it to broader clinical time series use cases.

References

1. Ann, K.-L., Douglas, M.S., Kathryn, H.B., Lawton, R.B., Linda, H.A.: Electronic health record adoption and nurse reports of usability and quality of care: the role of work environment. Appl. Clin. Inform. **10**(01), 129–139 (2019)
2. Krizhevsky, A., Sutskever, I., Hinton, G.: Imagenet classification with deep convolutional neural networks. In: NIPS (2012)
3. Ferrucci, D., Levas, A., Bagchi, S., Gondek, D., Mueller, E.T.: Watson: beyond jeopardy!. Artif. Intell. **199**, 93–105 (2013)
4. Amazon: Amazon Comprehend Medical (2018)
5. Google: System and Method for Predicting and Summarizing Medical Events from Electronic Health Records. US Patent. 2019/0034591 A1 (2019)
6. Fonarow, G.C., Adams, K.F., Abraham, W.T., Yancy, C.W., Boscardin, W.J.: ADHERE Scientific Advisory Committee, Study Group, and Investigators. Risk stratification for in-hospital mortality in acutely decompensated heart failure: classification and regression tree analysis. JAMA **293**, 572–580 (2005). https://doi.org/10.1001/jama.293.5.572.
7. Hochreiter, S., Schmidhuber, J.: Long short-term memory. Neural Comput. **9**(8), 1735–1780 (1997)
8. Lipton, Z.C., Kale, D.C., Elkan, C., Wetzell, R.: Learning to diagnose with LSTM recurrent neural networks. arXiv preprint arXiv:1511.03677 (2015)
9. Song, H., Rajan, D., Thiagarajan, J.J., Spanias, A.: Attend and diagnose: clinical time series analysis using attention models. In: AAAI (2018)
10. Chen, T., Guestrin, C.: Xgboost: a scalable tree boosting system. In: Proceedings of the 22nd ACM SIGKDD International Conference on Knowledge Discovery and Data Mining, pp. 785–794. ACM (2016)
11. Ke, G., Meng, Q., Finley, T., Wang, T., Chen, W., Ma, W., Ye, Q., Liu, T.-Y.: LightGBM: a highly efficient gradient boosting decision tree. Adv. Neural Inf. Process. Syst. **15**, 3149–3157 (2017)

12. Cho, K., van Merrienboer, B., Gulcehre, C., Bougares, F., Schwenk, H., Bengio, Y.: Learning phrase representations using RNN encoder-decoder for statistical machine translation. In: EMNLP (2014)
13. Harutyunyan, H., Khachatrian, H., Kale, D.C., Gal-styan, A.: Multitask learning and benchmarking with clinical time series data. arXiv preprint arXiv:1703.07771 (2017)
14. Johnson, A.E.W., Pollard, T.J., Shen, L., Lehman, L.H., Feng, M., Ghassemi, M., Moody, B., Szolovits, P., Celi, L.A., Mark, R.G.: MIMIC-III, a freely accessible critical care database. Sci. Data **3**, 160035 (2016)
15. Gers, F.A., Schmidhuber, J., Cummins, F.: Learning to forget: continual prediction with LSTM. Neural Comput. **12**(10), 2451–2471 (2000)
16. Chung, J., Güçehre, C., Cho, K., Bengio, Y.: Empirical evaluation of gated recurrent neural networks on sequence modeling. arXiv preprint arXiv:1412.3555 (2014)
17. Bergstra, J., Bengio, Y.: Random search for hyperparameter optimization. JMLR **13**(1), 281–305 (2012)
18. Bergstra, J., Yamins, D., Cox, D.D.: Making a science of model search: hyperparameter optimization in hundreds of dimensions for vision architectures. In: Proceedings of ICML, pp. 115–123 (2013)

Explaining Models by Propagating Shapley Values of Local Components

Hugh Chen, Scott Lundberg, and Su-In Lee

Abstract In healthcare, making the best possible predictions with complex models (e.g., neural networks, ensembles/stacks of different models) can impact patient welfare. In order to make these complex models explainable, we present DeepSHAP for mixed model types, a framework for layer wise propagation of Shapley values that builds upon DeepLIFT (an existing approach for explaining neural networks). We show that in addition to being able to explain neural networks, this new framework naturally enables attributions for stacks of mixed models (e.g., neural network feature extractor into a tree model) as well as attributions of the loss. Finally, we theoretically justify a method for obtaining attributions with respect to a background distribution (under a Shapley value framework).

1 Introduction

Neural networks and ensembles of models are currently used across many domains. For these complex models, explanations accounting for how features relate to predictions are often desirable and at times mandatory [1]. In medicine, explainable AI (XAI) is important for scientific discovery, transparency, and much more [2]

H. Chen (✉) · S.-I. Lee
Paul G. Allen School of CSE, University of Washington, Seattle, USA
e-mail: hughchen@cs.washington.edu

S.-I. Lee
e-mail: suinlee@cs.washington.edu

S. Lundberg
Microsoft Research, Albuquerque, USA
e-mail: scottmlundberg@gmail.com

© The Editor(s) (if applicable) and The Author(s), under exclusive license to Springer Nature Switzerland AG 2021

A. Shaban-Nejad et al. (eds.), *Explainable AI in Healthcare and Medicine*, Studies in Computational Intelligence 914,
https://doi.org/10.1007/978-3-030-53352-6_24

(examples include [3–5]). One popular class of XAI methods is per-sample feature attributions (i.e., importance values for each feature for a given prediction).

In this paper, we focus on SHAP values [6] – Shapley values [7] with a conditional expectation of the model prediction as the set function. Shapley values are the only additive feature attribution method that satisfies the desirable properties of local accuracy, missingness, and consistency. In order to approximate SHAP values for neural networks, we fix a problem in the original formulation of DeepSHAP [6] that previously used $E[x]$ as the reference by justifying a new method to create explanations relative to background distributions. Furthermore, we extend it to explain stacks of mixed model types as well as loss functions rather than margin outputs.

Popular model agnostic explanation methods that also aim to obtain SHAP values are Kernel SHAP [6] and IME [8]. Unfortunately, model agnostic methods are sampling based and consequently high variance or slow. Alternatively, local feature attributions targeted to deep networks have been addressed in numerous works: Occlusion [9], Saliency Maps [10], Layer-Wise Relevance Propagation [11], DeepLIFT, Integrated Gradients (IG) [12], and Generalized Integrated Gradients (GIG) [13].

Of these methods, IG and GIG are connected to Shapley Values. IG integrates gradients between a baseline and a sample being explained, approaching the Aumann-Shapley value. GIG is a generalization of IG to explain losses and mixed model types. IG and GIG have two downsides: 1.) integrating along a path can be expensive or imprecise and 2.) the Aumann-Shapley values fundamentally differ to the SHAP values we aim to approximate. Finally, DASP [14] is an approach that approximates SHAP values for deep networks. This approach works by replacing point activations at all layers by probability distributions and requires many more model evaluations than DeepSHAP. Furthermore, because DASP aims to obtain the same SHAP values as in DeepSHAP it is possible to use DASP as a part of the DeepSHAP framework.

2 Approach

2.1 *Propagating SHAP Values*

DeepSHAP builds upon DeepLIFT; in this section we aim to better understand how DeepLIFT's rules connect to SHAP values. This has been briefly touched upon in [15] and [6], but here we explicitly define the relationship.

DeepSHAP is a method that explains a sample (foreground sample), by setting features to be "missing". Missing features are set to corresponding values in a baseline sample (background sample). Note that DeepSHAP generally uses a background distribution, however focusing on a single background sample is sufficient because we can rewrite the SHAP values as an average over attributions with respect to a single background sample at a time (see next section for more details). In this section, we define a foreground sample to have features x f_{x_i} and neuron values f_h (obtained by

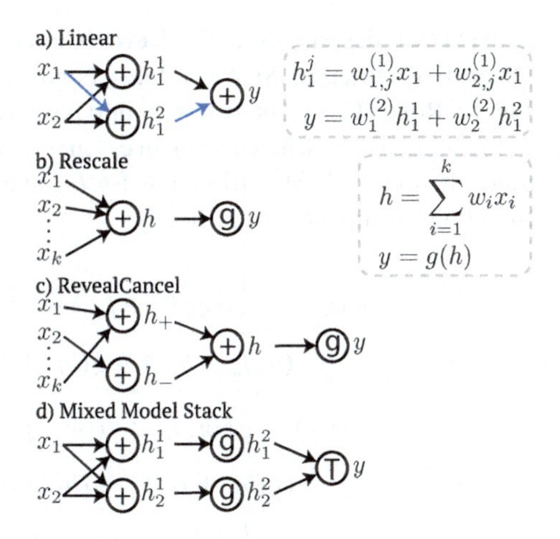

Fig. 1 *Visualization of models for understanding DeepLIFT's connection to SHAP values*. In the figure g is a non-linear function and T is a non-differentiable tree model

a forward pass) and a background sample to have b_{x_i} or b_h. Finally we define $\phi(\cdot)$ to be attribution values.

If our model is **fully linear** as in Fig. 1a, we can get the exact SHAP values for an input x_i by summing the attributions along all possible paths between that input x_i and the model's output y. Therefore, we can focus on a particular path (in blue). Furthermore, the path's contribution to $\phi(x_i)$ is exactly the product of the weights along the path and the difference in x_1: $w_2^{(2)}w_{1,2}^{(1)}(f_{x_1} - b_{x_1})$, because we can rewrite the layers of linear equations in 1a as a single linear equation. Note that we can derive the attribution for x_1 in terms of the attribution of intermediary nodes (as in the chain rule):

$$\begin{aligned}\phi(h_1^2) &= w_2^{(2)}(f_{h_1^2} - b_{h_1^2})\\ \phi(x_1) &= \frac{\phi(h_1^2)}{f_{h_1^2} - b_{h_1^2}} w_{1,2}^{(1)}(f_{x_1} - b_{x_1})\end{aligned} \tag{1}$$

Next, we move on to reinterpreting the two variants of DeepLIFT: the Rescale rule and the RevealCancel rule. First, a gradient based interpretation of the **Rescale rule** has been discussed in [16]. Here, we explicitly tie this interpretation to the SHAP values we hope to obtain.

For clarity, we consider the example in Fig. 1b. First, the attribution value for $\phi(h)$ is $g(f_h) - g(b_h)$ because SHAP values maintain local accuracy (sum of attributions equals $f_y - b_y$) and g is a function with a single input. Then, under the Rescale rule, $\phi(x_i) = \frac{\phi(h)}{f_h - b_h} w_i(f_{x_i} - b_{x_i})$ (note the resemblance to Eq. (1)). Under this formulation it is easy to see that the Rescale rule first computes the exact SHAP value for h and then propagates it back linearly. In other words, the non-linear and linear functions are treated as separate functions. Passing back nonlinear attributions linearly is clearly an approximation, but confers two benefits: 1.) fast computation on order of a backward pass and 2.) a guarantee of local accuracy.

Next, we describe how the **RevealCancel rule** (originally formulated to bring DeepLIFT closer to SHAP values) connects to SHAP values in the context of Fig. 1c. RevealCancel partitions x_i into positive and negative components based on if $w_i(f_{x_i} - b_{x_i}) < t$ (where $t = 0$), in essence forming nodes h_+ and h_-. This rule computes the exact SHAP attributions for h_+ and h_- and then propagates the resultant SHAP values linearly. Specifically:

$$\phi(g_+) = \frac{1}{2}((g(f_{h_+}+f_{h_-}) - g(b_{h_+}+f_{h_-}) + (g(f_{h_+}+b_{h_-}) - g(b_{h_+}+b_{h_-}))$$

$$\phi(g_+) = \frac{1}{2}((g(f_{h_+}+f_{h_-}) - g(f_{h_+}+b_{h_-}) + (g(b_{h_+}+f_{h_-}) - g(b_{h_+}+b_{h_-}))$$

$$\phi(x_i) = \begin{cases} \frac{\phi_{h+}}{f_{h+}-b_{h+}} w_i(f_{x_i} - b_{x_i}), & \text{if } w_i(f_{x_i} - b_{x_i}) > t \\ \frac{\phi_{h-}}{f_{h-}-b_{h-}} w_i(f_{x_i} - b_{x_i}), & \text{otherwise} \end{cases}$$

Under this formulation, we can see that in contrast to the Rescale rule that explains a linearity and nonlinearity by exactly explaining the nonlinearity and backpropagating, the RevealCancel rule exactly explains the nonlinearity and a partition of the inputs to the linearity as a single function prior to backpropagating. The RevealCancel rule incurs a higher computational cost in order to get a an estimate of $\phi(x_i)$ that is ideally closer to the SHAP values.

This reframing naturally motivates explanations for **stacks of mixed model types**. In particular, for Fig. 1d, we can take advantage of fast, exact methods for obtaining SHAP values for tree models to obtain $\phi(h_j^2)$ using Independent Tree SHAP [3]. Then, we can propagate these attributions to get $\phi(x_i)$ using either the Rescale or RevealCancel rule. This argument extends to explaining losses rather than output margins as well.

Although we consider specific examples here, the linear propagation described above will generalize to arbitrary networks if SHAP values can be computed or approximated for individual components.

2.2 *SHAP Values with a Background Distribution*

Note that many methods (Integrated Gradients, Occlusion) recommend the utilization of a single background/reference sample. In fact, DeepSHAP as previously described in [6] created attributions with respect to a single reference equal to the expected value of the inputs. However, in order to obtain SHAP values for a given background distribution, we prove that the correct approach is as follows: obtain SHAP values for each baseline in your background distribution and average over the resultant attributions. Although similar methodologies have been used heuristically [15, 17], we provide a theoretical justification in Theorem 1 in the context of SHAP values.

Theorem 1 *The average over single reference SHAP values approaches the true SHAP values for a given distribution.*

Proof Define D to be the data distribution, N to be the set of all features, and f to be the model being explained. Additionally, define $\mathcal{X}(x, x', S)$ to return a sample where the features in S are taken from x and the remaining features from x'. Define C to be all combinations of the set $N \setminus \{i\}$ and P to be all permutations of $N \setminus \{i\}$. Starting with the definition of SHAP values for a single feature: $\phi_i(x)$

$$
\begin{aligned}
&= \sum_{S \in C} W(|S|, |N|)(\mathbb{E}_D[f(X)|x_{S \cup \{i\}}] - \mathbb{E}_D[f(X)|x_S]) \\
&= \frac{1}{|P|} \sum_{S \subseteq P} \mathbb{E}_{\mathcal{D}}[f(x)|\text{do}(x_{S \cup \{i\}})] - \mathbb{E}_{\mathcal{D}}[\text{do}(f(x)|x_S)] \\
&= \frac{1}{|P|} \sum_{S \subseteq P} \frac{1}{|D|} \sum_{x' \in D} f(\mathcal{X}(x, x', S \cup \{i\})) - f(\mathcal{X}(x, x', S)) \\
&= \frac{1}{|D|} \sum_{x' \in D} \underbrace{\frac{1}{|P|} \sum_{S \subseteq P} f(\mathcal{X}(x, x', S \cup \{i\})) - f(\mathcal{X}(x, x', S))}_{\text{single reference SHAP value}}
\end{aligned}
$$

where the second step depends on an interventional conditional expectation [18] which is very close to Random Baseline Shapley in [19]). □

3 Experiments

Background Distributions Avoid Bias
We utilize the popular CIFAR10 dataset [20] to demonstrate that single references lead to bias in explanations. We train a CNN that achieves 75.56% test accuracy and evaluate it using either a zero baseline as in DeepLIFT or with a random set of 1000 baselines as in DeepSHAP.

In Fig. 2, we can see that for these images drawn from the CIFAR10 training set, DeepLIFT has a clear bias that results in low attributions for darker regions of the image. For DeepSHAP, having multiple references drawn from a background distribution solves this problem and we see attributions in sensical dark regions in the image.

Explaining Mortality Prediction
In this section, we validate DeepSHAP's explanations for an MLP predicting 15 year mortality (82.6% test accuracy). The dataset has 79 features for 14,407 individuals released by [3] based on NHANES I Epidemiologic Followup Study [21].

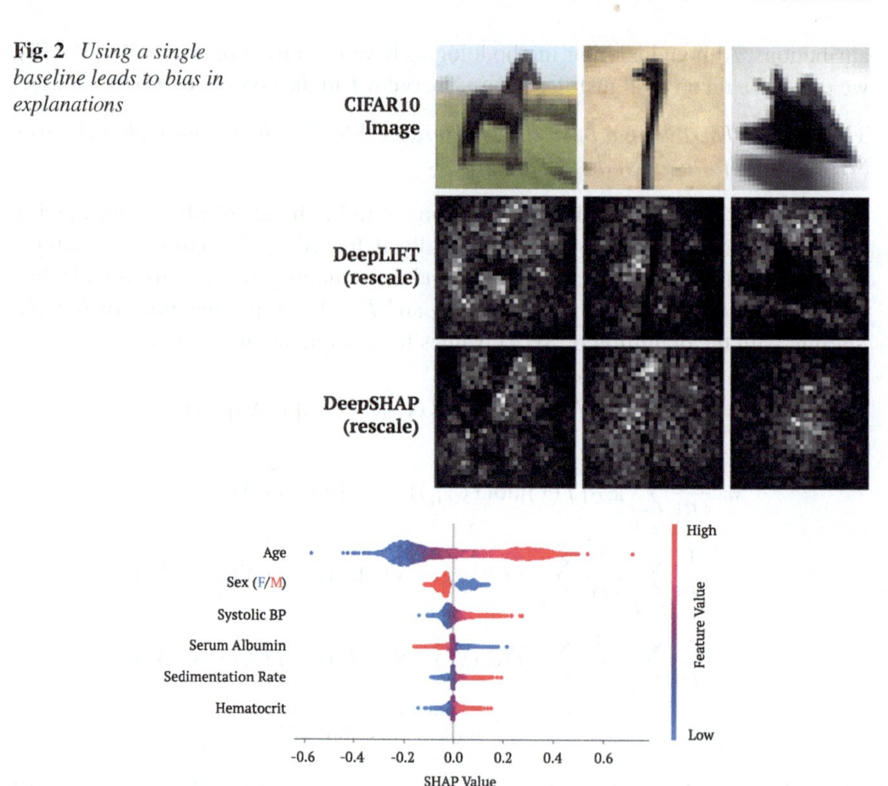

Fig. 2 *Using a single baseline leads to bias in explanations*

Fig. 3 *Summary plot of DeepSHAP attribution values.* Each point is the local feature attribution value, colored by feature value. For brevity, we only show the top 6 features

In Fig. 3, we plot a summary of DeepSHAP (with 1000 random background samples) attributions for all NHANES training samples (n = 8023) and notice a few trends. First, Age is predictably the most important and old age contributes to a positive mortality prediction (positive SHAP values). Second, the Sex feature validates a well-known difference in mortality [22]. Finally, the trends linking high systolic BP, low serum albumin, high sedimentation rate, and high hematocrit to mortality have been independently discovered [23–26].

Next, we show the benefits of specifying a background distribution. In Figure 4a, we see that explaining an individual's mortality prediction with respect to a general population emphasizes the individual's age and gender. However, in practice doctors are unlikely to compare a 67-year old male to a general population that includes much younger individuals. In Fig. 4b, being able to specify a background distribution allows us to compare our individual against a more relevant distribution of males over 60. In this case, gender and age are naturally no longer important, and the individual actually may not have cause for concern.

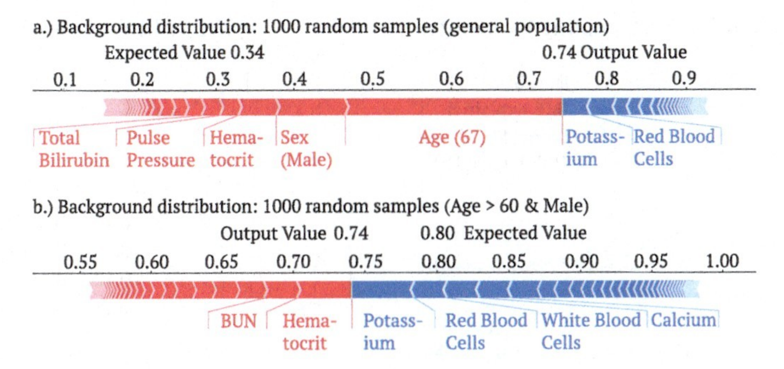

Fig. 4 *Explaining an individual's mortality prediction for different backgrounds distributions*

Interpreting a Stack of Mixed Model Types

Stacks and ensembles of models are increasingly popular [27–29]. In this section, our aim is to evaluate the efficacy of DeepSHAP for a neural network feature extractor fed into a tree model. For this experiment, we use the Rescale rule for simplicity and Independent TreeSHAP to explain the tree model [3]. The dataset is a simulated one called Corrgroups60. Features $X \in \mathbb{R}^{1000\times 60}$ have tight correlation between groups of features (x_i is feature i), where $\rho_{x_i,x_i}=1$, $\rho_{x_i,x_{i+1}}=\rho_{x_i,x_{i+2}}=\rho_{x_{i+1},x_{i+2}}=.99$ if $(i \bmod 3)=0$, and $\rho_{x_i,x_j}=0$ otherwise. The label $y \in \mathbb{R}^n$ is generated linearly as $y=X\beta+\epsilon$ where $\epsilon\sim\mathcal{N}_n(\mu=0,\sigma^2=10^{-4})$ and $\beta_i=1$ if $(i \bmod 3)=0$ and $\beta_i=0$ otherwise.

We evaluate DeepSHAP with an ablation metric (*keep absolute (mask)* [3]) that works as follows: 1) Obtain the feature attributions for all test samples 2) Mean impute all features (mask) 3) Introduce one feature at a time (unmask) from largest absolute attribution value to smallest for each sample and measure R^2. The R^2 should initially increase rapidly, because we introduce the "most important" features first.

We compare against two sampling-based methods (natural baselines for explaining mixed model stacks) that provide SHAP values in expectation: KernelSHAP and IME. In Fig. 5a, DeepSHAP (rescale) has no variability and requires a fixed number of model evaluations. IME and KernelSHAP, benefit from having more samples (and therefore more model evaluations). For the final comparison, we check the variability of the tenth largest attribution (absolute value) of the sampling based methods to determine "convergence" across different numbers of samples. Then, we use the number of samples at the point of "convergence" for the next figure.

In Fig. 5b, we can see that DeepSHAP has a slightly higher performance than model agnostic methods. Promisingly, all methods demonstrate initial steepness in their performance; this indicates that the most important features had higher attribution values. We hypothesize that KernelSHAP and IME's lower performance is due in part to noise in their estimates. This highlights an important point: model agnostic methods often have sampling variability that makes determining convergence difficult. For a fixed background distribution, DeepSHAP does not suffer from this variability and generally requires fewer model evaluations.

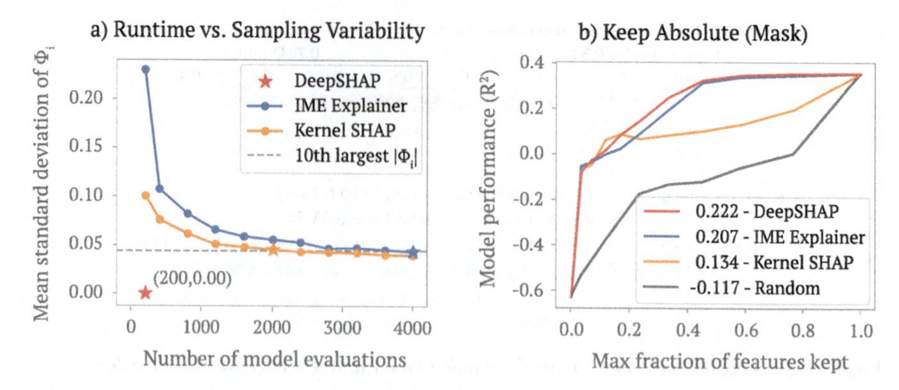

Fig. 5 *Ablation test for explaining an LSTM feature extractor fed into an XGB model.* All methods used background of 20 samples obtained via kmeans. [a.] Convergence of methods for a single explanation. [b.] Model performance versus # features kept for DeepSHAP (rescale), IME Explainer (4000 samples), KernelSHAP (2000 samples) and a baseline (Random) (AUC in the legend)

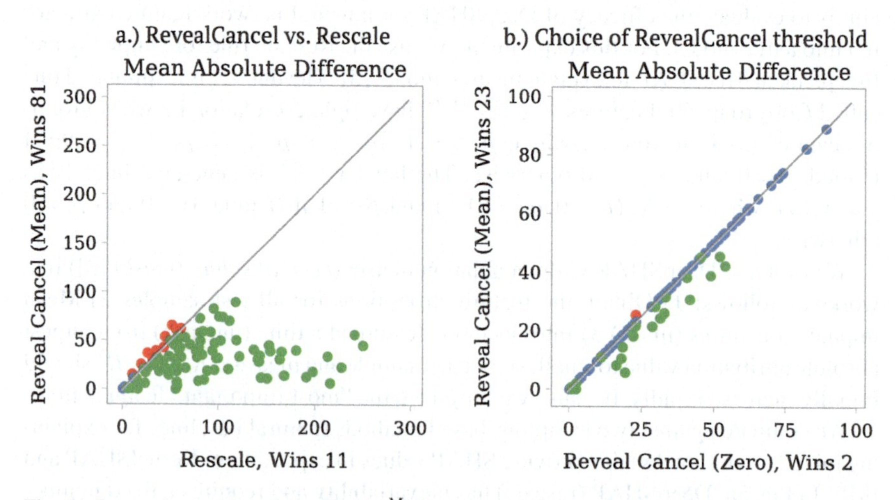

Fig. 6 *Comparison of new RevealCancel^{Mean} rule for estimating SHAP values on a toy example.* The axes correspond to mean absolute difference from the SHAP values (computed exactly). Green means RevealCancel^{Mean} wins and red means it loses

Improving the RevealCancel Rule

DeepLIFT's RevealCancel rule's connection to the SHAP values is touched upon in [15]. Our SHAP value framework explicitly defines this connection. In this section, we propose a simple improvement to the RevealCancel rule. In DeepLIFT's RevealCancel rule the threshold t is set to 0 (for splitting h_- and h_+). Our proposed rule RevealCancel^{Mean} sets the threshold to the mean value of $w_i(f_{x_i} - b_{x_i})$ across i. Intuitively, splitting by the mean better separates x_i nodes, resulting in a better approximation than splitting by zero.

We experimentally validate RevealCancel^{Mean} in Fig. 6, explaining a simple function: ReLU$(x_1 + x_2 + x_3 + x_4 + 100)$. We fix the background to zero: $b_{x_i} = 0$ and draw 100 foreground samples from a discrete uniform distribution: $f_{x_i} \sim U\{-1000, 1000\}$.

In Fig. 6a, we show that RevealCancel^{Mean} offers a large improvement for approximating SHAP values over the Rescale rule and a modest one over the original RevealCancel rule (at no additional asymptotic computational cost).

4 Conclusion

In this paper, we improve the original DeepSHAP formulation [6] in several ways: we 1.) provide a new theoretically justified way to provide attributions with a background distribution 2.) extend DeepSHAP to explain stacks of mixed model types 3.) present improvements of the RevealCancel rule. Future work includes more quantitative validation on different data sets and comparison to more interpretability methods. In addition, we primarily used Rescale rule for many of these evaluations, but more empirical evaluations of RevealCancel are also important.

References

1. Goodman, B., Flaxman, S.: AI Magazine **38**(3), 50 (2017)
2. Holzinger, A., Biemann, C., Pattichis, C.S., Kell, D.B., arXiv preprint arXiv:1712.09923 (2017)
3. Lundberg, S.M., Erion, G., Chen, H., DeGrave, A., Prutkin, J.M., Nair, B., Katz, R., Himmelfarb, J., Bansal, N., Lee, S.: CoRR **abs/1905.04610** http://arxiv.org/abs/1905.04610 (2018)
4. Lundberg, S.M., Nair, B., Vavilala, M.S., Horibe, M., Eisses, M.J., Adams, T., Liston, D.E., Low, D.K.W., Newman, S.F., Kim, J., Lee, S.I.: bioRxiv. 10.1101/206540. https://www.biorxiv.org/content/early/2017/10/21/206540 (2017)
5. Arcadu, F., Benmansour, F., Maunz, A., Willis, J., Haskova, Z., Prunotto, M.: NPJ Dig. Med. **2**(1), 1 (2019)
6. Lundberg, S.M., Lee, S.I.: Advances in Neural Information Processing Systems, pp. 4765–4774 (2017)
7. Shapley, L.S.: Contributions to the Theory of Games **2**(28), 307 (1953)
8. Štrumbelj, E., Kononenko, I.: Knowl. Inf. Syst. **41**(3), 647 (2014)
9. Zeiler, M.D., Fergus, R.: European Conference on Computer Vision, pp. 818–833. Springer (2014)
10. Simonyan, K., Vedaldi, A., Zisserman, A.: arXiv preprint arXiv:1312.6034 (2013)
11. Bach, S., Binder, A., Montavon, G., Klauschen, F., Müller, K.R., Samek, W.: PLoS One **10**(7), e0130140 (2015)
12. Sundararajan,, M., Taly, A., Yan, Q.: arXiv preprint arXiv:1703.01365 (2017)
13. Merrill, J., Ward, G., Kamkar, S., Budzik, J., Merrill, D.: CoRR **abs/1909.01869** http://arxiv.org/abs/1909.01869 (2019)
14. Ancona, M., Öztireli, C., Gross, M.: arXiv preprint arXiv:1903.10992 (2019)
15. Shrikumar, A., Greenside, P., Kundaje, A.: Proceedings of the 34th International Conference on Machine Learning, vol. 70 (JMLR.org), pp. 3145–3153 (2017)
16. Ancona, M., Ceolini, E., Oztireli, C., Gross, M.: 6th International Conference on Learning Representations (ICLR 2018) (2018)

17. Erion, G., Janizek, J.D., Sturmfels, P., Lundberg, S., Lee, S.I.: arXiv preprint arXiv:1906.10670 (2019)
18. Janzing, D., Minorics, L., Blöbaum, P.: arXiv preprint arXiv:1910.13413 (2019)
19. Sundararajan, M., Najmi, A.: arXiv preprint arXiv:1908.08474 (2019)
20. Krizhevsky, A., Hinton, G., et al.: Learning multiple layers of features from tiny images. Technical report, Citeseer (2009)
21. Cox, C.S., Feldman, J.J., Golden, C.D., Lane, M.A., Madans, J.H., Mussolino, M.E., Rothwell, S.T.: Vital and Health Statistics (1997)
22. Gjonça, A., Tomassini, C., Vaupel, J.W., et al.: Male-female differences in mortality in the developed world, Citeseer (1999)
23. Port, S., Demer, L., Jennrich, R., Walter, D., Garfinkel, A.: The Lancet **355**(9199), 175 (2000)
24. Goldwasser, P., Feldman, J.: J. Clin. Epidemiol. **50**(6), 693 (1997)
25. Paul, L., Jeemon, P., Hewitt, J., McCallum, L., Higgins, P., Walters, M., McClure, J., Dawson, J., Meredith, P., Jones, G.C., et al.: Hypertension **60**(3), 631 (2012)
26. Go, D.J., Lee, E.Y., Lee, E.B., Song, Y.W., Konig, M.F., Park, J.K.: J. Korean Med. Sci. **31**(3), 389 (2016)
27. Bao, X., Bergman, L., Thompson, R.: Proceedings of the Third ACM Conference on Recommender Systems, pp. 109–116. ACM (2009)
28. Güneş, F., Wolfinger, R., Tan, P.Y.: SAS Conference Proceedings (2017)
29. Zhai, B., Chen, J.: Sci. Total Environ. **635**, 644 (2018)

Controlling for Confounding Variables: Accounting for Dataset Bias in Classifying Patient-Provider Interactions

Kristen Howell, Megan Barnes, J. Randall Curtis, Ruth A. Engelberg, Robert Y. Lee, William B. Lober, James Sibley, and Trevor Cohen

Abstract Natural Language Processing (NLP) is a key enabling technology for re-use of information in free-text clinical notes. However, a barrier to deployment is the availability of labeled corpora for supervised machine learning, which are expensive to acquire as they must be annotated by experienced clinicians. Where corpora are available, they may be opportunistically collected and thus vulnerable to bias. Here we evaluate an approach for accounting for dataset bias in the context of identifying specific patient-provider interactions. In this context, bias is the result of a phenomenon being over or under-represented in a particular type of clinical note as a result of the way a dataset was curated. Using a clinical dataset which represents a great deal of variation in terms of author and setting, we control for confounding variables using a backdoor adjustment approach [1, 2], which to our knowledge has not been previously applied the clinical domain. This approach improves precision by up to 5% and the adjusted models' scores for false positives are generally lower,

---

K. Howell (✉) · M. Barnes
Department of Linguistics, University of Washington, Seattle, WA, USA
e-mail: kphowell@uw.edu

J. Randall Curtis · R. A. Engelberg · R. Y. Lee · W. B. Lober · J. Sibley
Cambia Palliative Care Center of Excellence, University of Washington, Seattle, WA, USA

J. Randall Curtis · R. A. Engelberg · R. Y. Lee
Division of Pulmonary, Critical Care, and Sleep Medicine, Department of Medicine, University of Washington, Seattle, WA, USA

J. Randall Curtis · W. B. Lober · J. Sibley
Department of Biobehavioral Nursing and Health Informatics, University of Washington, Seattle, WA, USA

W. B. Lober · J. Sibley · T. Cohen
Department of Biomedical Informatics and Medical Education, University of Washington, Seattle, WA, USA

© The Editor(s) (if applicable) and The Author(s), under exclusive license to Springer Nature Switzerland AG 2021

A. Shaban-Nejad et al. (eds.), *Explainable AI in Healthcare and Medicine*, Studies in Computational Intelligence 914,
https://doi.org/10.1007/978-3-030-53352-6_25

resulting in a more generalizable model with the potential to enhance the downstream utility of models trained using opportunistically collected clinical corpora.

1 Introduction

While natural language processing (NLP) techniques show promise for information extraction from the free text of clinical notes [3], the clinical domain presents some challenges to NLP. In particular, clinical datasets are difficult to share and produce due to patient privacy regulations and the need for annotators with medical expertise. As a result, datasets that represent a particular phenomenon tend to be small and may be opportunistically assembled from a variety of sources and represent a range of practice settings, authors and visit types [4] which can introduce bias into the data. A situation of particular concern for NLP applications involves biases present in the development set that vary from those at the point of deployment.

Eliminating sources of confounding bias can prove impractical during data collection and annotation, particularly in the low-resource setting of clinical NLP.[1] When data is scarce, removing sources of bias could result in a dataset that is too small to train a model. At the same time, failure to fully represent the contexts in which a phenomenon of interest may occur can result in models that do not generalize beyond the narrow range of practice settings represented in the training corpus.

We explore this issue in the context of a real-world clinical NLP problem: classifying notes as either containing or lacking documentation of a goals-of-care discussion. Because of the variety of practice settings in which such discussions may occur, our dataset comprises a variety of note types from various sources. Of particular concern for such studies is the issue of *confounding* bias, where an unseen variable (eg. note type) influences both the presence of a feature (eg. a word indicating a note type) and the outcome (eg. presence of a goals-of-care discussion), leading the model to the erroneous conclusion that this feature is predictive of the outcome in other contexts. Confounding bias may result in a model that weights features associated with a particular note type more heavily, due to the dataset-specific frequency of documented goals-of-care discussions within that category. While such an association may be a useful predictor for this dataset, the prevalence of such discussions in particular note types is likely to differ at the point of deployment, where other note types beyond those in the development set may also be considered. It would be preferable for a model to identify features that are correlated with a goals-of-care discussion regardless of the type of note it is found in. Such a model should generalize to other datasets more readily and reduce false-positive predictions made on account of language that indicates an over-represented note type rather than a goals-of-care discussion.

Recent work has confronted the problem of dataset bias using the framework of causal reasoning [1, 2]. This has shown promise as a way to address bias in

[1] In this context, low-resource refers to settings in which datasets are relatively small from a machine learning perspective.

general-domain datasets, but to our knowledge has not been applied to biomedical or clinical datasets. We adopt the backdoor adjustment approach [5], which involves identifying sources of confounding bias and controlling for them at training time by treating them as confounding variables in the model. This method corrects for the influence of a potential confounding variable by treating the presence and absence of the confounder as independent binary features at training time and takes an outcome-frequency weighted sum across both possible values of the confounder at test time.

We evaluate the utility of this approach on a small clinical dataset drawn from diverse practice settings. This approach accounts for confounding bias so that we do not have to reduce the size of our dataset and performs better overall. Because confounders can be dataset specific, we examine the effect of controlling for a range of potential confounders by producing a separate model for each and comparing the feature vectors in the resulting models and analyzing their performance. We evaluate in the context of a classification task on patient-provider interactions and show that this method can effectively correct for confounding bias in a clinical training set.

2 Previous Work

This work is contextualized by the literature on dataset bias, confounding variables, and sublanguages, which we will proceed to discuss in turn.

Dataset bias is well studied in NLP and is often addressed with data augmentation and data re-balancing (see inter alia [6, 7]). These approaches are generally applied to correct class imbalance, while we are looking at intra-confounder class imbalance, the imbalance between certain confounders and the class, making augmentation more complex. Furthermore, effective data augmentation requires a degree of domain expertise, which is particularly expensive in the clinical domain. Data re-balancing requires a sufficient set of examples to over-sample in the training set while still having examples for evaluation, making small clinical datasets poor candidates.

A particularly relevant source of bias is *confounding bias* – a circumstance in which a third, variable influences both a predictive feature and the outcome of interest. This may lead a model to overestimate the strength with which this feature indicates the outcome in the absence of the confounding variable. While the importance of confounding bias has long been recognized in the causal modeling literature [5], recent work has attempted to address this issue in NLP. Pryzant et al. (2018) address confounding bias by inducing a lexicon that is uncorrelated with confounding variables [8], while Landeiro and Culotta (2016, 2018) propose an approach in which they add confounding variables to the feature set and adjust the coefficients of features that are correlated with the confound [1, 2]. As they do not require modification of the dataset, these approaches are more readily applicable to clinical NLP.

In many cases, the confounding variable in a training set is an extra-linguistic feature which is correlated with linguistic features. Hovy (2015) finds that including demographic variables that may be related to linguistic differences help in text classification [9]. Instead of demographic variables, we consider the setting in which a note was written, the *sublanguage* or *genre* [10–12], associated with a clinical setting. Patterson and Hurdle (2011) identify sub-genres in clinical notes, which we consider as potential confounders in our dataset [12].

3 Task: Classifying Goals-of-Care

Our task is situated within the larger goal of improving patient and family outcomes by evaluating provider-patient interactions. To achieve this, we need to identify the documentation of different conversations in clinical notes. In general, but especially in the context of life-limiting illness, goals-of-care discussions warrant documentation, as these discussions provide patients the opportunity to express their treatment preferences in advance, facilitating proactive care planning. These conversations can be documented by doctors, nurses and social workers in admission, progress and discharge notes, but the vast majority are found in progress notes written by doctors.

In an ideal scenario, we could train a classifier on a dataset that is evenly balanced for goals-of-care discussions across potential confounders such as author, note type and source database. However, in practice curating a dataset that is representative of the phenomena of interest, large enough for a machine learning task and evenly balanced is not feasible. In a more realistic scenario, a medical group develops a dataset from the electronic health records (EHR) that are available to represent as wide a range of examples as possible. Using such a dataset, we demonstrate that confounding variables can be accounted for to reduce their effect in the model.

Identifying goals-of-care discussions is a scenario in which precision is especially important. If one is identifying documentation of these discussions for the purpose of directing providers to a prior discussion to inform clinical care, false positives would deter providers from using the results of the technology, while false negatives would have somewhat less impact, so we prioritize precision over recall.

4 Dataset

We use a dataset which contains 3,173 clinical notes written in English and annotated for the presence (GOC+) or absence (GOC-) of a goals-of-care discussion [13]. We split the notes in the dataset randomly into train (80%), dev (10%) and test (10%), ensuring only that each set would contain the same proportion of GOC+ and GOC- notes. As a result the distribution of notes across each of the categories in Table 1 is random, but representative of the distribution of the larger dataset.

The dataset contains clinical notes from three sources: FCS, a clinical trial of inpatients [14]; PICSI, a clinical trial of outpatients [15]; and PCC, a purposive sample of EHR notes (both inpatient and outpatient) from patients with serious illness. The PCC sample includes both GOC+ notes written by palliative care specialists and

Table 1 Distribution of goals-of-care discussions across note types, source datasets and EHR categories (Variables that are mutually exclusive with each other are shown in the same font style)

Variable	Train		Dev		Test		Total	
	GOC+	GOC-	GOC+	GOC-	GOC+	GOC-	GOC+	GOC-
Inpatient	373	782	47	90	48	95	468	967
Outpatient	177	1212	22	159	22	156	221	1527
PCC Dataset	**336**	**1135**	**46**	**143**	**42**	**146**	**424**	**1424**
PICSI Dataset	**170**	**859**	**20**	**106**	**22**	**105**	**212**	**1070**
FCS Dataset	**44**	**0**	**3**	**0**	**6**	**0**	**53**	**0**
Admit	*1*	*31*	*0*	*4*	*0*	*5*	*1*	*40*
Progress	*542*	*995*	*68*	*122*	*70*	*121*	*680*	*1238*
Discharge Summary	*0*	*28*	*0*	*3*	*0*	*4*	*0*	*35*
Other Summary	*0*	*18*	*1*	*1*	*0*	*1*	*1*	*20*
Emergency Department	*0*	*56*	*0*	*6*	*0*	*7*	*0*	*69*
Code Status	*0*	*7*	*0*	*0*	*0*	*0*	*0*	*7*
Nursing	*4*	*58*	*0*	*6*	*0*	*6*	*4*	*70*
Social Work	*1*	*24*	*0*	*5*	*0*	*7*	*1*	*36*
Other[a]	2	*777*	*0*	*102*	*0*	*100*	2	*979*
All	550	1994	69	249	60	251	679	2494

[a]This miscellaneous category includes notes such as immunization records, mental health visits, and anesthesia reports

GOC- notes selected randomly from a cohort of patients with any of nine chronic life-limiting conditions which are used to study end of life care [16, 17].

In addition to the source dataset, metadata for each note specifies if it is inpatient or outpatient and what category it falls into within the EHR. Table 1 shows the distribution of GOC+ and GOC- notes for each of these categories. All notes are labeled with a note type (inpatient/outpatient), source dataset (PCC, PICSI, FCS) and note category . Nursing and social work notes inherently indicate the type of author, while all other notes are authored by doctors. In the following experiment we will consider each of these categories as a potential confounder.

5 Feature Extraction and Backdoor Adjustment

We pre-processed the dataset by splitting each note into sentences using the NLTK sentence tokenizer [18], and annotating each sentence using the National Center for Biomedical Ontology (NCBO) concept annotator [19]. This process identifies phrases that map to known medical concepts while normalizing variant expressions of the same concept.[2] We counted each concept once per note and created a binary feature vector for each note.

[2]We used the "match longest only" and "exclude numbers" settings for concept annotation.

Our baseline model is a Logistic Regression model with L2 regularization from the scikit-learn Python library [20], trained on the features described above and the GOC + or - label for each note. Our adjusted models use the same pipeline, before being supplemented with features for the confounding variable as follows.

Landeiro and Culotta propose a methodology for accounting for confounding variables in a logistic regression model [1, 2], based on Pearl's *backdoor adjustment*[3] [5]. We are interested in the causal effect of a term feature vector generated from the text in a clinical note (x) on the class label GOC (y), where a causal effect is the likelihood that intervening on x results in y. If there is an influence between the the note type (z) and the features extracted from note text (x) and between the note type (z) and the GOC label (y), z is a confounding variable and the causal effect of x on y taking into account both possible values z of a confounder Z can be estimated as:

$$p(y|do(x)) = \sum_{z \in Z} p(y|x, z)p(z) \tag{1}$$

Landeiro and Culotta apply this effect to text classification with the goal of predicting y based on the term feature vector x while controlling for z. The effect of z on x and y is illustrated in Fig. 1. $P(y|x)$ is affected by z through both $P(x|z)$ and $\mathrm{P}(y|z)$. We would like to ascertain the degree to which x influences y irrespective of the value of z, which may be unknown at the point of deployment or misleading because $P(y|z)$ differs from the training set. Using the equation in (1) to classify each test example is robust to cases where $P(y|z)$ changes between training and testing data, and assumes that z is known at training time, but not at testing time. Whereas (1) *measures* causal effect, our primary interest is in *controlling* for confounding effects.[4] To do so, we use an open source implementation of the backdoor adjustment method[5] to supplement a logistic regression model with two additional features indicating presence and absence of the confounder (for example $inpatient = 0$ and $\neg inpatient = 1$, or $inpatient = 1$ and $\neg inpatient = 0$). These features, which are known at training time effectively reduce the weight of the features they are correlated with in the

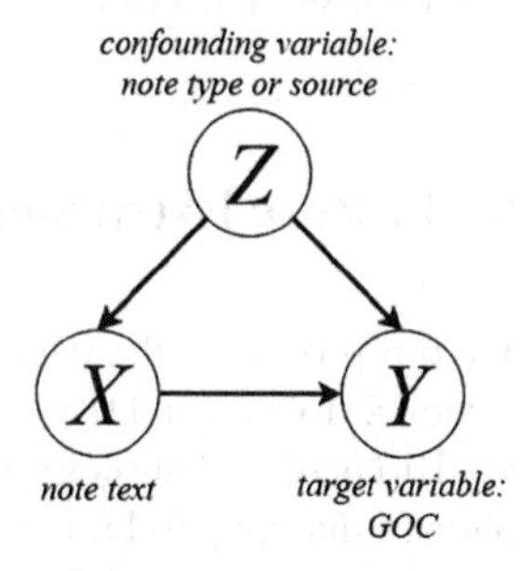

Fig. 1 Directed graphical model depicting confounder variable Z influencing both term vector X and class variable Y. (Adapted from [1])

[3] Also called *covariate adjustment*.

[4] For an approach that uses text classifiers to analyze causal effects, see [21].

[5] https://github.com/tapilab/aaai-2016-robust [1].

model. At testing time when the confounder is not known, we sum out these features using Eq. (1).

6 Results and Analysis

We use the backdoor adjustment approach to produce a separate model that accounts for each of the source datasets and note types described in Sect. 4,[6] and evaluate these models over the test data.[7] The precision, recall, F1-scores and F0.5-scores are given in Fig. 2. The F0.5-score was included as a metric because it emphasises precision over recall, which aligns with the priorities of this task.

Using backdoor adjustment to control for the source dataset improves precision of the positive class over the baseline by 4.9% for the PCC dataset and 2.4% for PICSI. We also see some improvement when we control for progress notes. We observe a trade-off, such that the models that improve most in precision, regress in recall; however, the overall improvement in PCC and PICSI models is borne out in the F1-score and particularly in the F0.5-score, as it favors precision.

We used 10-fold cross-validation over the full dataset to verify the robustness of these results. The mean scores (Fig. 3) affirm the improvements in precision for the PCC and PICSI models, which outperformed the baseline by 4.0% and 3.3%, and also show marginal improvements for some of the other models. Lower recall for the models with improved precision results in lower F1 scores, but the F0.5 scores

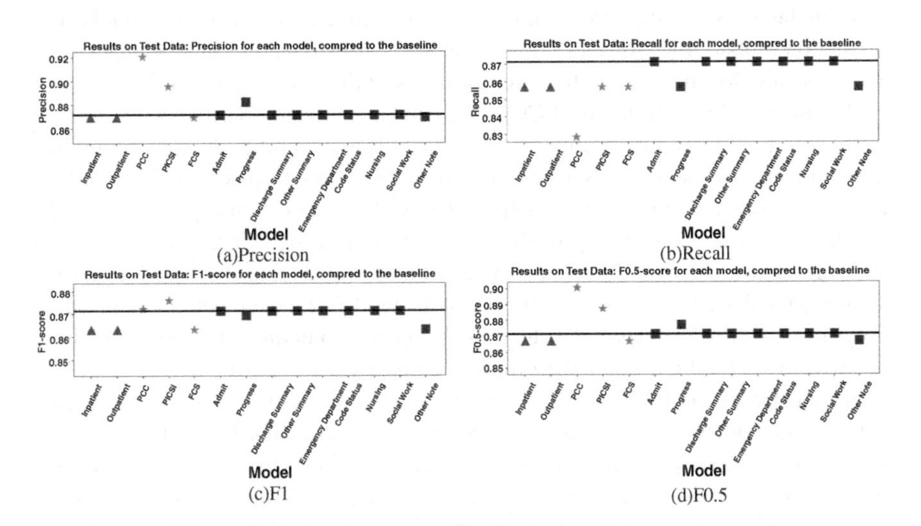

(a)Precision (b)Recall (c)F1 (d)F0.5

Fig. 2 Results for each model compared with the baseline on the test set

[6]Each model is trained on the entire dataset, but we adjust for only one variable per model.

[7]The code for this experiment is available at https://github.com/uwcirg/cambia-palliative-nlp-public.

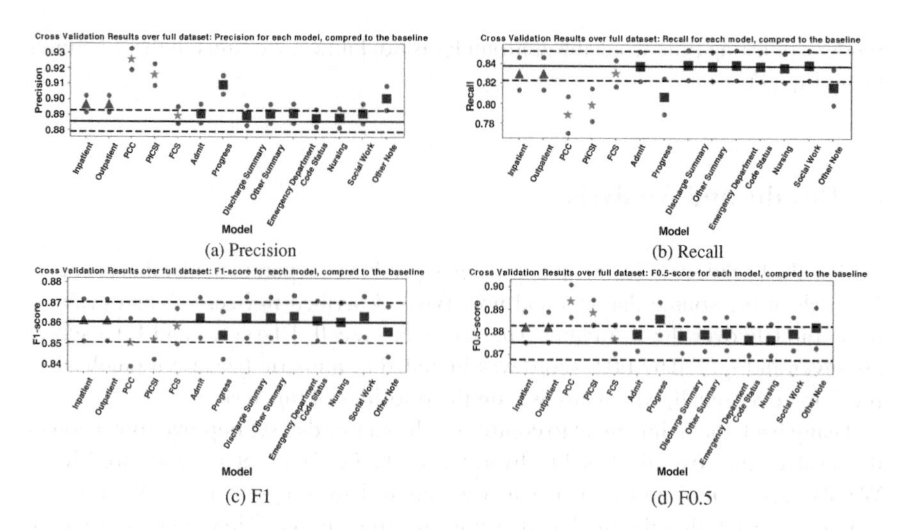

(a) Precision (b) Recall

(c) F1 (d) F0.5

Fig. 3 Mean (indicated with blue, green or purple) and standard error (red) over 10 folds for each model, compared with the baseline mean and its standard error

demonstrate some overall improvement in the PCC and PICSI models.

Error Analysis on False Positives

We can further evaluate our models by examining the false positives. Of the 251 GOC- notes in the test set, only 10 were misclassified as GOC+ by any of the models. The prediction probabilities for these false positives is illustrated in Fig. 4, which shows 10 notes and plots the probabilities for each model that misclassified it.[8]

The models that adjusted for PCC, PICSI and progress notes predicted false positives with the lowest probability, in some cases less than 60%, demonstrating that these models are considerably less confident in their false positive predictions. Furthermore, even though the other adjusted models had the same precision as the baseline on the test set, their probabilities for false positives were generally lower, showing measurable improvement between those models and the baseline. Of the ten notes labeled as false positives by any of the models, nine are outpatient progress notes from the PICSI dataset. The tenth is an inpatient, nursing, PCC note. The false positives from the PCC and PICSI models comprise a subset of seven notes, all of which are outpatient, PICSI notes. Two of these notes refer to a goals-of-care discussion (without documenting it directly) and one describes counseling, which involves many of the same terms as a goals-of-care documentation. The false positives for the progress notes model were a subset of these seven notes, plus one nursing note from PCC describing a patient in recovery and their long-term goals and outlooks. Though this is not a goals-of-care discussion, it contains similar terms. These marginal cases can be particularly challenging for a classifier trained on limited data. Nonetheless,

[8]Predictions for models that correctly classified these notes as GOC- are not shown.

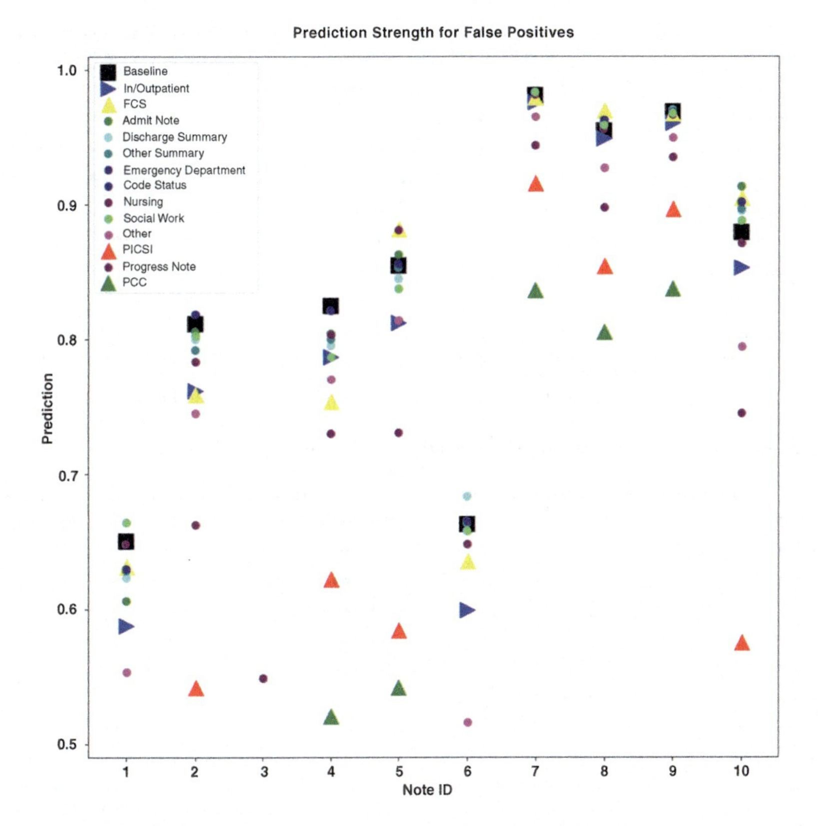

Fig. 4 Probability of false positive predictions across models

the adjusted models predict false positives for marginal cases with lower probability.

Features Undergoing Confounding Adjustment
Examining model feature coefficients provides insight into its generalizability [1]. The magnitude of the absolute value of a coefficient indicates importance of a given feature to the model. Using backdoor adjustment, we expect features correlated with the confounder to become weaker (smaller magnitude than the baseline model) and features that are independently correlated with goals-of-care discussions to become stronger (larger magnitude than the baseline). Figure 5 shows the features that were most weakened when we adjusted for PCC and PICSI. Most of these features ('saw', 'months', 'neuro', 'minutes', 'team', 'perl', 'discharge', 'tablet', 'additional', 'medications', 'time') have nothing to do with goals-of-care,[9] so trending towards zero indicates an improvement. Others could be part of a goals-of-care discussion: 'daughter',

[9]Arguably features having to do with time could be indicative of a goals-of-care discussion, but they are also indicative of many other things.

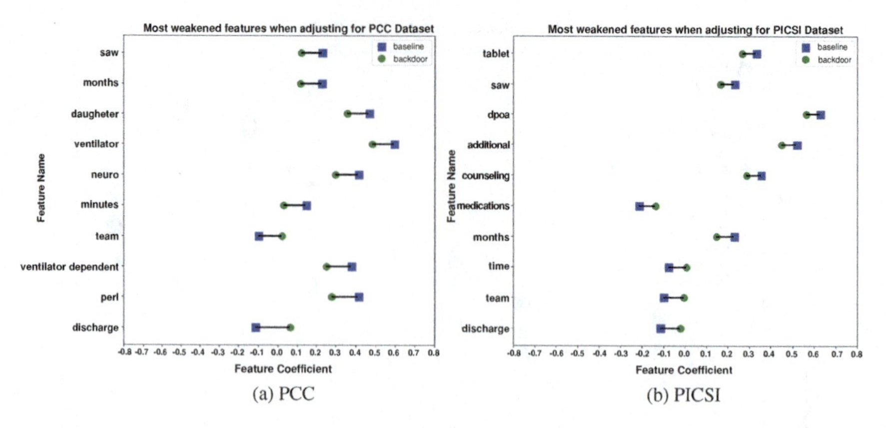

(a) PCC (b) PICSI

Fig. 5 Features that shift closer to 0 between the baseline and adjusted model

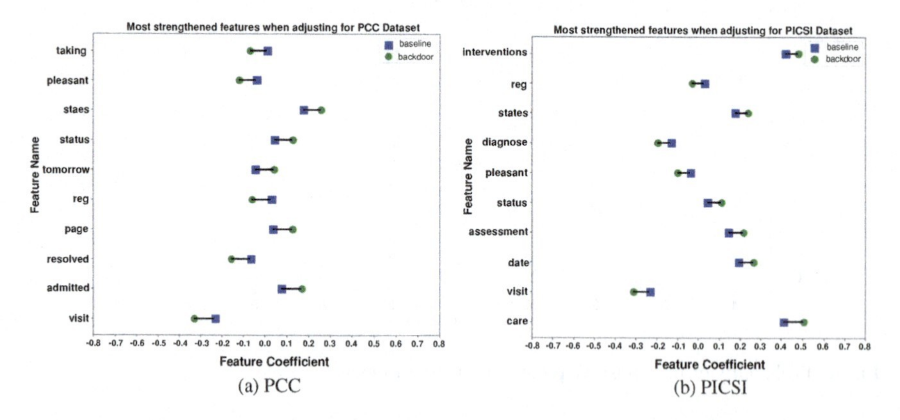

(a) PCC (b) PICSI

Fig. 6 Features that shift farther from 0 between the baseline and adjusted model

'ventilator', 'ventilator dependent', 'dpoa',[10] and 'counseling'), but may be strongly correlated with the confounder. 'dpoa' and 'counseling' are likely over-represented in the PICSI dataset, which came from a clinical trial targeting provider-patient communication, while terms related to specific serious illness scenarios such as 'ventilator' and 'ventilator dependent' are more likely to be used by palliative care specialists who wrote a significant portion of the PCC notes. Figure 6 illustrates the features that strengthened most when controlling for source dataset. These features are more noisy and many do not seem particularly indicative of goals-of-care discussions. 'tomorrow' and 'reg' simply shift from one side of zero to the other, but do not move very far from zero. Other features like 'page', 'status', 'interventions' and 'care' are likely to be part of a goals-of-care discussion so in becoming stronger, they represent an improvement in these models.

[10]Durable Power of Attorney.

Limitations

Our experiment looked closely at models that considered confounders individually. Future work will examine models with multiple confounders. Furthermore, we only consider confounders as binary features. However, one could conceive of a model that accounts for dataset as a non-binary feature, containing three values: PCC, PICSI, and FCS, for example. Examining models that account for multiple confounders or non-binary confounders has the potential to further improve these classifiers. Finally, we applied backdoor adjustment to a logistic regression model, following Landiero and Culotta (2016, 2018). In principle this approach could be applied to any discriminative classifier, but future work is required to test its effect.

7 Conclusion

We find that controlling for the clinical settings in which note types originated improves a logistic regression classifier for goals of care discussion. In the clinical domain datasets may be small and heterogeneous—controlling for such confounding variables results in models that have higher precision and features more likely to generalize to categories of notes that are under-represented in the training data.

Acknowledgements This material is based upon work supported by the Cambia Health Foundation. Any opinions, findings, and conclusions or recommendations expressed in this material are those of the author(s) and do not necessarily reflect the views of the Cambia Health Foundation.

References

1. Landeiro, V., Culotta, A.: Robust text classification in the presence of confounding bias. In: Thirtieth AAAI Conference on Artificial Intelligence. (2016)
2. Landeiro, V., Culotta, A.: Robust text classification under confounding shift. Journal of Artificial Intelligence Research. **63**, 391–419 (2018)
3. Demner-Fushman, D., Chapman, W.W., McDonald, C.J.: What can natural language processing do for clinical decision support? Journal of biomedical informatics. **42**, 760–772 (2009)
4. Cios, K.J., Moore, G.W.: Uniqueness of medical data mining. Artificial Intelligence in Medicine. **26**(1–2), 1–24 (2002)
5. Pearl, J.: Causality: models, reasoning and inference. In: Economic Theory. 675–685 (2003)
6. Zhao, J.m Wang, T., Yatskar, M., Ordonez, V., Chang, K.: Gender Bias in Coreference Resolution: Evaluation and Debiasing Methods. In: The 2018 Conference of the North American Chapter of the Association for Computational Linguistics. (2018)
7. Calders, T., Kamiran, F., Pechenizkiy, M.: Building classifiers with independency constraints. In: International Conference on Data Mining Workshops. (2009)
8. Pryzant, R., Shen, K., Jurafsky, D., Wagner, S.: Deconfounded lexicon induction for interpretable social science. In: The 2018 Conference of the North American Chapter of the Association for Computational Linguistics: Human Language Technologies. (2018)

9. Hovy, D.: Demographic factors improve classification performance. In: The 53rd Annual Meeting of the Association for Computational Linguistics and the 7th International Joint Conference on Natural Language Processing. (2015)
10. Biber, D., Finegan, E.: Sociolinguistic perspectives on register. Oxford University Press (1994)
11. Friedman, C., Kra, P., Rzhetsky, A.: Two biomedical sublanguages: a description based on the theories of Zellig Harris. Journal of Biomedical Informatics. **35**(4), 222–235 (2002)
12. Patterson, O., Hurdle, J.F.: Document clustering of clinical narratives: a systematic study of clinical sublanguages. In: American Medical Informatics Association Annual Symposium. (2011)
13. Lee, R.Y., Lober, W.B., Sibley, J.K., Kross, E.K., Engelberg, R.A., Curtis, J.R.: Identifying Goals-of-Care Conversations in the Electronic Health Record Using Machine Learning and Natural Language Processing. In: American Thoracic Society 2019 International Conference. (2019)
14. Curtis, J.R., Treece, P.D., Nielsen, E.L., Gold, J., Ciechanowski, P.S., Shannon, S.E., Khandelwal, N., Young, J.P., Engelberg, R.A.: Randomized Trial of Communication Facilitators to Reduce Family Distress and Intensity of End-of-life Care. American Journal of Respiratory and Critical Care Medicine. **193**(2), 154–162 (2016)
15. Curtis, J.R., Downey, L., Back, A.L., Nielsen, E.L., Paul, S., Lahdya, A.Z., Treece, P.D., Armstrong, P., Peck, R., Engelberg, R.A.: Effect of a patient and clinician communication-priming intervention on patient-reported goals-of-care discussions between patients with serious illness and clinicians: a randomized clinical trial. JAMA Internal Medicine. **178**, 930–940 (2018)
16. Iezzoni, L.I., Heeren, T., Foley, S.M., Daley, J., Hughes, J., Coffman, G.A.: Chronic conditions and risk of in-hospital death. Health Services Research **29**(4), 435–460 (1994)
17. Goodman, D.C., Esty, A.R., Fisher, E.S.: Trends and Variation in End-of-Life Care for Medicare Beneficiaries with Severe Chronic Illness: A Report of the Dartmouth Atlas Project. The Dartmouth Institute for Health Policy and Clinical Practice. (2011)
18. Bird, S., Loper, E., Klein, E.: Natural language processing with Python. O'Reilly Media, Inc. (2009)
19. Musen, M.A., Noy, N.F., Shah, N.H., Whetzel, P.L., Chute, C.G., Story, M., Smith, B.arry, and NCBO team: The national center for biomedical ontology. Journal of the American Medical Informatics Association. **19:2**, 190–195 (2011)
20. Pedregosa, F., Varoquaux, G., Gramfort, A., Michel, V., Thirion, B., Grisel, O., Blondel, M., Prettenhofer, P., Weiss, R., Dubourg, V. and others: Scikit-learn: Machine learning in Python. Journal of Machine Learning Research. **12**, 2825–2830 (2011)
21. Wood-Doughty, Z., Shpitser, I., Dredze, M.: Challenges of Using Text Classifiers for Causal Inference. In: Conference on Empirical Methods in Natural Language Processing. (2018)

Learning Representations to Augment Statistical Analysis of Drug Effects on Nerve Tissues

Hamid R. Karimian, Kevin Pollard, Michael J. Moore, and Parisa Kordjamshidi

Abstract We learn representations for classifying electrophysiological waveforms in a novel application, that is, analysis of electrical impulse conduction in microengineered peripheral nerve tissues. The goal is to understand the influence of distinct neuropathic conditions on the properties of electrophysiological waveforms produced by such tissues treated with known neurotoxic compounds and healthy controls. We show that statistical data analysis provides insight to the design of deep neural networks and the trained neural networks provide more informative representations compared to using pure statistical techniques. Based on this analysis, we design deep learning architectures that learn interpretable representations of the signals jointly with classification models. The proposed architecture provides new smooth representations of the signals that highlight the important points and patterns necessary for recognizing the electrophysiological effects of neurotoxic drug treatments.

This project is supported in part by Michigan State University, AxoSim Inc., and a grant from the NIH (R42-TR001270).

H. R. Karimian (✉) · P. Kordjamshidi
Michigan State University, East Lansing, MI 48824, USA
e-mail: karimian@msu.edu

P. Kordjamshidi
e-mail: kordjams@msu.edu

K. Pollard · M. J. Moore
Tulane University, New Orleans, LA 70118, USA
e-mail: kpollar@tulane.edu

M. J. Moore
e-mail: mooremj@tulane.edu

M. J. Moore
Tulane Brain Institute; and AxoSim Inc., New Orleans, LA 70112, USA

© The Editor(s) (if applicable) and The Author(s), under exclusive license to Springer Nature Switzerland AG 2021

A. Shaban-Nejad et al. (eds.), *Explainable AI in Healthcare and Medicine*,
Studies in Computational Intelligence 914,
https://doi.org/10.1007/978-3-030-53352-6_26

Keywords Deep learning · Signal processing · Neurotoxic drug treatments · Peripheral nerve tissue · Electrophysiological waveforms · PCA

1 Introduction

Inadequate preclinical identification of peripheral nerve toxicity frequently leads to failure of emerging pharmaceutical compounds at the clinical or post-market stages [8, 9, 13]. As a result, the cost to bring a new drug to market has become unacceptably large, reaching $2.6B [4]. Despite this, the FDA offers limited guidelines for correlation of *in vitro* toxicity screening with clinical outcomes, limiting their recommendations to identification of the relationship between drug dissolution and bioavailability [5]. The ability to identify peripheral nerve toxicity *in vitro*, before the initiation of animal studies and long before clinical trials, would significantly reduce the cost to bring a drug to market. However, conventional *in vitro* culture methods have not proven to be powerful predictors of *in vivo* toxicity [1]. Our lab has developed "nerve-on-a-chip" systems more capable of replicating structure and physiological function of *in vivo* nerve tissue [7]. Using these constructs, peripheral nerve health can be assessed through both direct, high-resolution imaging of nerve tissue and recording of the electrophysiological waveforms produced when the nerve fibers are electrically stimulated. These physiological outcomes are more analogous to clinical outcomes than traditional cell culture measures. Treatment with various neurotoxic drugs changes both the structure of nerve tissue described through high-resolution imaging and the properties of the electrophysiological waveforms produced when the tissue is electrically stimulated. Analysis of tissue architecture and electrophysiological function is conventionally only possible through use of costly and time-consuming animal models. The ability to generate similar, physiologically-relevant data from *in vitro* microphysiological systems will reduce both the time and cost associated with drug discovery pipelines and basic peripheral nerve research.

Approach: We build a prediction model capable of classifying specific drug-induced neuropathies based on electrophysiological waveforms alone. We design an AutoEncoder module [2, 3, 11, 12, 14, 15] to obtain an informative representation of the signals, and train this jointly [6] with a classification module to exploit the labels supervision in learning the representations. Both AutoEncoder and the classifier use long short-term memory (LSTM) layers and fully connected convolutional layers respectively. To design the layers of the NN model we use Principal component analysis (PCA) [10] to make informed decisions about the hyper-parameters of the network and the size of the latent layer. The PCA guides us to determine the lowest number of the neurons in the latent layer of the AutoEncoder. The data includes electrophysiological waveforms obtained from cultures treated the neurotoxic compounds bortezomib, forskolin, cisplatin, vincristine, and paclitaxel as well as healthy, vehicle-treated controls. The trained model maps the electrophysiological waveforms to the neuropathic classification of the tissue with a reasonable accuracy.

2 Network Architecture

Our first step is to attain an expressive representation of the electrophysiological waveforms to help neuropathic classification of the nerve constructs.

To train an ideal representation of the signals, we apply an AutoEncoder network adapted with LSTM layers. We evaluate the performance of the AutoEncoders in two ways: a) feeding the outputs of the latent layer of the AutoEncoder to a convolutional classifier; b) analyzing the outputs of the AutoEncoders using bioinformatic statistical methods. In our best model we use principled component analysis to determine the size of the intermediate layer of the AutoEncoder. This analysis helps to avoid losing information caused by choosing low data dimensions and also to avoid unnecessarily high dimensions that will require more data for training. Our experiments show that the neural network training provides more informative representations compared to using pure PCA. The modules of the proposed architecture are shown in Fig. 1. **Classifier:** The classifier is a point-wise network. It is composed of four blocks labeled "dense layers", "BatchNormalization", "Dropout", and "Activation layers". The classifier has a "Softmax" activation in its final layer. This architecture, using Dense layer, decreases the dimensions of the signals into the number of classes and the above mentioned layers prevents the models from over-fitting on the training data. We train the model by "RMSprop" optimizer along with Categorical CrossEntropy (CCE) loss function. The six treatments are class labels encoded with one hot representation. **Encoder and Decoder:** These two modules are used to build an AutoEncoder network. We use PCA method on the whole signals to determine the dimension of the connecting layer of the encoding and decoding modules. This dimension is 800 and obtained from the graph of cumulative explained variance versus the number of components. We also use this size to connect the classifier and encoder. The AutoEncoder is a sequence to sequence (LSTM) network including of six LSTM layers (three decoding and three encoding) followed by Dropout and dense layers in a time-distributed fashion. We train the model by "RMSprop" optimizer with mean squared error (MSE) as loss function.

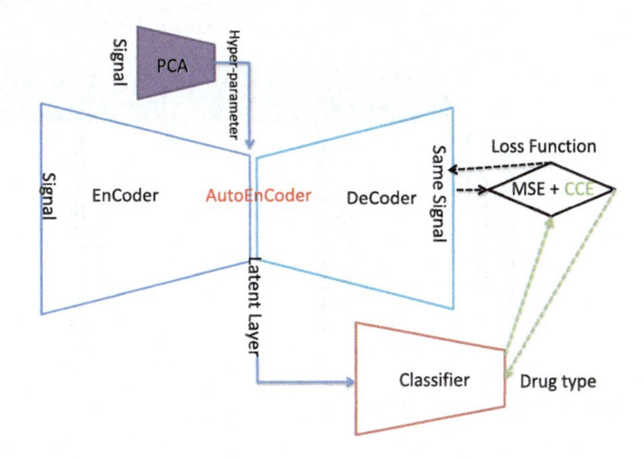

Fig. 1 Modules of the proposed model

3 Experiments

Data Collection. To obtain electrophysiological waveforms, the stimulating electrode is placed in the region of axonal outgrowth while the recording electrode is placed in the cell body region. Figure 2 depicts a fully mature "nerve-on-a-chip" construct. The stimulating electrode delivers an electrical impulse to the nerve tissue which, in response, generates its own electrical signal. The nerve fibers actively propagate this bioelectric signal towards the recording site where it is recorded as an electrophysiological waveform.

Experimental setup. The signals have a high distortion level; to smooth out the signals, we apply an AutoEncoder with LSTM Layers on the whole data-set with 1296 signals. Each signal has 1980 data points. We split the signals into 30 sequences of 66 data points. To evaluate the LSTM AutoEncoder, the output signals are given to a Bioinformatic Statistical model and a Classifier.

We split the data set into test and training subsets in setting of 85% as training and 15% testing set, we use Keras for implementation.

3.1 Deep Neural Networks

We implement several models using aforementioned encoder, decoder and classifier modules (see Fig. 1). $\mathbf{M_1}$: It uses the encoder and the classifier modules. The training happens end-to-end by receiving supervision from signals class labels; $\mathbf{M_2}$: It uses

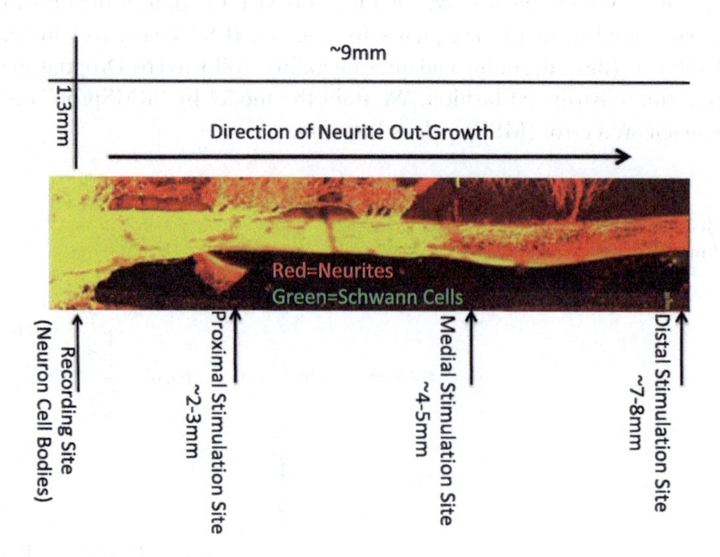

Fig. 2 A constructed Nerve

the encoder and decoder modules which form an AutoEncoder, in addition to the classifier module. The AutoEncoder and the classifier are trained jointly by receiving supervision from the label-loss as well as AutoEncoder loss; $\mathbf{M_3}$: The encoder and classifier trained jointly. This model uses class-label loss and the another loss –we call PCA-loss. The PCA-loss tries to make the representation of the encoder closer to the PCA representation; $\mathbf{M_4}$: This model is a combination of the $M1$, M_2 and M_3 that includes three different losses in a joint training setting. Those are the class-label loss, the AutoEncoder loss and the PCA-loss; $\mathbf{M_5}$: The AutoEncoder is trained separately to learn the representations. The obtained representations are given to the classifier in a pipeline. The classifier is trained separately; $\mathbf{M_6}$: Same as M_5, but here we obtain the input of the classifier by using PCA as a preprocessing step instead of the AutoEncoder representation; $\mathbf{M_7}$: We train the AutoEncoder separately. Then we use the pretrained encoder part of it and retrain it jointly with the classifier using the class-label supervision. In training this model, instead of raw inputs, the input representation obtained from the pretrained AutoEncoder is used. This is our best performing model (see Fig. 1).

Table 1 shows that the M_7 model yields the highest accuracy. The PCA helps to determine the best size for the number of the neurons in the latent layer to convey the information in the signals from encoder part into the decoder part of the AutoEncoder. Based on the PCA analysis we choose 800 as the final number of the neurons. We experimented with different number of layers but the results of the classifier dropped dramatically. For all aforementioned models when we use PCA-loss, the accuracy drops. Moreover, using the PCA representation of the signals as input of the classifier, the accuracy drops down dramatically to 33.5%. This means while PCA analysis is very helpful in designing the hyper-parameters of the model, using it directly does not perform well. We also implemented encoder and decoder modules adapted with Convolutional layers and built all aforementioned models based that. The results show that models with LSTM encoder/decoder modules provide better results. The peak detection was more effective when using AutoEncoder with LSTM layers compared to Convolutional layer (See Bioinformatic Statistical Analysis subsection). As Table 1 shows, the best obtained accuracy is 61.66% which is reasonable but has room for improvement. One source of difficulty is that the treatments on the tissues and the applied voltages in the lab experiments influence the status of the tissues (labeled) and make them indistinguishable from each other when different treatments are applied. One future plan is to use our machine learning models to discover the best lab conditions for engineering the tissues.

Table 1 The total accuracy of classifier from all models

Models	M_1	M_2	M_3	M_4	M_5	M_6	M_7
Accuracy	52.66%	60.5%	56.33%	57.5%	54.83%	33.5%	**61.66%**

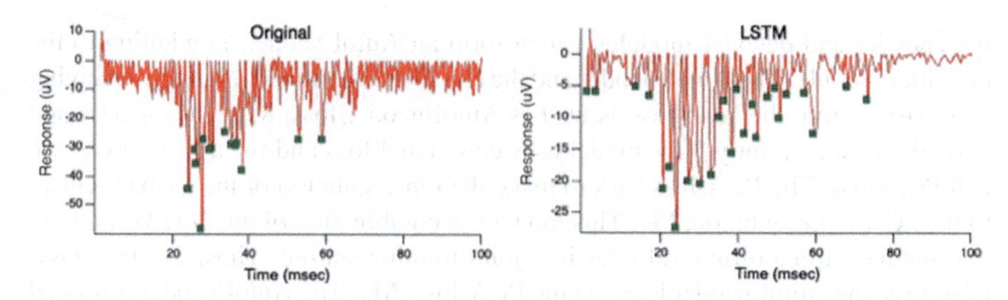

Fig. 3 Original trace (left) and smoothed trace (right). Green dots are distinct peaks identified by the peak-finding algorithm

3.2 Bioinformatic Statistical Analysis

Descriptive metrics are extracted from trace data using a custom analysis algorithm written in Igor Pro (WaveMetrics, Inc). The baseline-subtracted trace and results of the peak-picking algorithm for a raw example trace and the learned representation version of the same trace are shown in Fig. 3. For each data set, the maximum amplitude, integrated area under the curve (AUC), gross nerve conduction velocity (NCV), and peak count was extracted from the waveforms generated at each recording distance of each nerve construct and the values were averaged across all samples in the same treatment group.

Improved Peak Detection: Statistically significant drug-induced differences are identified with a two-way mixed model ANOVA using Prism 7 (Graphpad Software, Inc). The between-subjects factor "treatment" consisted of six levels labeled "vehicle", "vincristine", "cisplatin", "bortezomib", "forskolin", and "paclitaxel" while the within-subjects factor "recording distance" consisted of three levels labeled "distal", "medial", and "proximal". This analysis was performed on the descriptive metric dataset obtained from both the raw and learned-smoothed traces and any differences in statistical conclusions were noted. The smoothing procedures of our proposed neural architecture did not significantly affect the analysis of gross electrophysiological parameters of maximum amplitude, AUC, or NCV. Interestingly, the neural learning procedure significantly reduced the amplitude of background noise and therefore resulted in detection of significantly more smaller peaks.

As shown in Table 2, bioinformatic statistical analysis of the raw stat set identified a significant main effect of "recording distance" and "treatment" on total peak count but no significant interaction, while analysis of the data set that was smoothed using the proposed architecture identified a significant interaction between "recording distance" and "treatment" in addition to the main effects of each variable alone.

Table 2 ANOVA table

	Raw signal		Smoothed signal	
ANOVA table	F(DFn, DFd)	P value	F(DFn, DFd)	P value
Recording distance	$F(10, 60) = 0.9576$	$P = 0.4891$	$F(10, 60) = 3.33$	$P = 0.0017$
Treatment	$F(5, 30) = 9.596$	$P < 0.0001$	$F(5, 30) = 20.26$	$P < 0.0001$
Location	$F(2, 60) = 10.71$	$P = 0.0001$	$F(2, 60) = 17.64$	$P < 0.0001$
Subjects	$F(30, 60) = 1.276$	$P = 0.2084$	$F(30, 60) = 1.537$	$P = 0.0787$

4 Conclusion

We designed a predictive model for learning interpretable representations and classification of electrophysiological waveforms. Our best model is a joint training one that uses insights form statistical analysis, that is, PCA for determining the most appropriate hyper-parameters of the network. Our results show that the trained representations of our best model are consistently interpretable by human experts and are informative for the type of classical analysis that is usually performed based on the statistical signal processing using qualitative features such as peaks and frequency of peaks in the signals. The results of this research will largely impact the costs and the effectiveness of the drug design procedure which will impact the treatment of neoropathic health issues in general.

References

1. Astashkina, A., Grainger, D.W.: Critical analysis of 3-D organoid in vitro cell culture models for high-throughput drug candidate toxicity assessments. Adv. Drug Delivery Rev. **69-70**, 1–18 (2014), innovative tissue models for drug discovery and development
2. Baldi, P., Hornik, K.: Neural networks and principal component analysis: learning from examples without local minima. Neural Netw. **2**(1), 53–58 (1989). Jan
3. Chen, M., Weinberger, K., Sha, F., Bengio, Y.: Marginalized denoising auto-encoders for non-linear representations. In: Xing, E.P., Jebara, T. (eds.) Proceedings of the 31st International Conference on Machine Learning. Proceedings of Machine Learning Research, vol. 32, pp. 1476–1484. PMLR, Bejing, China (22–24 June 2014)
4. DiMasi, J.A., Grabowski, H.G., Hansen, R.W.: Cost to develop and win marketing approval for a new drug is $2.6 billion. Tufts Center for the Study of Drug Development (2014)
5. Emami, J., et al.: In vitro-in vivo correlation: from theory to applications. J. Pharm Pharm. Sci. **9**(2), 169–189 (2006)
6. Gogna, A., Majumdar, A., Ward, R.K.: Semi-supervised stacked label consistent autoencoder for reconstruction and analysis of biomedical signals. IEEE Trans. Biomed. Eng. **64**(9), 2196–2205 (2017)
7. Huval, R.M., Miller, O.H., Curley, J.L., Fan, Y., Hall, B.J., Moore, M.J.: Microengineered peripheral nerve-on-a-chip for preclinical physiological testing. Lab Chip **15**, 2221–2232 (2015)
8. Kola, I., Landis, J.: Can the pharmaceutical industry reduce attrition rates? Nat. Rev. Drug Disc. **3**, 711–716 (2004)

9. Li, A.P.: Accurate prediction of human drug toxicity: a major challenge in drug development. Chemico-Biological Interactions **150**(1), 3–7 (2004). accurate Prediction of Human Drug Toxicity
10. Pearson, K.: On lines and planes of closest fit to systems of points in space. Philosoph. Mag. **2**, 559–572 (1901)
11. Rifai, S., Vincent, P., Muller, X., Glorot, X., Bengio, Y.: Contractive auto-encoders: explicit invariance during feature extraction. In: ICML (2011)
12. Rumelhart, D.E., Hinton, G.E., Williams, R.J.: Learning representations by back-propagating errors. Nature **323**, 533–536 (1986)
13. Schuster, D., Laggner, C., Langer, T.: Why drugs fail-a study on side effects in new chemical entities. Curr. Pharmaceutical Des. **11**(27), 3545–59 (2005)
14. Vincent, P., Larochelle, H., Lajoie, I., Bengio, Y., Manzagol, P.A.: Stacked denoising autoencoders: learning useful representations in a deep network with a local denoising criterion. J. Mach. Learn. Res. **11**, 3371–3408 (2010). Dec
15. Wen, T., Zhang, Z.: Deep convolution neural network and autoencoders-based unsupervised feature learning of eeg signals. IEEE Access **6**, 25399–25410 (2018)

Automatic Segregation and Classification of Inclusion and Exclusion Criteria of Clinical Trials to Improve Patient Eligibility Matching

Tirthankar Dasgupta, Ishani Mondal, Abir Naskar, and Lipika Dey

Abstract Clinical trials are aimed to observe the effectiveness of a new intervention. For every clinical trial the eligibility requirements of a patient are specified in the form of *inclusion* or *exclusion* criteria. However, the process of eligibility determination is extremely challenging and time-consuming. Such a process typically involves repeated manual reading followed by matching of the trial descriptions mentioning the eligibility criteria and patient's electronic health record (EHR) for multiple trials across every visit. Thus, the number of patients to be evaluated gets reduced. In this work, we have focused on a small but important step towards automatic segregation and classification of inclusion-exclusion criteria of clinical trials to improve patient eligibility matching. Accordingly, we have proposed an attention aware CNN-Bi-LSTM model. We evaluate our model with different word and character level embeddings as input over two different openly available datasets. Experimental results demonstrate our proposed model surpasses the performance of the existing baseline models. Furthermore, we have observed that character level information along with word embeddings boost up the predictive performance of criteria classification (in terms of F1-Scores), which is a promising direction for further research.

T. Dasgupta (✉) · A. Naskar · L. Dey
TCS Research and Innovation, Kolkata, India
e-mail: dasgupta.tirthankar@tcs.com

A. Naskar
e-mail: abir.naskar@tcs.com

L. Dey
e-mail: lipika.dey@tcs.com

I. Mondal
IIT Kharagpur, Kharagpur, India
e-mail: ishani340@gmail.com

© The Editor(s) (if applicable) and The Author(s), under exclusive license to Springer Nature Switzerland AG 2021

A. Shaban-Nejad et al. (eds.), *Explainable AI in Healthcare and Medicine*, Studies in Computational Intelligence 914,
https://doi.org/10.1007/978-3-030-53352-6_27

Keywords Inclusion and exclusion criteria · Clinical trials · Criteria classification

1 Introduction

Clinical trials (CTs) are research studies that are aimed at evaluating a medical, surgical, or behavioral intervention [11]. Through such trials researchers aim to find out whether a new treatment, like a new drug or diet or medical device is more effective than the existing treatments for a particular ailment [2, 8].

A successful completion of a trial is dependent on achieving a significant sample size of patients enrolled for the trial within a limited time period. Therefore, it is important to "search" for patients eligible for a given trial, which is done by retrieving Electronic Health Records (EHR) of patients and manually matching them against eligibility criteria specified in trials in the form of specified in the form of *inclusion* and *exclusion* conditions.

The complexity of eligibility criteria in terms of volume, medical jargon, complex language structure and presentation format (semistructured and unstructured (see Table 1)) makes it tedious for manual search [5, 6]. Therefore, screening of clinical trials to automatically segregate and classify inclusion/exclusion criteria is important.

Table 1 Different formats of eligibility criteria. Texts in italics represent inclusion, texts in bold are exclusion

Format-1: Unstructured form
Myopic volunteers, ages 20 to 35, who had not worn contact lenses were eligible to participate in the study if they were free of ocular disease, were in good physical health, and are not taking systemic medications that could have ocular side effects
In addition, eligibility was limited to persons with corneal curvature between 40.50 and 47.00 D (flatter keratometry reading), corrected visual acuity of 6/6 (20/20)or better in each eye, astigmatism less than 0.75 D, anisometropia less than 1D, and myopia between 1 and 4D.
Format-2: Semi-structured form
Eligibility: Gender: All Age: 14 Years to 35 Year Criteria: Inclusion criteria: - diagnosed with Congenital Adrenal Hyperplasia (CAH) - normal ECG during baseline evaluation Exclusion criteria: - history of liver disease, or elevated liver function tests - history of cardiovascular disease

We formulate the problem as a supervised binary classification task, in which given the eligibility criteria, the aim is to classify whether a given sequence of words talks about an *inclusion* or *exclusion* criteria. Accordingly, we have proposed an attention aware CNN-Bi-LSTM model for the classification of word sequences into *inclusion* or *exclusion* criteria. The model have been evaluated using the TREC 2018 Precision medicine track and the Kaggle-2017 data challenge task. Our preliminary investigation shows that our proposed attention aware CNN-BiLSTM model surpasses the existing state-of-the-art neural network architecture including the Clinical-BERT-base model.

2 Attention Aware CNN-BiLSTM Model for Criteria Classification

We have formulated the problem of criteria classification of clinical trials as a binary text classification task. Given a criteria sentence S consisting of n words denoted by w_1, w_2, w_3, $w_4 .... w_n$, the task is to predict the label of S in which the target label $t \in$ {inclusion, exclusion}. Accordingly, we have applied the CNN-BiLSTM based model for the classification task. In addition to this we have also incorporated a relation aware self attention mechanism between the pairs of words in order to capture the contextual information more accurately. The architecture of the proposed model is depicted in Fig. 1.

Word Embeddings. For the present work we have used three type of word embedding models namely, word2vec (W) [3], FastText (FT) [4] and context specific embedding

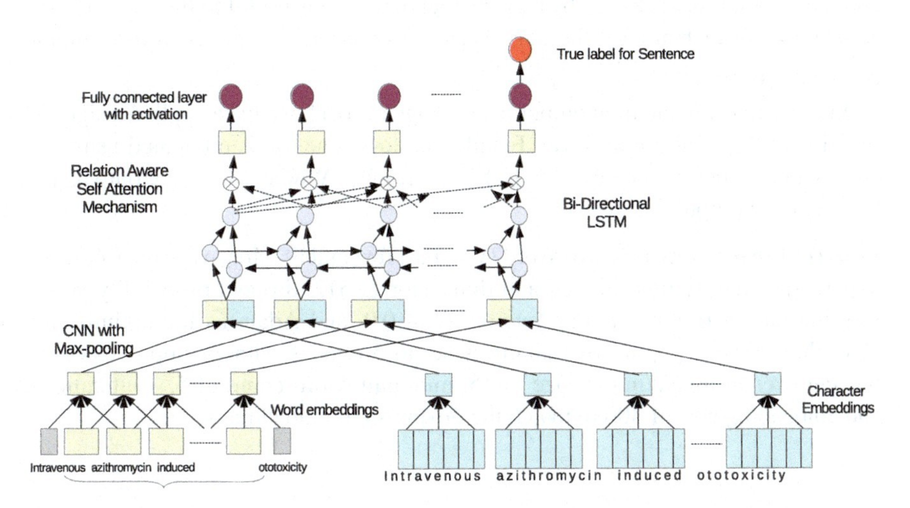

Fig. 1 The proposed relation aware self-attention based CNN-BiLSTM based model for criteria classification

like ELMo (E) [7]. All the embeddings were constructed over the entire clinical trial dataset of TREC 2018 Precision Medicine Track[1] and Kaggle Challenge.[2] All words that occur at least five times in the corpus are included.

In our training dataset, we have encountered many out of vocabulary words which are specific to the given clinical domain for which no word-level embeddings were not available (e.g. METHOTREXATE or FLUOXETINE or dosages like *'10mg'*, *'12 cc'*,). To compensate for the loss we have also included the character-level embeddings.

Bidirectional LSTM + CNN Layer. Each of the convolution network assumes a pre-defined sequence of n-gram words as input and perform a series of linear transformation. We represent every sentence using our joint CNN and LSTM architecture. The CNN is able to learn the local features from words to phrases from the text, while the LSTM learns the long-term dependencies of the text. The main goal of this step is to learn good intermediate semantic representations of the sentences. In order to ensure a uniform dimension over all sequences of input and output vectors, we perform a zero padding.

The Attention Network. We have used the relation aware self attention model as proposed in the literature [9] and [10]. Here instead of considering the traditional self attention network with the hidden layers, we consider a pair wise relation network between all possible pairs of input vectors. Thus, forming a labeled, fully-connected graph. The edge between input words x_i and x_j is represented by vectors $a_{ij}^V, a_{ij}^K \in R^{d_a}$, where V and K are the hyper-parameters to be learned. These representations can be shared across attention heads. Therefore, the existing self-attention representation can be enhanced by propagating the edge information to the sub-layer output as: $Z_i = \sum_{j=1}^{N} \alpha_{ij}(x_j.W^V + a_{ij}^V)$ This modification is important for tasks where information about the edge types selected by a given attention head is useful to the classification engine. We further modify the edge vectors to consider edges when determining compatibility as: $e_{ij} = \frac{x_i W^Q (x_j W^K + a_{ij}^K)^T}{\sqrt{d_z}}$.

In order to bring the final output in the range of [0.1], we have applied a sigmoid function at the final output layer. Finally, the loss function is computed using the cross-entropy loss defined by $L = -\sum_{i=1}^{2} \bar{y}_i log(y_i)$ Where $\bar{y}$ is actual output, y is the predicted output.

Fine Tuning the Network. For Word2Vec, FastText and ELMo we use the following hyperparameters setting: hidden unit dimension at 512, dropout probability at 0.5, learning rate at 0.001, learning rate decay at 0.9, and Adam as the optimization algorithm. Early stopping of training is set to 5 epochs without improvement to prevent over-fitting. Window size of 15, minimum word count of 5, 15 iterations, and embedding size of 300 to match the off-the-shelf embeddings.

[1] http://www.trec-cds.org/2018.html.

[2] https://www.kaggle.com/auriml/eligibilityforcancerclinicaltrials.

3 Experiment and Results

Dataset1: We have used the 241000 TREC 2018 precision medicine task Clinical Trials to train, validate and test our proposed models [12].

Dataset2: Kaggle has made available a total of 6186572 labeled clinical statements specifically on cancer trials were extracted from 49201 interventional protocols[3].

We have explored a number of other deep neural network-based models as a baseline system. Each of the above neural network model were individually trained and tested with Domain-specific Word2Vec embedding, FastText, ELMo and BERT embeddings. Results of each of the experiment are reported in Table 2. We also compare the performance of the aforementioned models with the state-of-the-art model proposed in [1]. In all our experiments, 10-fold cross validation was used for the purpose of fair evaluation on the datasets.

We observe that in terms of F1-score improves significantly in comparison to the output of the individual CNN or Bi-LSTM alone. Thus, the local filters used by CNN do not significantly boost up the predictive performance of the model in criteria classification. We further observe that the relation aware self attention mechanism ($C - BiLSTM - Re_{att}$) works better than the standard self-attention model ($C - BiLSTM - S_{att}$). This also justifies the long range dependencies between the word sequences that are captured by the fully-connected pairwise attention mechanism.

Finally, we have compared the performance of the proposed model with the BERT based classification architecture. We observe that for Dataset-I $C - BiLSTM - S_{att}$ clearly outperforms the BERT based architecture even when BERT is fine-tuned over the specific TREC dataset. On the other hand, BERT when fine-tuned over Dataset-II gives a slight better performance then $C - BiLSTM - Re_{att}$ architecture. However, a detailed analysis of the results are yet to be done which remains a future scope of this work.

Table 2 Results demonstrating F1 scores for the classification of Dataset-I and Dataset-II for different word embeddings, word2Vec (W), FastText(FT) and ELMo(E). Models with * implies use of character embedding along with word embedding

Word embedding	Dataset-I (TREC-2019)						Dataset-II (Kaggle)					
	Inclusion			Exclusion			Inclusion			Exclusion		
	W	FT	E	W	FT	E	W	FT	E	W	FT	E
BiLSTM	0.70	0.74	0.78	0.72	0.72	0.76	0.70	0.73	0.80	0.71	0.75	0.76
BiLSTM*	0.73	0.76	0.80	0.74	0.75	0.78	0.72	0.77	0.84	0.73	0.76	0.78
CNN	0.68	0.67	0.68	0.65	0.65	0.70	0.71	0.73	0.62	0.61	0.65	0.70
CNN*	0.69	0.71	0.73	0.67	0.66	0.74	0.79	0.77	0.67	0.63	0.68	0.71
C-BiLSTM	0.70	0.72	0.76	0.71	0.71	0.74	0.73	0.74	0.81	0.70	0.72	0.79
C-BiLSTM*	0.77	0.75	0.81	0.73	0.73	0.78	0.75	0.77	0.86	0.72	0.75	0.80
$C - BiLSTM - S_{att}$	0.75	0.78	0.82	0.72	0.73	0.80	0.82	0.83	0.89	0.78	0.85	0.88
$C - BiLSTM - S_{att}*$	0.80	0.82	0.83	0.78	0.76	0.81	0.84	0.86	0.90	0.82	0.88	0.90
$C - BiLSTM - Re_{att}$	0.81	0.79	0.84	0.79	0.80	0.81	0.85	0.87	0.91	0.86	0.85	0.91
$C - BiLSTM - Re_{att}*$	0.84	0.84	0.89	0.82	0.81	0.85	0.87	0.87	0.95	0.88	0.90	0.93
BERT-base	Dataset-I						Dataset-II					
	Inclusion			Exclusion			Inclusion			Exclusion		
BERT-base	0.86			0.80			0.91			0.92		

[3]https://clinicaltrials.gov/ct2/results?term=neoplasmtype=Intrshowdow.

4 Conclusion

In this work we have explored different deep neural network based architectures to automatically segregate and classify inclusion and exclusion criteria from clinical trials. The proposed model combines the strength of both LSTM, CNN and relation aware self attention along with context specific word embeddings. We have also observed a significant improvement of the model when word level representations are combined with character level ones. We have systematically compared the performance of aforementioned models. Overall, we have observed that our proposed model performs better than all the existing baseline models. However, a fine grained analysis of the experiments and results are yet to be done as a future work.

References

1. Bustos, A., Pertusa, A.: Learning eligibility in cancer clinical trials using deep neural networks. Appl. Sci. **8**(7), 1206 (2018)
2. Cao, Y., Liu, F., Simpson, P., Antieau, L., Bennett, A., Cimino, J.J., Ely, J., Yu, H.: Askhermes: an online question answering system for complex clinical questions. J. Biomed. Inform. **44**(2), 277–288 (2011)
3. Goldberg, Y., Levy, O.: Word2vec explained: deriving Mikolov et al.'s negative-sampling word-embedding method. arXiv preprint arXiv:1402.3722 (2014)
4. Joulin, A., Grave, E., Bojanowski, P., Douze, M., Jégou, H., Mikolov, T.: FastText. zip: compressing text classification models. arXiv preprint arXiv:1612.03651 (2016)
5. Penberthy, L., Brown, R., Puma, F., Dahman, B.: Automated matching software for clinical trials eligibility: measuring efficiency and flexibility. Contemp. Clin. Trials **31**(3), 207–217 (2010)
6. Penberthy, L.T., Dahman, B.A., Petkov, V.I., DeShazo, J.P.: Effort required in eligibility screening for clinical trials. J. Oncol. Pract. **8**(6), 365–370 (2012)
7. Peters, M.E., Neumann, M., Zettlemoyer, L., Yih, W.T.: Dissecting contextual word embeddings: architecture and representation. arXiv preprint arXiv:1808.08949 (2018)
8. Samson, C.W., Tu, W., Sim, I., Richesson, R.: Formal representation of eligibility criteria: a literature review. J. Biomed. Inform. **43**(3), 451–467 (2010). https://doi.org/10.1016/j.jbi.2009.12.004. http://www.sciencedirect.com/science/article/pii/S1532046409001592
9. Shaw, P., Uszkoreit, J., Vaswani, A.: Self-attention with relative position representations. arXiv preprint arXiv:1803.02155 (2018)
10. Shido, Y., Kobayashi, Y., Yamamoto, A., Miyamoto, A., Matsumura, T.: Automatic source code summarization with extended tree-LSTM. arXiv preprint arXiv:1906.08094 (2019)
11. Shivade, C., Hebert, C., Lopetegui, M., de Marneffe, M.C., Fosler-Lussier, E., Lai, A.M.: Textual inference for eligibility criteria resolution in clinical trials. J. Biomed. Inform. **58**, 211–218 (2015). https://doi.org/10.1016/j.jbi.2015.09.008. http://www.sciencedirect.com/science/article/pii/S1532046415002014. Proceedings of the 2014 i2b2/UTHealth Shared-Tasks and Workshop on Challenges in Natural Language Processing for Clinical Data
12. Voorhees, E.M.: The TREC medical records track. In: Proceedings of the International Conference on Bioinformatics, Computational Biology and Biomedical Informatics, p. 239. ACM (2013)

Tell Me About Your Day: Designing a Conversational Agent for Time and Stress Management

Libby Ferland, Jude Sauve, Michael Lucke, Runpeng Nie, Malik Khadar, Serguei Pakhomov, and Maria Gini

Abstract Growing interest in applications of AI in healthcare has led to a similarly elevated interest in fully integrated smart systems in which disparate technologies, such as biometric sensors and conversational agents, are combined to address health problems like medical event detection and response. Here we describe an ongoing project to develop a supportive health technology for stress detection and intervention and discuss a pilot application for one component, the conversational agent.

1 Introduction

Stress is a universal constant in modern life, for both young and old alike. Stress has heavy societal impact–since stress plays a major role in physical and mental well-being, people repeatedly experiencing stress without appropriate coping mechanisms cannot achieve their full potential in any number of situations [15, 17]. A variety of mitigation and coping techniques have been proposed to address stress, notably effective time management and planning skills, but these can be difficult to learn without proper tools [10, 17].

Stress management is a key component of lifestyle modification in the treatment of many conditions among elderly people [4, 16]. The increased availability and ease of use of intelligent technologies such as conversational agents (CAs) gives them

L. Ferland (✉) · J. Sauve · M. Lucke · R. Nie · M. Khadar · M. Gini
Department of Computer Science and Engineering, University of Minnesota, Minneapolis, MN 55455, USA
e-mail: ferla006@umn.edu

S. Pakhomov
Department of Pharmaceutical Care and Health Systems, University of Minnesota, Minneapolis, MN 55455, USA

© The Editor(s) (if applicable) and The Author(s), under exclusive license to Springer Nature Switzerland AG 2021

A. Shaban-Nejad et al. (eds.), *Explainable AI in Healthcare and Medicine*, Studies in Computational Intelligence 914,
https://doi.org/10.1007/978-3-030-53352-6_28

great potential as supplements for human caregivers in supporting older populations and for more effective stress management in general.

Here we describe the first phases of an ongoing project, "SMARTHUGS," which is a multidisciplinary effort to create a system designed to detect stress in users and deliver a multimodal touch- and speech-based intervention when users experience high stress. The end product is a prototype of a compression garment, such as a vest, with an integrated unobtrusive physiological signal monitoring using a wrist-worn device (Empatica E4) and a CA interface. The monitor and the CA in tandem will be able to detect potential stressors in real time and flag them, activating the garment to provide therapeutic compression similar to a hug. The CA component is the focus of this paper, specifically the design of its architecture and the domain as initial steps towards addressing open challenges in the field as informed by real user data.

2 Background and Related Work

The key motivating concepts to this work include the effects of stress on health and the usefulness of CAs as support mechanisms.

2.1 Stress and Health

Many studies have shown the effects of stress on conditions like heart disease, depression, and sleep disorders, to name a few [11]. It has also been shown that stress has negative effects on disease progression and quality of life, and managing stress can significantly affect disease management [16]. Stress also impacts mental health, which appropriate interventions can mitigate or reverse [4]. This is notably true for cognitive functioning, which is salient for age-related cognitive decline [7, 11]. Stress management and coping skills are important aspects of treatment plans and can be used as cognitive support [6].

2.2 Time Management and Coping

Time management is closely related to stress. The development of effective time management skills has been linked to stress mediation and anxiety reduction in many situations and age groups [10, 15, 17]. Time management skills lead to a perception of control over time, which in turn increases reported satisfaction levels in measures such as academic/job performance and overall life satisfaction [15]. Time management skills are also important for older adults, especially those experiencing cognitive decline [6].

2.3 Conversational AI and Support

CA systems have been used in many applications, such as changing habits, patient education, and as social support for older users [8]. Health applications are sensitive [14] and so arguably benefit most from data-driven ground-up design. However, the cycle of prototyping and user testing is costly [3], making the use of user-centered design in many support domains as difficult and uncommon as it is necessary, even despite its desirability [12].

3 Proposed Approach

Our prototype system incorporates three separate applications. The tasks done by the system consist of two separate yet similar workflows. In a *prospective* use case, the CA talks with a user about their upcoming schedule and stress levels for the day. In the *retrospective* use case, the CA acts like a diary and talks with a user about the events of their day and the stress they actually experienced.

The key challenges reflect open challenges with CAs and NLP in general: (1) interpreting ambiguous or complex temporal expressions [18], (2) appropriately responding to stress-related disclosures, and (3) doing the same for other affective states and self disclosure [14]. These challenges informed our architecture choices (Sect. 3.1) and were directly drawn from observations in human experiments (Sect. 3.2).

3.1 System Components

3.1.1 MindMeld

MindMeld is an open-source conversational AI platform recently released by Cisco [2], which neatly packages all the NLP modules necessary to create a fully-integrated dialogue system. We chose MindMeld for this pilot due to its robustness and flexibility, which will be needed to further address challenges 2 and 3. Unlike other frameworks which require a complex set of dialogue rules, MindMeld fully exploits machine learning methods to train an agent given only a set of domains, entities, and intents within the domain.

3.1.2 SUTime

SUTime, a library developed by the Stanford NLP group, is a robust deterministic temporal expression parser [1]. SUTime uses an extensive rule base to parse temporal expressions into the TIMEX3 format. We chose SUTime as an alternate pipeline for temporal information because Mindmeld does not handle some types of temporal expressions (eg., discussion of recurring or conditional events) natively, so more support is needed to effectively address challenge 2. SUTime also allows us to reduce the amount of training data required; the application only needs to recognize an entity, not fully understand its contents.

3.1.3 Google Calendar API

We included an external calendar service in our system as a way to increase system performance. With an external API, we can store, retrieve, and modify events without slowing the application down. We chose the Google Calendar API for ease of use and familiarity by developers and users [5].

3.2 Implementation

3.2.1 Data-Driven Dialogue Development

The dialogue flow for our CA is based on scripted interactions with participants in a Wizard of Oz protocol (WOZ) [3]. We sought to answer two questions: When faced with a machine, how do users actually talk about the events in their day? How do they talk about stress and time, and how comfortable are they when discussing these topics? To answer those questions, participants talked to a realistic "prototype" about their daily schedule and their stress levels and mood. Using real user data as a basis for our CA allows us to create a more natural dialogue flow and provides us examples of possible inputs. User data are also reserved as a test set for the CA. Figure 1 shows two dialogue segments from this dataset with rich features.

(1): Response to schedule prompt, with complex temporal expressions and habit disclosure

System: What's next on your schedule after reading and studying?
Participant: Um...Around that time...once I'm done with that it should be about maybe 4:00 or something. I usually just wait until {PERSON} gets home or when {PERSON} needs to get picked up and then I'll go do that but it'll be around maybe 6:00 or 8:00 depending on what days of the week it is.

(2): Response to stress prompt with indirect voluntary preference disclosure

System: How stressful do you think that will be?
Participant: Um, probably around a two, because it's cold outside.

Fig. 1 Sample real-world dialogue fragments

3.2.2 Domain Modeling

Our implementation began with developing a training dataset of calendar entries provided by other researchers. We then developed our ontology of calendar events and annotation scheme. The key entity in our application is an event, which contains these entities: (1) event name, (2) location, (3) start date/time, (4) end date/time, (5) duration, (6) frequency/frequency duration, and (7) stress level. Of these, only 3, 4, and 5 were predefined in the Mindmeld platform. In the first training set, 300 calendar entries were annotated by two independent raters.

3.2.3 Conversational Agent

Our pilot CA is designed to exploit learned domain models rather than relying on a rule base. Users need not fill in event information in any particular order, and can mostly move freely between different domains and intents. The only rule enforced is that users must provide all necessary details for each event. However, we built some fail-safes into the application to minimize repetitiveness and user frustration. First, the application fills missing slots automatically using information already provided by the user. Second, in the case of entity resolution failure, the system will re-prompt for the desired information twice before switching to a rule base. Third, in the case of major failure, such as a user attempting an unknown workflow, the agent will try a redirect and then switch to a rule base.

Figure 2 shows a sample dialogue with the system when adding a calendar event.

User: I have a meeting with Paul and Ryan at one.
System: Where will that be?
User: in my office
System: How long do you think it will take?
User: Hopefully about an hour
System: How stressful do you think it will be?
User: I'm not sure.
System: How would you rate that on a scale from one to seven?
User: Maybe a 5
System: That sounds stressful. Please tell me how it goes later.
User: Okay.
System: Thank you. I'll add an event called meeting with Paul and Ryan in your office to your calendar for today. I'll schedule it to start at 1:00 PM and end at 2:00 PM. Does that sound good?
User: yep
System: Great. What's the next item on your schedule?
User: That's it
System: Goodbye.

Fig. 2 Sample dialogue for event entry during a prospective interaction, with features such as automated slot filling and ability to switch between intents and domains

4 Conclusions and Future Work

This work is in early stages and presents many opportunities for improvement. In the short term, we will increase data privacy and flexibility by both connecting this CA to an in-house speech recognition server and developing a simple calendar graphical interface to be hosted internally. We will also expand our dataset and validate our annotation scheme with naïve raters. Long-term questions to explore include:

4.1 Designing Conversational Interventions for Stress

CAs in mental health applications face a number of unique challenges and pitfalls. How does an agent appropriately detect affective state from conversation? What constitutes an appropriate reply to self-disclosure and different user states? The answers to these questions are vital and difficult, considering the sensitivity of mental health issues [14]. CAs used to address mental health issues must be carefully designed to be appropriate and sensitive to user needs, while not being artificial or repetitive.

4.2 Improved Handling of Complex Temporal Expressions

Humans talk about time in daily conversation in a way which is often coarse-grained and ambiguous. Fully understanding temporal expressions encompasses problems such as commonsense reasoning and temporal resolution [18, 19]. For instance, our prototype currently struggles to interpret expressions with long tails and conditional expressions ("Remind me to take the garbage out every Wednesday, except when it's after a holiday."), which it should understand to converse naturally.

4.3 Enhancing User Modeling with Self Disclosure

The need for better user modeling in CAs is not a new problem, but it is both complex and a notable source of dissatisfaction in users [9, 13]. Though user-dependent, our WOZ testing has shown that users display at least a moderate degree of self-disclosure. A history of a user's schedule, including patterns, could be used as part of a more complex user model.

Acknowledgements Work supported in part by the University of Minnesota Grand Challenges Research, NSF I/UCRC (IIP-1439728) and NSF EAGER (IIS 1927190) grants. The authors would like to thank Anja Wiesner and Sarah Schmoller for their assistance in model and data development.

References

1. Chang, A.X., Manning, C.D.: SUTime: a library for recognizing and normalizing time expressions. In: Proceedings of 8th International Conference on Language Resources and Evaluation (LREC), pp. 3735–3740. European Language Resources Association (2012)
2. Cisco: The Conversational AI Playbook (2019). https://www.mindmeld.com/docs/. Accessed 19 Nov 2019
3. Dahlbäck, N., Jönsson, A., Ahrenberg, L.: Wizard of Oz studies–why and how. Knowl.-Based Syst. **6**(4), 258–266 (1993)
4. George, L.K.: Social factors, depression, and aging. In: Handbook of Aging and the Social Sciences, pp. 149–162. Elsevier (2011)
5. Google: Calendar API (2019). https://developers.google.com/calendar. Accessed 17 Nov 2019
6. Jean, L., Bergeron, M., Thivierge, S., Simard, M.: Cognitive intervention programs for individuals with mild cognitive impairment: systematic review of the literature. Am. J. Geriatr. Psychiatry **18**(4), 281–296 (2010)
7. Justice, N.J.: The relationship between stress and Alzheimer's disease. Neurobiol. Stress **8**, 127–133 (2018)
8. Laranjo, L., Dunn, A.G., Tong, H.L., Kocaballi, A.B., Chen, J., Bashir, R., Surian, D., Gallego, B., Magrabi, F., Lau, A.Y., et al.: Conversational agents in healthcare: a systematic review. J. Am. Med. Inform. Assoc. **25**(9), 1248–1258 (2018)
9. Luger, E., Sellen, A.: Like having a really bad PA: the gulf between user expectation and experience of conversational agents. In: Proceedings of CHI Conference on Human Factors in Computing Systems (CHI), pp. 5286–5297 (2016)
10. Macan, T.H., Shahani, C., Dipboye, R.L., Phillips, A.P.: College students' time management: correlations with academic performance and stress. J. Educ. Psychol. **82**(4), 760–768 (1990)
11. McEwen, B.S.: Central effects of stress hormones in health and disease: understanding the protective and damaging effects of stress and stress mediators. Eur. J. Pharmacol. **583**(2–3), 174–185 (2008)
12. Meiland, F., Innes, A., Mountain, G., Robinson, L., van der Roest, H., García-Casal, J.A., Gove, D., Thyrian, J.R., Evans, S., Dröes, R.M., et al.: Technologies to support community-dwelling persons with dementia: a position paper on issues regarding development, usability, effectiveness and cost-effectiveness, deployment, and ethics. JMIR Rehabil. Assist. Technol. **4**(e1), 1 (2017)
13. Milhorat, P., Schlogl, S., Chollet, G., Boudy, J., Esposito, A., Pelosi, G.: Building the next generation of personal digital assistants. In: Proceedings of 1st International Conference on Advanced Technologies for Signal and Image Processing (ATSIP), pp. 458–463. IEEE (2014)
14. Miner, A., Chow, A., Adler, S., Zaitsev, I., Tero, P., Darcy, A., Paepcke, A.: Conversational agents and mental health: theory-informed assessment of language and affect. In: Proceedings of 4th International Conference on Human-Agent Interaction (HAI), pp. 123–130. ACM (2016)
15. Monat, A., Lazarus, R.S.: Stress and Coping: An Anthology. Columbia University Press, New York (1991)
16. Newman, S., Steed, L., Mulligan, K.: Self-management interventions for chronic illness. Lancet **364**(9444), 1523–1537 (2004)
17. Prenda, K.M., Lachman, M.E.: Planning for the future: a life management strategy for increasing control and life satisfaction in adulthood. Psychol. Aging **16**(2), 206–216 (2001)
18. Rong, X., Fourney, A., Brewer, R.N., Morris, M.R., Bennett, P.N.: Managing uncertainty in time expressions for virtual assistants. In: Proceedings of CHI Conference on Human Factors in Computing Systems (CHI), pp. 568–579. ACM (2017)
19. Zhou, B., Khashabi, D., Ning, Q., Roth, D.: "Going on a vacation" takes longer than "Going for a walk": a study of temporal commonsense understanding. In: Proceedings of the 2019 Conference on Empirical Methods in Natural Language Processing and the 9th International Joint Conference on Natural Language Processing (EMNLP-IJCNLP), pp. 3354–3360. ACL (2019)

Accelerating Psychometric Screening Tests with Prior Information

Trevor Larsen, Gustavo Malkomes, and Dennis Barbour

Abstract Classical methods for psychometric function estimation either require excessive measurements or produce only a low-resolution approximation of the target psychometric function. In this paper, we propose solutions for rapid high-resolution approximation of the psychometric function of a patient given her or his prior exam. We develop a rapid screening algorithm for a change in the psychometric function estimation of a patient. We use Bayesian active model selection to perform an automated pure-tone audiometry test with the goal of quickly finding if the current estimation will be different from the previous one. We validate our methods using audiometric data from the National Institute for Occupational Safety and Health (NIOSH). Initial results indicate that with a few tones we can (i) detect if the patient's audiometric function has changed between the two test sessions with high confidence, and (ii) learn high-resolution approximations of the target psychometric function.

1 Introduction

Machine learning has great potential to improve healthcare, with such applications as personalized medicine and automated robotic surgery [7, 11, 16]. In particular,

---

T. Larsen · D. Barbour
Washington University in St. Louis, St. Louis, MO 63130, USA
e-mail: trevorlarsen@wustl.edu

D. Barbour
e-mail: dbarbour@wustl.edu

G. Malkomes (✉)
SigOpt, San Francisco, CA 94108, USA
e-mail: gustavo@sigopt.com

© The Editor(s) (if applicable) and The Author(s), under exclusive license to Springer Nature Switzerland AG 2021

A. Shaban-Nejad et al. (eds.), *Explainable AI in Healthcare and Medicine*,
Studies in Computational Intelligence 914,
https://doi.org/10.1007/978-3-030-53352-6_29

machine learning can be use to aid in the judicious application of healthcare resources in resource-poor settings. An example of this is implementing the most appropriate use of expensive or intrusive diagnostic procedures.

Perceptual testing to diagnose disorders of hearing and vision requires many repeated stimulus presentations to patients in order to determine their condition. By carefully controlling the experiment conditions (*e.g.* the strength, duration, or other characteristics of the stimulus), clinicians estimate the subject's perception. Precisely, the clinician control the experiment conditions in order to learn a psychometric function, an inference model mapping features of the physical stimulus to the patient response. We focus on audiometry tests, but our methodology generalizes to any psychometric test reflecting perceptual or cognitive phenomena.

Traditional audiometry tests, such as the modified Hughson-Westlake procedure [2], are low-resolution and time-consuming. Clinicians present a series of tones at various frequencies (corresponding to pitch) and intensities (corresponding to loudness) to the patients and record their response (*i.e.* whether he or she hears the tone). A standard audiometry test requires 15–30 min, and it only estimates the hearing threshold—the softest sound level one can hear—at a few discrete frequencies. This labor-intensive approach scales poorly to large populations. Crucially, some disorders such as noise-induced hearing loss can be entirely preventable with a sensitive, early diagnostic. Unfortunately, the standard methodology greatly hinders rapid screening at high-resolution.

Alternative tests have been investigated, and special interest has been given to procedures using Bayesian active learning [12, 15]. In this framework, it is possible to leverage audiologist's expertise to construct automated audiometry tests. Active learning strategies for estimating a patient's hearing threshold were studied in [5, 15, 17]. [4] further applied Bayesian active model selection to rapid screening for noise-induced hearing loss.

We extend previous approaches on Bayesian active learning for audiometry tests by mathematically incorporating prior information about the patients. Our framework tackles clinically relevant psychometric tests. We propose a rapid screening method for changes in the patient's clinical condition. Given a previous audiometry test, our approach automatically delivers a sequence of stimuli to quickly determine whether the patient's hearing thresholds differ from a reference exam, such as a previous test in the same patient or a population average. Initial results show that our methods can result in fewer tone deliveries (less than 20) when compared to previously methods such as [17], which delivers 80 tones, or the conventional modified Hughson-Westlake method, which requires more than 90 tones.

2 Bayesian Active Differential Inference

Consider supervised learning problems defined on an input space $\mathcal{X}$ and an output space $\mathcal{Y}$. We are given a set of observations $\mathcal{D} = (\mathbf{X}, \mathbf{y})$, where $\mathbf{X}$ represents the design matrix of independent variables $\mathbf{x}_i \in \mathcal{X}$ and $\mathbf{y}$ the associated vector of dependent variables $y_i = y(\mathbf{x}_i) \in \mathcal{Y}$. We assume that these data were generated via a latent function $f: \mathcal{X} \to \mathbb{R}$ with a known observation model $p(\mathbf{y} \mid \mathbf{f})$, where $f_i = f(\mathbf{x}_i)$. In this context the latent function f is the psychometric function, and the initial data were obtained during a previous exam in the same patient.

Suppose that, after some undetermined period of time following the first exam, we wish to collect a new set of observations $\mathcal{D}'$ from the same phenomenon f, *e.g.*, the psychometric function in the same individual. Our goal is to perform measurements—select $\mathbf{x}^* \in \mathcal{X}$ and observe $y^* = y(\mathbf{x}^*)$—to quickly (i) learn the target function or (ii) distinguish whether or not the latent function f has changed. In our medical application, this translates to either a faster audiometry pure-tone test (i.e. it converges with fewer tones) or rapid screening for a different clinical condition.

We begin by modeling this as a two-task *active learning* problem[1]. We define a new input space by augmenting $\mathcal{X}$ with a feature representing which task (or test) the data points come from, *i.e.*, $\mathcal{X}' : \mathcal{X} \times \mathcal{T}$, where $\mathcal{T} = \{1, 2\}$. For all prior observations we have $\mathcal{D} = ([\mathbf{X}, \mathbf{1}], \mathbf{y})$ and each new observation will be from the new task, $\mathbf{x}^* = [\mathbf{x}^T, 2]^T, \mathbf{x} \in \mathcal{X}$. Next, we hypothesize that the data can be explained by one of two probabilistic models: $\mathcal{M}_f$, which assumes that $\mathcal{D}$ and $\mathcal{D}'$ come from the same underlying function f; and $\mathcal{M}_g$, which offers a different explanation for the most recent set of observations $\mathcal{D}'$. Under these assumptions we are interested in selecting candidate locations $\mathbf{x}^*$ to quickly differentiate these two models. We pursue these goals motivated by ideas from information theory, which were successfully applied in a series of active-learning papers [4–6, 8–10].

Specifically, for rapid diagnostic of a clinical condition, we select $\mathbf{x}^*$ maximizing the mutual information between the observation y^* and the unknown model class:

$$I(y^*; \mathcal{M} \mid \mathbf{x}^*, \mathcal{D} \cup \mathcal{D}') = H[y^* \mid \mathbf{x}^*, \mathcal{D} \cup \mathcal{D}'] - \mathbb{E}_{\mathcal{M}}[H[y^* \mid \mathbf{x}^*, \mathcal{D} \cup \mathcal{D}', \mathcal{M}]] \quad (1)$$

We use the (more-tractable) formulation of mutual information [4] that requires only computing the model-conditional predictive distributions: $\{p(y^* \mid \mathbf{x}^*, \mathcal{D} \cup \mathcal{D}', \mathcal{M}_i)\}$, the differential entropy of each of these distributions, and the differential entropy of the model-marginal predictive distribution:

$$p(y^* \mid \mathbf{x}^*, \mathcal{D} \cup \mathcal{D}') = \sum_j p(y^* \mid \mathbf{x}^*, \mathcal{D} \cup \mathcal{D}', \mathcal{M}_j)\, p(\mathcal{M}_j \mid \mathcal{D} \cup \mathcal{D}') \quad (2)$$

2.1 *Bayesian Active Differential Inference for Gaussian Processes*

Following a series of work on active learning for audiometry, we use a GP to model the psychometric function [1, 3–5, 15, 17, 18]. A Gaussian Process (GP) is completely

[1] We use active learning in its broader sense of intelligently selecting observations to achieve *any* goal, as opposed to restricting it to *learning* predictors with few training samples.

defined by its first two moments, a mean function $\mu: \mathcal{X}' \to \mathbb{R}$ and a covariance or kernel function $K: \mathcal{X}'^2 \to \mathbb{R}$. For further details on GPs see [14].

Audiometric Function Model $\mathcal{M}_f$. We use a constant prior mean function $\mu_f(.) = c$ to model a frequency-independent natural threshold. While audiograms do not necessary have a constant mean, previous research has shown that a constant mean function is sufficient for modeling audiograms, as the covariance function captures the shape of the psychometric function in the posterior distribution [1, 5, 17]. For the covariance function, we use a linear kernel in intensity and a squared-exponential covariance in frequency as proposed by [4, 5]. Let $[i, \phi]$ represent a tone stimulus, with i representing its intensity and ϕ its frequency, and set $\mathbf{x} = [i, \phi, t] \in \mathcal{X}'$, where t is the task-related feature. The covariance function of our first model is independent of the task and given by:

$$K_f([i, \phi, t], [i', \phi', t']) = K_{[i,\phi]}([i, \phi], [i', \phi']) = \alpha i i' + \beta \exp(-\frac{1}{2\ell^2}|\phi - \phi'|^2) \quad (3)$$

where $\alpha, \beta > 0$ weight each component and $\ell > 0$ is a length scale of frequency-dependent random deviations from a constant hearing threshold. This kernel enforces the idea that (i) hearing is monotonic as a function of intensity, and (ii) the audiometric function should be continuous and smooth across the frequency domain because nearby locations in the cochlea are physically coupled [5].

Our $\mathcal{M}_f$ is a GP model, $f = \mathcal{GP}(\mu_f, K_f)$, with cumulative Gaussian likelihood (probit regression), where y takes a Bernoulli distribution with probability $\Phi(f(\mathbf{x}))$, and Φ is the Gaussian CDF. Computing the exact form of the latent posterior distribution is intractable because of the probit likelihood function. Thus, the posterior must be approximated. For this model, we use the *expectation propagation* (EP) approximation [13, 14]. Additionally, we perform inference by maximizing the hyperparameter posterior, finding the maximum *a posteriori* (MAP) hyperparameters. Notice that we only perform inference after observing $\mathcal{D}$. During the active-learning procedure for constructing $\mathcal{D}'$, we fix the MAP hyperparameters, due to our hypothesis that the new observations $\mathcal{D}'$ should not drastically change our belief about f.

Differential Model $\mathcal{M}_g$. The previous model assumes that both $\mathcal{D}$ and $\mathcal{D}'$ can be explained by the same probabilistic model. Now, we present our differential model, which has a less restrictive assumption. We assume that the first set of observations $\mathcal{D}$ was generated by a latent function f and that the new observations $(\mathbf{x}^*, y^*) \in \mathcal{D}'$ are generated by a different latent function $g: \mathcal{X}' \to \mathbb{R}$. Specifically, we use the following kernel to capture the correlation between both tasks:

$$K_g([i, \phi, t], [i', \phi', t']) = K_t(t, t') \times K_{[i,\phi]}([i, \phi], [i', \phi']) \quad (4)$$

where $K_{[i,\phi]}$ is the same as in (3) and the task or conjoint kernel is defined as: 1 if $t = t'$ and ρ_t, if $t \neq t'$. The new parameter ρ_t can be interpreted as the correlation between tasks and is referred to as the task or conjoint correlation. We learn the hyperparameters of this model for both mean and covariance functions. Hyperparameters associated exclusively to the first task are fixed, similarly to previous model. For the input locations related to the second task $t = 2$, we optimize $\theta_g = (c_g, \alpha_g, \beta_g, \ell_g)$ after obtaining each new observation. The exception is the conjoint correlation parameter ρ_t, which is marginalized by sampling 50 linearly uniformly spaced points in $[-1, 1]$.

2.1.1 Related Work

Prior work using GPs for audiogram estimation have proposed a variant of uncertainty sampling that selected points with maximum variance [17]. This work increased both the speed and accuracy of the standard threshold audiogram estimation by at least an order of magnitude. In [4], a framework was proposed to quickly distinguish which of a number of predetermined models best explains a function being actively observed. Both of the aforementioned papers explored the use of machine learning audiometry in one ear, this approach was extended to exploit the shared variance between ears using a conjoint audiogram [1, 3]. Learning the audiogram for both ears was shown to be as fast as, or even faster than, learning a single ear individually.

Our work is complementary to these methodologies. Instead of ignoring prior information from the individual, we proposed starting our current test acknowledging a previous audiometric test might be available to reference. The underlying psychometric function that the audiogram is meant to estimate is correlated with previous audiometric tests that the subject has taken. In the same way that conjoint audiogram estimation utilizes covariance between ears to reduce the number of tones needed to estimate the underlying function, we propose a novel approach for determining whether two audiograms represent the same or a different psychometric function.

3 Discussion

3.1 Related Work

We compare BADS against three different active learning strategies. Our first baseline is similar to how we collected the data from the previous exam, using Bayesian active learning by disagreement BALD that computes $I(y^*;\ f)$. This simulates a procedure that ignores previous information; hence during the decision-making part we simply compute the strategy with respect to the most recent data, $I(y^*;\ f \mid \mathcal{D}')$. Our second baseline is an adaptation of uncertainty sampling (US) considering both models and all data. The last baseline is random sampling (RND). Figure 1 shows results for three selected experiments. BADS results were averaged across 10 experiments during 15 iterations, and the baselines ran for more iterations and were averaged across 5 experiments. These initial results indicate that BADS outperforms the baselines.

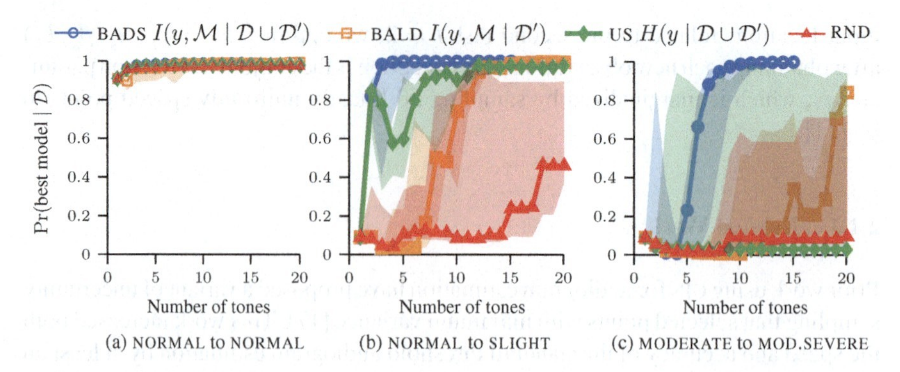

Fig. 1 Posterior probability of best model as function of number of iterations (tones delivered). Solid lines represent the median. Upper and lower quartiles displayed by the shaded regions

4 Conclusion

We proposed a novel framework for quickly determining whether a complex physiological system is in a different state than a reference. Previously, complete models of both systems would need to be estimated before their similarity could be evaluated. BADS enables queries of the new system that are most informative in order to answer the question of whether it is the same as or different from the reference system.

References

1. Barbour, D.L., DiLorenzo, J.C., Sukesan, K.A., Song, X.D., Chen, J.Y., Degen, E.A., Heisey, K.L., Garnett, R.: Conjoint psychometric field estimation for bilateral audiometry. Behav. Res. Methods **51**(3), 1271–1285 (2019)
2. Carhart, R., Jerger, J.F.: Preferred method for clinical determination of pure-tone thresholds. J. Speech Hear. Disord. **24**(4), 330–345 (1959)
3. Di Lorenzo, J.C.: Conjoint audiogram estimation via Gaussian process classification. In: Master of Science thesis (2017)
4. Gardner, J., Malkomes, G., Garnett, R., Weinberger, K.Q., Barbour, D., Cunningham, J.P.: Bayesian active model selection with an application to automated audiometry. Advances in Neural Information Processing Systems **28**, 2386–2394 (2015)
5. Gardner, J.R., Song, X., Weinberger, K.Q., Barbour, D., Cunningham, J.P.: Psychophysical detection testing with Bayesian active learning. In: UAI (2015)
6. Garnett, R., Osborne, M.A., Hennig, P.: Active learning of linear embeddings for Gaussian processes. In: UAI, pp. 230–239 (2014)
7. Hamburg, M.A., Collins, F.S.: The path to personalized medicine. New Engl. J. Med. **363**(4), 301–304 (2010). pMID: 20551152
8. Hernández-Lobato, J.M., Hoffman, M.W., Ghahramani, Z.: Predictive entropy search for efficient global optimization of black-box functions. In: NIPS, pp. 918–926 (2014)
9. Houlsby, N., Hernández-Lobato, J.M., Ghahramani, Z.: Cold-start active learning with robust ordinal matrix factorization. In: ICML, pp. 766–774 (2014)
10. Houlsby, N., Huszar, F., Ghahramani, Z., Hernández-Lobato, J.M.: Collaborative Gaussian processes for preference learning. In: NIPS, pp. 2096–2104 (2012)

11. Kononenko, I.: Machine learning for medical diagnosis: history, state of the art and perspective. Artif. Intell. Med. **23**(1), 89–109 (2001)
12. Leek, M.R.: Adaptive procedures in psychophysical research. Percept. Psychophys. **63**(8), 1279–1292 (2001). Nov
13. Minka, T.P.: Expectation propagation for approximate Bayesian inference. In: Proceedings of the Seventeenth Conference on Uncertainty in Artificial Intelligence, pp. 362–369. Morgan Kaufmann Publishers Inc. (2001)
14. Rasmussen, C.E., Williams, C.K.I.: Gaussian Processes for Machine Learning. MIT Press, Cambridge (2006)
15. Schlittenlacher, J., Turner, R.E., Moore, B.C.J.: Audiogram estimation using Bayesian active learning. J. Acoust. Soc. Am. **144**(1), 421–430 (2018)
16. Shademan, A., Decker, R.S., Opfermann, J.D., Leonard, S., Krieger, A., Kim, P.C.W.: Supervised autonomous robotic soft tissue surgery. Sci. Transl. Med. **8**(337), 337ra64 (2016)
17. Song, X.D., Wallace, B.M., Gardner, J.R., Ledbetter, N.M., Weinberger, K.Q., Barbour, D.L.: Fast, continuous audiogram estimation using machine learning. Ear Hear. **36**, 326–335 (2015)
18. Song, X.D., Garnett, R., Barbour, D.L.: Psychometric function estimation by probabilistic classification. J. Acoust. Soc. Am. **141**(4), 2513–2525 (2017)

Can an Algorithm Be My Healthcare Proxy?

Duncan C. McElfresh, Samuel Dooley, Yuan Cui, Kendra Griesman, Weiqin Wang, Tyler Will, Neil Sehgal, and John P. Dickerson

Abstract Planning for death is not a process in which everyone participates. Yet a lack of planning can severely impact a patient's well-being, the well-being of her family, and the medical community as a whole. Advance Care Planning (ACP) has been a field in the United States for a half-century, and often using short surveys or questionnaires to help patients consider future end of life (EOL) care decisions. Recent web-based tools promise to increase ACP participation rates; modern techniques from artificial intelligence (AI) could further improve and personalize these tools. We discuss two hypothetical AI-based apps and their potential implications. We hope that this paper will encourage thought about appropriate applications of AI in ACP as well as implementation of AI to ensure patient intentions are honored.

Keywords Artificial intelligence · Machine learning · Advance Care Planning

D. C. McElfresh (✉) · S. Dooley · N. Sehgal · J. P. Dickerson
University of Maryland, College Park, USA
e-mail: dmcelfre@umd.edu

Y. Cui
Oberlin College, Oberlin, USA

K. Griesman
Haverford College, Haverford, USA

W. Wang
Pennsylvania State University, State College, USA

T. Will
Michigan State University, East Lansing, USA

© The Editor(s) (if applicable) and The Author(s), under exclusive license to Springer Nature Switzerland AG 2021
A. Shaban-Nejad et al. (eds.), *Explainable AI in Healthcare and Medicine*, Studies in Computational Intelligence 914,
https://doi.org/10.1007/978-3-030-53352-6_30

1 Introduction

Modern medicine is exceptional at prolonging life. New devices, drugs, and medical procedures can stave off death in cases that–until recently–were uniformly fatal. Legal and ethical norms dictate that each patient may decide *for themselves* whether or not to receive these life-sustaining treatments (LST); however patients in need of LST and/or EOL care are often physically or cognitively unable to make these decisions. In anticipation of this, some people use Advance Care Planning (ACP) in order to better understand their care options and ensure they receive only the care they want. Medical decision-making is challenging in general, and is complicated by each patient's unique values and health conditions. ACP is often an iterative process with many conversations–involving the patient's family, medical professionals, and sometimes ethicists, lawyers, and religious representatives. In the best case, ACP culminates in a written description of the patient's wishes, such as an Advance Directive (AD) or Physician Orders for Life Sustaining Treatment (POLST); these documents are intended to represent the patient's wishes to their family and medical professionals, when the patient is unable to do so.

However, most people do not participate in ACP; in the US, document completion rates are especially low among less-educated, poorer, and non-white people [13]. Without ACP, EOL care decisions are often left to a proxy decision maker–usually a family member or friend. Proxies often make decisions that contradict their patient's wishes–usually resulting in prolonged LST until it becomes medically or financially impossible. Insufficient ACP has ethical, legal, and financial implications; LST is often inconsistent with patients' authentic wishes, expensive, and emotionally and morally burdensome for the proxy decision maker. Some of these problems persist even *with* ACP since existing forms do not address all EOL care scenarios, and may not lead to preferred outcomes. With Artificial Intelligence (AI) being adopted in various medical and health fields, we pose the question: what role *can* and *should* AI play in ACP[1]?

1.1 A Very Brief History of Advance Care Planning

Modern ADs include several components developed over the last fifty years. These include a living will (LW)–a patient's exact wishes for when to receive LST; a *medical power of attorney* (MPOA)[2] in which a patient designates someone (often a family member or friend) to make care decisions on their behalf when they are unable to do so themselves; Do Not Resuscitate (DNR) and Do Not Intubate (DNI) orders; and POLST: a document translating a patient's goals and wishes into concrete medical orders [3].

---

[1]Please see our full paper on arXiv for a more in-depth discussion.

[2]Nomenclature varies by state and may be phrased as Durable Power of Attorney for Health Care, Health Care Proxy, Surrogate Decision Maker, or other terms.

Two challenges persist: (1) it is difficult to translate patients' unique preferences and goals into care decisions, and (2) few people actually participate in ACP (less than one-third of adults in the US [13]). Recent web-based tools such as PREPARE[3] [16] and MyDirectives[4] [6] can potentially increase ACP participation rates, but they still do not capture patients' nuanced values and wishes, and often fail to influence care decisions [14]. With these challenges in mind, we now turn to the future of AI and ACP.

2 Looking Forward: An ACP Decision Aid

AI techniques have been shaping the medical profession in myriad ways. Early examples include tools for diagnosis and treatment planning [9]; modern ML methods have been broadly applied to detect cancers [10]; automated alerts are widely used to detect drug interactions and sepsis [15].

Future applications for ACP will likely take the form of *decision aids*, designed to help care providers make better decisions for their patients. Computerized decision aids have been developed for a variety of purposes including diagnosis, prevention, disease management, and prescription management; however their impact on patient outcomes is not always clear [7]. These applications leverage expert knowledge to make predictions (e.g., a diagnosis) or recommendations (e.g., a treatment) to a care provider. In ACP, the decision question is often whether or not to provide LST, and for how long and in what forms. Answering these questions is not a matter of expert knowledge, but rather one of patient goals and wishes. And patient preferences are not simple: they depend on a variety of factors, such as personal values, religious beliefs, goals of care, cultural background, and family preferences [17].

An effective ACP decision aid would understand, and accurately represent, a patient's wishes for care. This is a complicated task, but recent advancements in computational methods for preference elicitation and recommender systems could provide innovation. We briefly describe these fields here.

Preference Elicitation is the study of people's preferences, usually by learning a *utility function*. This field has its roots in marketing and economics though recent applications include healthcare [11] and public policy [1]. The AI community has developed a wealth of elicitation methods which can be applied to a huge variety of scenarios.

Recommender Systems share a similar lineage with preference elicitation. Used mostly in commercial settings, recommender systems use consumer data to suggest products (e.g., Amazon), content (e.g., Netflix), or social connections (e.g., Facebook) to users; for a review, see [12].

---

[3]https://prepareforyourcare.org.

[4]https://mydirectives.com.

We anticipate that an effective ACP decision aid would apply these fields (among others) to learn a patient's care preferences. Of course every AI application has risks–in healthcare, such applications can threaten patient autonomy, privacy, and can worsen existing healthcare disparities [8]. Our goal is to anticipate these applications and their potential impact on patient outcomes. Next, to spark discussion, we outline two hypothetical AI-based ACP applications.

2.1 Applications

We present two hypothetical applications, each designed to improve adherence to patient wishes through ACP. While these applications are similar in appearance and function, their implications are quite different.

Application 1: First we consider CAREDECIDER, an AI-driven healthcare proxy. CAREDECIDER consists of two different web interfaces: one for patients, and one for care providers. Patients create a profile and enter basic demographic and health information; they might be prompted to answer some follow-up questions resembling those on ADs. After answering these questions, each user is instructed to print and sign an MPOA, designating CAREDECIDER as their proxy (in the US this is a matter of state law). Patients would provide copies to their family and care providers, as they would with (human) MPOAs. Most importantly, the CAREDECIDER MPOA provides a unique *patient login code*, which care providers use to access a patient's CAREDECIDER profile. When a provider logs in, they answer several questions regarding the patient's state, prognosis, and the treatment options being considered. Upon entering this data, CAREDECIDER provides an estimate of whether or not the patient would want to receive aggressive LST.

Behind the scenes, CAREDECIDER resembles a standard ML application–predicting patient care preferences using training data. In this case, training data might consist of care decisions made by other patients in the past (from hospital records, or the application itself). Engineers might encode these decisions into numerical vectors representing each patient's state and the treatment option considered (i.e., the *input variables*), and whether they decided to receive the treatment or not (i.e., the *output variables*). Using modern ML methods, CAREDECIDER's engineers train a model that infers patient care preferences (e.g., they learn a *utility function* for the patient). Upon creating a profile, each use is prompted with additional questions to provide personalized training data.

While apocryphal, CAREDECIDER represents a simple and technically feasible use of AI in healthcare. Deploying this application would only require sufficient training data, and a web interface to initialize and query patient's personalized ML model. From a ML perspective, CAREDECIDER allows care providers to directly query the (approximation of) a patient's utility function. Indeed this is the general structure of hypothetical applications proposed for a similar purpose–predicting patient mortality to facilitate EOL care discussions [2].

In the best case, CAREDECIDER might represent patient goals and wishes more accurately than their AD or (human) proxy. In the worst case, CAREDECIDER might misrepresent a patient's wishes, or violate patient autonomy (e.g., making decisions that the patient never considered). Perhaps the greatest risk is that CAREDECIDER will encode human biases into its predictions, and will disproportionally impact already-disadvantaged groups (e.g. people without health insurance or access to a doctor) [5]. We leave further discussion to future work. While some of these risks are inherent to any AI-based system, others are a matter of design. We leave further discussion to future work.

Next we consider an ACP application that–using very similar methods to CAREDECIDER–avoids (or perhaps conceals) some of the apparent risk.

Application 2: We now consider PREFLIST, which closely resembles existing web-based applications such as MyDirectives. However unlike these applications, PREFLIST uses an AI back-end to create a unique set of questions for each patient to answer. The patient-facing interface of PREFLIST is identical to that of CAREDECIDER: patients enter demographic and health information, followed by some additional questions. However the provider-facing interface is far simpler: it displays only the information provided by the patient, including their answers to any AD-style questions. In this way, PREFLIST fits into the existing field of AD tools, while also leveraging modern AI methods.

Behind the scenes, PREFLIST may be identical to CAREDECIDER. However unlike CAREDECIDER, the patient's personalized ML model is only used to select questions–and is not directly available to care providers. Yet this ML model *guides* the interaction between patient and questioner. Methods from active learning, preference elicitation, and recommender systems can still be applied in this setting, with a less-direct impact on patient treatment decisions.

The benefits of PREFLIST seem clear: patients can create personalized ACP documents (similar to a POLST), without assistance from a medical professional. Furthermore, the AI-selected questions might be tailored to each patient, given their unique goals and wishes. PREFLIST doesn't appear to raise any risk to patient autonomy, at least not compared to CAREDECIDER. However many risks inherent to AI might still be present, and less apparent, in PREFLIST.

In particular, issues of bias would be less apparent, yet still present in an elicitation-focused application such as PREFLIST. For example, if training data includes only decisions and health conditions of (say) older white men, the utility of this application will be questionable for other groups. Using a biased model, PREFLIST might guide patients to answer uninformative questions.

Yet another concern is raised by any interaction of AI systems with humans. It is not immediately clear how a doctor would perceive or react to a description from a computer about a human's desires. For instance, does the doctor trust the computer more over time and eventually stop questioning its authority? There is some initial evidence that *stated* accuracy–not just observed accuracy–impacts a human's trust in an AI system [18], which might also lead to questions of manipulation. On the patient side, interacting with an AI system with (apparent) authority might influence her preferences; recent work finds that even spurious AI-generated advice can influence peoples' expressed preferences in a moral decision-making scenario [4].

3 Discussion

ACP, by nature, is an interdisciplinary social problem, and it is natural to ask about the appropriate use of new technologies in this domain. Indeed AI researchers have already proposed applications to improve EOL care decisions, and web-based apps promise to increase ACP participation rates. But to immediately apply AI without regard for potential implications is unwise. For the purposes of discussion, we propose some hypothetical applications that leverage existing technologies and platforms; we encourage the ACP community, medical ethicists, and AI researchers to discuss their appropriateness.

Before any technological intervention in peoples' lives, particularly in the medical field, there should be considered attention given to whether that intervention is appropriate. To facilitate this dialog, there should be more formal discussion between AI researchers and medical ethicists where collaboration occurs. Possible questions to be asked here include *whether* a computer should/could make *health* decisions on behalf of a human, and if so, how will conflicts between a patient's computer-assisted AD and a human MPOA be resolved?

Acknowledgements Dickerson, Dooley, and McElfresh were supported in part by NSF CAREER Award IIS-1846237 and by a generous gift from Google. Cui, Griesman, Wang, and Will were supported via an REU grant, NSF CCF-1852352, and were advised by Dickerson at the University of Maryland. Many thanks to Patty Mayer, who provided thoughtful guidance from the perspective of clinical ethics.

References

1. Álvarez-Farizo, B., Hanley, N.: Using conjoint analysis to quantify public preferences over the environmental impacts of wind farms. An example from Spain. Energy policy **30**(2), 107–116 (2002)
2. Avati, A., Jung, K., Harman, S., Downing, L., Ng, A., Shah, N.H.: Improving palliative care with deep learning. BMC Med. Inform. Decis. Mak. **18**(4), 122 (2018)
3. Bomba, P.A., Kemp, M., Black, J.S.: POLST: an improvement over traditional advance directives. Cleve Clin. J. Med. **79**(7), 457–464 (2012)
4. Chan, L., Doyle, K., McElfresh, D.C., Conitzer, V., Dickerson, J.P., Borg, J.S., Sinnott-Armstrong, W.: Artificial artificial intelligence: measuring influence of AI 'assessments' on moral decision-making. In: Proceedings of the 2020 AAAI/ACM Conference on AI, Ethics, and Society, AIES 2020. ACM (2020)
5. Crawford, K., Calo, R.: There is a blind spot in AI research. Nat. News **538**(7625), 311 (2016)
6. Fine, R., Yang, Z., Spivey, C., Boardman, B., Courtney, M.: Early experience with digital advance care planning and directives, a novel consumer-driven program. Baylor Univ. Med. Cent. Proc. **29**(3), 263–267 (2016)

7. Garg, A.X., Adhikari, N.K., McDonald, H., Rosas-Arellano, M.P., Devereaux, P., Beyene, J., Sam, J., Haynes, R.B.: Effects of computerized clinical decision support systems on practitioner performance and patient outcomes: a systematic review. Jama **293**(10), 1223–1238 (2005)
8. Gianfrancesco, M.A., Tamang, S., Yazdany, J., Schmajuk, G.: Potential biases in machine learning algorithms using electronic health record data. JAMA Intern. Med. **178**(11), 1544–1547 (2018)
9. Kahn, C.: Decision aids in radiology. Radiol. Clin. N. Am. **34**(3), 607–628 (1996)
10. Kourou, K., Exarchos, T.P., Exarchos, K.P., Karamouzis, M.V., Fotiadis, D.I.: Machine learning applications in cancer prognosis and prediction. Comput. Struct. Biotechnol. J. **13**, 8–17 (2015)
11. Llewellyn-Thomas, H.A., Crump, R.T.: Decision support for patients: values clarification and preference elicitation. Med. Care Res. Rev. **70**, 50S–79S (2013)
12. Lu, J., Wu, D., Mao, M., Wang, W., Zhang, G.: Recommender system application developments: a survey. Decis. Support Syst. **74**, 12–32 (2015)
13. Rao, J.K., Anderson, L.A., Lin, F.C., Laux, J.P.: Completion of advance directives among us consumers. Am. J. Prev. Med. **46**(1), 65–70 (2014)
14. Sabatino, C.: Advance care planning tools that educate, engage, and empower. Public Policy Aging Rep. **24**(3), 107–111 (2014)
15. Sawyer, A.M., Deal, E.N., Labelle, A.J., Witt, C., Thiel, S.W., Heard, K., Reichley, R.M., Micek, S.T., Kollef, M.H.: Implementation of a real-time computerized sepsis alert in nonintensive care unit patients. Crit. Care Med. **39**(3), 469–473 (2011)
16. Sudore, R.L., Schillinger, D., Katen, M.T., Shi, Y., Boscardin, W.J., Osua, S., Barnes, D.E.: Engaging diverse English-and Spanish-speaking older adults in advance care planning: the prepare randomized clinical trial. JAMA Intern. Med. **178**(12), 1616–1625 (2018)
17. Winter, L.: Patient values and preferences for end-of-life treatments: are values better predictors than a living will? J. Palliat. Med. **16**(4), 362–368 (2013)
18. Yin, M., Wortman Vaughan, J., Wallach, H.: Understanding the effect of accuracy on trust in machine learning models. In: Proceedings of the 2019 CHI Conference on Human Factors in Computing Systems, p. 279. ACM (2019)

Predicting Mortality in Liver Transplant Candidates

Jonathon Byrd, Sivaraman Balakrishnan, Xiaoqian Jiang, and Zachary C. Lipton

Abstract Donated livers are assigned to eligible matches among patients on the transplant list according to a sickest-first policy, which ranks patients by their score via the Model for End-stage Liver Disease (MELD). While the MELD score is indeed predictive of mortality on the transplant list, the score was fit with just three features, for a different task (outcomes from a shunt insertion procedure), and on a potentially un-representative cohort. These facts motivate us to investigate the MELD score, assessing its predictive performance compared to modern ML techniques and the fairness of the allocations vis-a-vis demographics such as gender and race. We demonstrate that assessing the quality of the MELD score is not straightforward: waitlist mortality is only observed for those patients who remain on the list (and don't receive transplants). Interestingly, we find that MELD performs comparably to a linear model fit on the same features and optimized directly to predict same-day mortality. Using a wider set of available covariates, gradient-boosted decision trees achieve .926 AUC (compared to .867 for MELD-Na). However, some of the additional covariates might be problematic, either from a standpoint of procedural fairness, or because they might expose the process to possible gaming due to manipulability by doctors.

J. Byrd (✉) · S. Balakrishnan · Z. C. Lipton
Carnegie Mellon University, Pittsburgh, PA 15213, USA
e-mail: jabyrd@cmu.edu

S. Balakrishnan
e-mail: siva@stat.cmu.edu

Z. C. Lipton
e-mail: zlipton@cmu.edu

X. Jiang
University of Texas Health Science Center at Houston, Houston, TX 77030, USA
e-mail: Xiaoqian.Jiang@uth.tmc.edu

© The Editor(s) (if applicable) and The Author(s), under exclusive license to Springer Nature Switzerland AG 2021
A. Shaban-Nejad et al. (eds.), *Explainable AI in Healthcare and Medicine*, Studies in Computational Intelligence 914,
https://doi.org/10.1007/978-3-030-53352-6_31

1 Introduction

Machine learning algorithms are increasingly being used to drive allocative decisions in applications with potential for high social impact, such as allocating bank loans, ranking job applicants, or matching organs to patients in need of transplants. The use of machine learning systems in consequential domains such as recidivism prediction in criminal justice, or the aforementioned problems, raises concerns of how to audit the quality and equitability of algorithmic decisions. Currently, the ML-based tools used to drive these decisions are classifiers, and perhaps the most mature set of practical tools for analyzing such data-driven decisions are those used for evaluating classifiers. Typically, we observe sample covariates, X, on which we make predictions $\widehat{Y}$, evaluated against observed outcomes, Y. Possibly, we also observe some protected feature Z, used to assess the fairness of classification decisions. Model performance is evaluated by comparing metrics based on true/false positives/negatives such as precision and recall, or comparing metrics based on output score such as calibration by group or ranking metrics like area under the receiving operator characteristic (ROC) curve.

However, this static view of examples and labels paints an incomplete picture of the situation in several real-world tasks. Firstly, we only observe outcomes under the decision made for each candidate. Potential outcomes under alternative policies are unobserved. Secondly, we are concerned with comparing the outcomes of allocating resources among eligible candidates. But aggregation across time periods and locations pools together incomparable candidates. (It is not possible to send a resource obtained in 2013 back through time to a recipient in 2006). To asses the quality of allocative decisions, we must look at the outcomes of each individual decision among the eligible recipients of that resource.

As of November 12, 2019, 12, 965 patients with end-stage liver disease are listed on the Organ Procurement and Transplantation Network (OPTN) waiting list for liver transplantation. In 2018, over 12, 700 patients were added to the list, while only 8, 250 liver transplantations were performed, necessitating a decision policy to select patients to receive available organs. 609 patients died while waiting for an organ, while another 627 were removed from the list due to being too sick for the procedure. In the United States, donor livers are roughly allocated according to a *sickest-first* policy, where compatible transplant candidates are ranked according to disease severity. First, candidates incompatible with the given donor liver are filtered out according to blood type, size, and geography to lower the risks of graft failure. The remaining candidates are ranked according to their Model for End-Stage Liver Disease (MELD) score [18], a simple model calculated from three blood measurements/tests: total bilirubin (g/dL), serum creatinine (g/dL), and international normalized ratio (INR) of prothrombin time. Currently, the MELD-Na score [3] is used, which also incorporates serum sodium.

Although MELD was originally developed to predict post-treatment survival in the transjugular intrahepatic portosystemic shunt (TIPS) procedure [12], it was re-purposed by the OPTN to rank liver transplant candidates, after being demonstrated to be a capable general predictor of mortality in patients with chronic liver disease [9]. The MELD-based allocation system was immediately successful, leading to the first ever reduction in the number of waiting list candidates and a 15% reduction in mortality among those on the waiting list [8]. However, the model is not without its drawbacks. In addition to being fit on a different cohort for a different task, the model thresholds the log-transformed values of its three features at 1.0 to avoid negative values. This is problematic, as a large percentage waiting list candidates possess serum creatinine levels below this threshold, and values below this threshold can reflect very different levels of kidney function [17].

Furthermore, as expected from such a simple model, the correlation between MELD and outcome is not equally strong for all patients. For some patients, MELD may not accurately reflect the severity of their condition. Other patients become ineligible for transplant before their condition deteriorates to the point of producing the higher MELD scores needed to be prioritized for transplantation. For example, in patients with Hepatocellular Carcinoma (HCC), as their tumors become larger and more numerous, post-transplant regrowth in the new liver becomes likely. Such patients are granted MELD exception scores which replace their calculated MELD score when being ranked for organ allocation. Due to the high difference in MELD scores among different regions, most MELD exception scores are based on the median MELD at transplant time of recent transplant recipients in the same region. 18.9% of candidates on the waiting list on December 31, 2017 have active MELD exceptions, 13.3% of which are for HCC. While the final rule prescribes disease severity as the measure with which to prioritize transplant candidates, the MELD system also prioritizes patients who may soon develop high risks for graft failure or disease recurrence. However, this is done in a post-hoc manner via heuristics, which prompts the question: why not estimate both mortality risk and transplant ineligibility risk in a statistically principled way?

In this paper, we evaluate the performance of modern machine learning methods on the task of mortality prediction for waiting list candidates. We train and evaluate models on predicting both same-day and 3-month mortality. We find that gradient boosting ensembles outperform MELD and MELD-Na in terms of area under the ROC curve by a wide margin on both the same-day and 90-day prediction tasks—0.900 (gradient-boosting) vs 0.831 (MELD-Na) for same-day prediction and 0.926 vs 0.867 for 3-month prediction. Removing demographic features including race, gender, education, etc, as well as more subjective features including ascites, encephelopathy, and diagnosis, does not have a large effect on model performance. Both our model and MELD-Na slightly underestimate mortality in female patients as compared to male patients, but we find no similar trends when comparing scores across ethnicities.

2 Related Work

Since the implementation of the MELD-based liver allocation system in 2002, many analyses and validations of the model's performance have been performed, and many modifications to the original MELD formula have been proposed. Merion et al. and Bambha et al. examine prediction using differences in MELD scores updated over time to incorporate information regarding a patient's change in condition over time, rather than just MELD at waiting list registration [1, 13]. Many authors [3, 11] have shown that serum sodium levels are predictive of waiting list mortality, ultimately leading to the adoption of the MELD-Na score for ranking transplant candidates. Sharma et al. refit MELD coefficients using a time-dependent Cox model on liver transplant waiting list patients, and showed that their updated model better ranks patients by waiting list mortality [17]. Leise et al. use a generalized additive form of Cox regression with smoothing splines to propose new cutoff values for MELD features [11]. Myers et al. propose the 5-variable MELD model which incorporates serum sodium and albumin in addition to the three MELD features [14]. 5-variable MELD is also derived using Cox regression, and was found to outperform MELD in prediction of 3-month mortality.

While there have been many attempts to predict post-treatment outcomes in liver transplant patients, comparatively little work has been done in predicting waiting list survival. Cuchetti et al. train a multi-layer perceptron to predict 3-month mortality in a cohort of 188 patients using age, sex, treatment indication, and a set of 10 laboratory values as features [6]. Recently, classification trees trained using mixed-integer optimization techniques were shown to reduce combined waitlist deaths/removals and post-transplant deaths by 17.6% as compared to MELD in simulation [2]. Their model, termed, "Optimized Predictor of Mortality" (OPOM) also outperformed MELD in ranking waiting list patients for 3-month survival. OPOM is comprised of two models, one trained on non-HCC candidates, and the other on HCC candidates. They train on data from the OPTN Standard Transplant Analysis and Research (STAR) dataset, treating every patient check-in as an example, and removing examples for which the patients receive treatment within three months (although they also explore imputing observations for treatment-censored patients).

MELD and other proposed patient-scoring methods have largely been created by fitting Cox regression models to patient features at the time of transplant listing. These models are then validated primarily using area under the ROC curves (ROC AUC) on a separate holdout set of patients. There are several potential issues with this procedure: 1) It is not clear that the proportional hazards assumption should hold among patients with different liver transplant indications. 2) Using standard Cox regression ignores feature updates from check-ins after patients are added to the waiting list (although some papers use time-dependent models [17]). As we would like models to perform well on both new waiting list candidates as well as candidates remaining on the list, it is important to include post-listing observations when evaluating models. When this has been done [2], each check-in update is treated as an example. However, this approach weights our evaluation metric based on the number of measurements of a patient. 3) Cox models assume uninformative censoring, which clearly does not hold in this application, because treatment is allocated according to patient features. Furthermore, choosing not to consider any patients treated within 3 months for model evaluation, biases metrics to deprioritize patients with more severe conditions, who tend to either quickly receive treatment or quickly die. This is especially problematic when the end goal is to identify patients with the highest mortality risk.

Much work has also been done in predicting graft failure following liver transplants, either as a function of the recipient, or the donor-recipient pair. The SOFT (Survival Outcomes Following Liver Transplantation) score identifies 19 factors as significant predictors of recipient mortality following transplantation, and provides a logistic regression model to estimate recipient mortality [16]. Delen et al. use a support vector machine with a Gaussian kernel, a multi-layer perceptron (MLP), and an M5-based regression tree to select features for a Cox survival model for post-transplant outcomes, demonstrating their method to be superior to traditional feature selection methods on this task [7]. More recently, machine learning methods including random forests and MLP's have been applied to predicting recipient mortality following transplantation [4, 10]. Perez-Ortiz et al. augment a dataset of liver transplant outcomes with recent unlabeled transplants and virtual donor-recipient pairs to combat class imbalance, and use a semi-supervised label propagation method to train support vector classifiers on the augmented data [15]. They then propose an organ allocation policy based on the model.

3 Dataset

Our data is composed of OPTN waiting list histories and Transplant Candidate Registration (TCR) form data from the STAR (Standard Transplant Analysis and Research) file for adult transplant candidates registered on the OPTN liver waiting list. Pre-2016 waiting list histories for candidates added to the waiting list after June 30, 2004 are divided into in-sample training, validation, and test sets using a random respective 50-25-25% split on individual patients. When deploying models to make real-world decisions, we are making decisions regarding future patients using data from past patients. Changes in society, patient care, and healthcare policy may cause to distribution shift among patients seen over time. For this reason, models are also evaluated on an out-of-sample test set of waiting list histories for a randomly selected 50% of patients added to the waiting list between January 1, 2016 and June 30, 2018.

Each day that a patient is on the waiting list (a patient-day) is treated as an example. As we wish to rank patient's chance of survival without treatment, patient-days on which the patient receives treatment or is removed from the waiting list for any reason other than death are excluded for the same-day mortality prediction task. For 3-month mortality prediction, all patient-days for which the patient is removed from the list for any non-death reason within 3 months are excluded, with the exception of patients removed from the list due to condition improvement such that they no longer require transplantation. These patient-days remain in the dataset as nega-

Table 1 Composition of train and test sets for same-day and 3-month prediction tasks

	Patient-days	Patients	Positive examples	Updates
Same-day mortality				
Training set	27.79M	61.0K	7886 (0.03%)	758K
In-sample test set	13.78M	30.5K	3943 (0.03%)	374K
Out-of-sample test set	3.02M	15.2K	991 (0.03%)	142K
3-month mortality				
Training Set	24.66M	61.0K	791K (2.6%)	513K
In-sample test set	12.23M	30.5K	397K (2.6%)	252K
Out-of-sample test set	32.45M	15.2K	90K (2.2%)	85K

tive examples. Table 1 breaks down the numbers of examples and patients in each dataset for the two prediction tasks. Demographic information is presented in Table 2. Figures 1, 2, and 3 examine MELD-Na scores for patients removed from the waiting list for different reasons.

Our models make use of 50 features, 31 of which are known at waiting list registration, while the remaining 19 are updated over time while a candidate remains on the list. Categorical features are encoded as dummy variables, and numerical features are standardized to have zero mean and unit variance in the training set. MELD and MELD-Na values are calculated using their respective formulas rather than read directly from the dataset. This results in a different value being used for 4% of observations, almost all of which have a difference of one point. The match MELD (value used for ranking patients that incorporates MELD exceptions) for inactive patient-days is set as the match MELD pre-inactivity. Missing values for numerical time-series features are forward-filled using the last known value for that feature. Missing values that cannot be forward-filled and missing values for numerical non-updated features are imputed using the feature median from the training set, and a corresponding missing value indicator feature is created for each numerical feature with missing values. When serum sodium values are missing and cannot be forward-filled, MELD-Na is calculated using the original MELD formula. Patient-days missing features to calculate lab MELD are removed from the training set if those values cannot be forward-filled. Numerical features are clipped to within four standard deviations from the mean. After pre-processing, the data has dimensionality 241.

Table 2 Demographic breakdown of train and validation sets. Percentages of patients and patient-days that fall into each category are shown as well as transplant rates, mortality rates, and rates of removal from the list for becoming too sick to transplant for each category. Age and diagnosis categories reflect values at waiting list registration time

Feature	Category	Patients	Patient-days	Transplant	Mortality	Removed: sick
	All Patients	100.0%	100.0%	55.6%	13.0%	10.6%
Age	18–34	6.0%	5.6%	57.5%	8.7%	5.6%
	35–49	20.2%	22.1%	56.0%	11.9%	7.4%
	50–64	60.9%	61.5%	56.0%	13.6%	11.1%
	65+	12.9%	10.8%	52.6%	13.8%	15.4%
Gender	Male	64.7%	62.8%	58.1%	12.3%	10.1%
	Female	35.3%	37.2%	51.0%	14.2%	11.4%
Ethnicity	White	70.6%	70.3%	55.9%	12.9%	10.2%
	Black	9.0%	7.5%	60.9%	11.5%	10.5%
	Hispanic	14.6%	16.5%	50.6%	15.1%	12.1%
	Asian	4.5%	4.5%	56.4%	9.7%	10.9%
	Other	1.3%	1.2%	54.7%	14.0%	11.1%
Diagnosis	Non-Chol. Cirrhosis	76.1%	79.4%	54.7%	13.4%	10.9%
	Chol. Liver Dis-ease/Cirr.	7.9%	8.6%	58.3%	12.0%	8.6%
	Biliary Artesia	0.3%	0.4%	61.1%	10.0%	6.1%
	Acute Hepatic Necrosis	4.7%	3.1%	54.4%	11.8%	10.4%
	Metabolic Diseases	2.2%	1.7%	68.2%	11.2%	6.7%
	Malignant Neoplasms	18.5%	12.6%	68.5%	6.2%	12.1%
	Other	1.9%	1.8%	61.6%	11.2%	7.1%

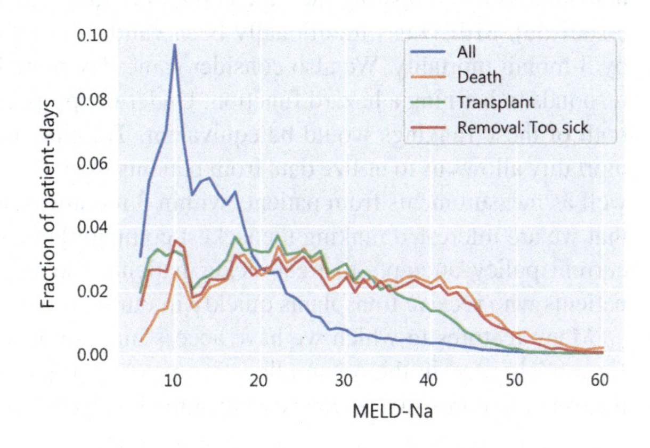

Fig. 1 Fraction of patient-days (y-axis) with different MELD-Na scores (x-axis) in the training and validation sets normalized by group. Lines for removal due to death, transplant, or too sick for transplant show MELD-Na scores at removal time

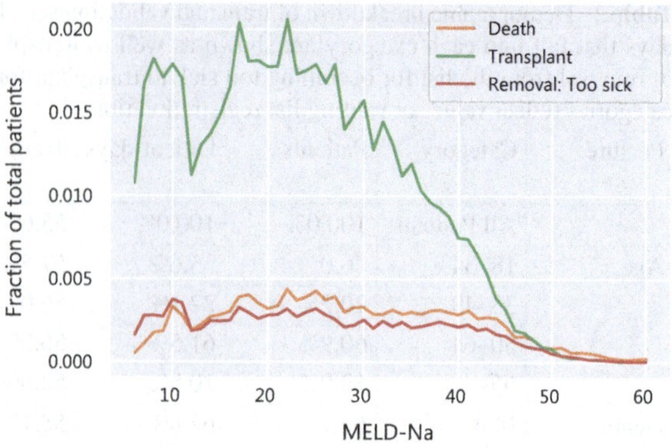

Fig. 2 MELD-Na scores (x-axis) of patients at waiting list removal due to death, transplant, or becoming too sick for transplant. Each point shows the fraction of total patients with corresponding outcome and MELD-Na score. Patients in the training and validation sets are shown

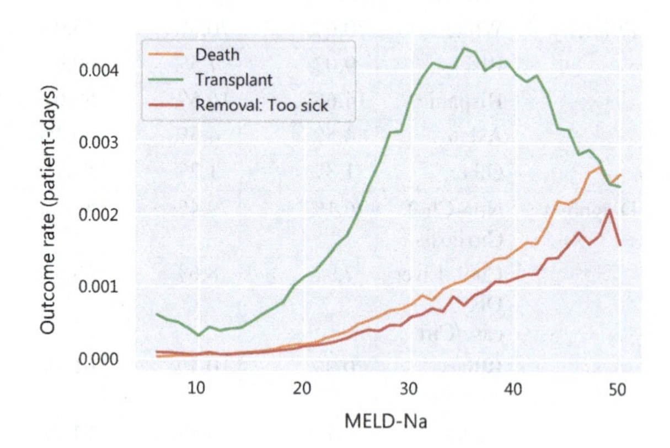

Fig. 3 Outcome rates among patient-days (y-axis) with different MELD-Na scores (x-axis) for removal due to death, transplant, or becoming too sick for transplant

4 Predicting Mortality

When allocating livers, we are much more concerned with giving livers to the patients who most need them, rather than accurately predicting mortality risks. Thus, we evaluate models by comparing their area under the ROC curve (ROC AUC, also termed c-statistic). MELD has traditionally been validated on the task of ranking patients by 3-month mortality. We also consider same-day mortality prediction, which corresponds to learning a hazard function. Under the proportional hazards assumption, both of these rankings would be equivalent. Training and evaluating on same-day mortality allows us to utilize data from patients treated within 3 months of listing, as well as measurements from patients within 3 months of transplantation date. Given that we are interested ranking the sickest compatible patients above others, and the current policy attempts to treat sicker patients sooner, it is important to include patients who receive transplants quickly in our datasets.

Many features to which we have access may not be appropriate to use as input for models used in the organ allocation process. Equitability concerns discourage the use of features such as race or education level, even if they were to be associated with waiting list outcomes. While knowledge of gender may be useful in interpreting how serum creatinine values reflect kidney function, naively feeding this feature to models invites discrimination on a protected class. We present results for models trained without such features.

Additionally, many features are the result of subjective judgements from physicians. End-stage liver disease symptoms such as ascites and encephelopathy are not judged identically by every physician, and their subjectivity was a primary criticism of the Child-Turcotte-Pugh score [5]. Primary and secondary diagnoses can be especially dependent on the discretion of the physician when multiple indications for liver transplant are present. These features do not only raise concerns due to their variability, but also do to their manipulability. Knowledge of the allocation system may influence physicians when measuring attributes or making decisions that could influence the patient's transplantation ranking. We identify four features with higher measurement subjectivity, and present results from models trained excluding these features.

In total, we train models using four different feature sets. The first contains all available features. The second set excludes the following demographic features: citizenship, education, gender, whether or not patient works for income, ethnicity, blood type, and donor service area (age, height, and weight are kept as features). The third set further excludes the following features which we believe have a higher degree of measurement subjectivity: ascites, encephelopathy, diagnosis, and functional status. The final set includes only features used in MELD-Na: bilirubin, international normalized ratio of prothrombin time (INR), serum creatinine, serum sodium, and dialysis twice within prior week.

We fit logistic regression and gradient-boosting ensembles with decision trees as base classifiers on each feature set and prediction task. Hyperparameters were selected by iterative grid searches on the validation set. Logistic regression models use an L2-penalty of 1. Gradient-boosting ensembles are trained with a learning rate of 0.1. Ensembles of 2000 trees with max tree depth of 4, minimum of 60 examples required to create a split, and a minimum of 7 examples required to create a leaf, are trained for the first three feature sets. On the MELD-Na feature set, we use ensembles of 500 trees with max tree depth of 3, minimum of 6 examples required to create a split, and a minimum of 5 examples required to create a leaf.

Table 3 shows ROC AUC on the test set for the same-day mortality and 3-month mortality prediction tasks respectively. Results are shown for MELD, MELD-Na, and our models across the four feature sets. Results are given for both holdout test sets containing patients from different time periods. We also compare the equitability of scoring models by examining mortality rates of protected classes with similar model scores (Fig. 4).

Table 3 ROC AUC scores for ranking patient-days by same-day and 3-month mortality. The first number in each cell is the performance on the in-sample holdout set which includes patients added to the waiting list between July 8, 2004 and December 31, 2015. The second number is the performance on the out-of-sample holdout set which includes patients added to the waiting list between January 1, 2016 and June 30, 2018. Results are shown for ranking by MELD, MELD-Na, match MELD, and logistic regression and gradient boosting models trained on four different feature sets. Selected features refers to feature set excluding both non-demographic features and four additional features with greater measurement subjectivity. Same-day and 3-month versions of logistic regression and gradient boosting models refer to the target the model was trained to estimate

	All features	Non-demographic features	Selected features	MELD-Na features
Same-day mortality				
MELD	N/A	N/A	N/A	0.825, 0.791
MELD-Na	N/A	N/A	N/A	0.831, 0.793
Match MELD	N/A	N/A	N/A	0.750, 0.729
Logistic Regression (same-day)	0.888, 0.867	0.886, 0.864	0.876, 0.855	0.817, 0.782
Gradient Boosting (same-day)	**0.935, 0.920**	0.931, 0.918	0.873, 0.857	0.793, 0.735
Logistic Regression (3-month)	0.881, 0.851	0.880, 0.849	0.872, 0.839	0.820, 0.774
Gradient Boosting (3-month)	0.902, 0.873	0.901, 0.873	0.894, 0.864	0.832, 0.796
3-month mortality				
MELD	N/A	N/A	N/A	0.715, 0.674
MELD-Na	N/A	N/A	N/A	0.730, 0.686
Match MELD	N/A	N/A	N/A	0.685, 0.651
Logistic Regression (same-day)	0.786, 0.756	0.786, 0.752	0.778, 0.745	0.700, 0.662
Gradient Boosting (same-day)	0.783, 0.767	0.781, 0.765	0.808, 0.781	0.731, 0.690
Logistic Regression (3-month)	0.820, 0.772	0.818, 0.770	0.809, 0.759	0.734, 0.687
Gradient Boosting (3-month)	**0.834, 0.800**	0.832, 0.798	0.827, 0.789	0.734, 0.696

Fig. 4 Mortality rates (y-axis) for rank quantiles (x-axis) by gender (a) and ethnicity (b). The x-axis shows the quantile for patient days ranked by our model (solid lines) or by MELD-Na (dashed lines). The y-axis shows the mortality rates for patient days in the corresponding demographic group. Each bin is 5 percentiles wide, and we only show the top 50% rank days, because lower-ranked patient-days are unlikely to be selected for transplantation. The 'GBoost' model corresponds to the gradient-boosting model trained to predict same-day mortality using the non-demographic feature set

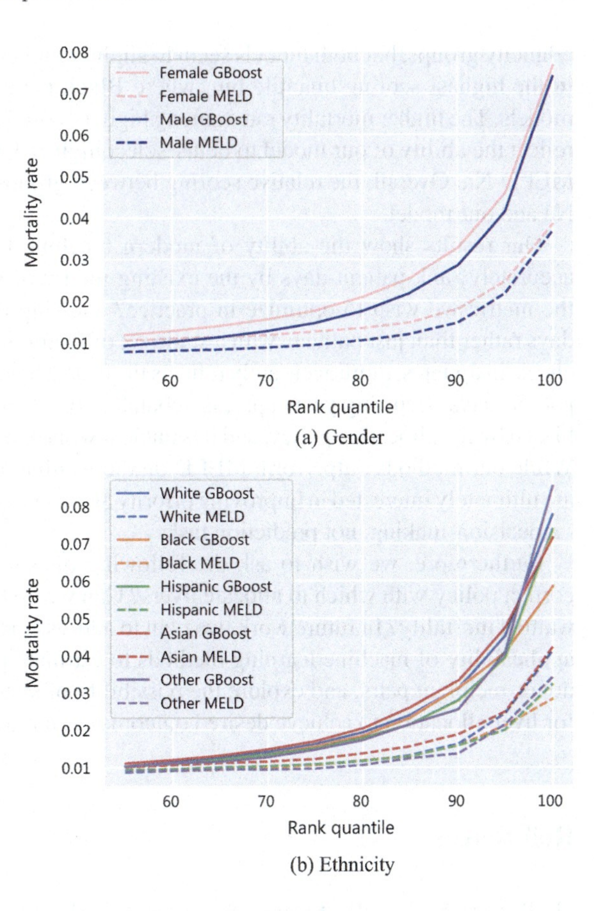

(a) Gender

(b) Ethnicity

5 Discussion

Firstly, we find that both logistic regression and gradient boosting models trained on a larger feature set outperform MELD and MELD-Na by a wide margin in terms of ROC AUC for both same-day and 3-month mortality prediction tasks (Table 3). These results show that there is potential for a large margin of improvement in ranking patients by mortality risk for the current MELD-based system. However, we do find that when limiting ourselves to the feature set used by MELD-Na, it is hard to perform too much better, and MELD-Na outperforms linear classifiers limited to the MELD-Na feature set. Larger feature sets may provide more of a performance benefit than more complex models. Eliminating demographic features such as gender, race, and location, as well as features with a more measurement subjectivity such as ascites and encephelopathy, does not sacrifice much test performance. These models may still perform well when restricted to more practical feature sets.

Both our model and MELD-Na consistently underestimate mortality for female patients by a small margin (Fig. 4). We do not find many notable trends in scoring ethnicity groups, but both models seem to slightly under-score Asian patients, except in the highest scoring quantile bin, where Black patients are over-scored by both models. The higher mortality rates among higher-scored patient-days from our model reflect the ability of our model to better select high-risk patient-days as compared to MELD-Na. Overall, the relative scoring between groups is similar between MELD-Na and our model.

Our results show the ability of modern machine learning techniques to more accurately rank patient-days by the existing metric of 90-day mortality, but is this the metric we wish to optimize in practice? Viewing the data in terms of patient-days rather than just patients with a static set of features at registration, or a series of check-in updates, more accurately reflects the task at hand, yet the merit of comparing patient-days occurring years apart is debatable, the distribution of test examples are biased by the allocation policy, and it is unclear what exactly this metric is estimating. While our methods outperform MELD on the waitlist mortality prediction task, we are ultimately interested in improving priority assignment within the matching system - a decision-making, not prediction task.

Furthermore, we wish to ask the following questions: is sickest first even the correct policy with which to allocate livers? Can we do better by directly optimizing waitlist mortality? In future work, we plan to address these questions, investigate the applicability of machine learning methods to estimate post-treatment outcomes for donor-recipient pairs, and explore the possibility of statistically-principled methods for liver allocation to achieve desired outcomes in a practical and equitable manner.

References

1. Bambha, K., Kim, W., Kremers, W., Therneau, T., Kamath, P., Wiesner, R., Thostenson, J., Benson, J., Dickson, E.: Predicting survival among patients listed for liver transplantation: an assessment of serial meld measurements. American Journal of Transplantation (2004)
2. Bertsimas, D., Kung, J., Trichakis, N., Wang, Y., Hirose, R., Vagefi, P.: Development and validation of an optimized prediction of mortality for candidates awaiting liver transplantation. Am. J. Transplantat. **19**, 1109–1118 (2018)
3. Biggins, S., Kim, W., Terrault, N., Saab, S., Balan, V., Schiano, T., Benson, J., Therneau, T., Kremers, W., Wiesner, R., Kamath, P., Klintmalm, G.: Evidence-based incorporation of serum sodium concentration into meld. Gastroenterology **130**, 1652–1660 (2006)
4. Chandra, V.: Graft survival prediction in liver transplantation using artificial neural networks. J. Health Med. Inform. **16**, 72–78 (2016)
5. Child, C.: The liver and portal hypertension. In: Annals of Internal Medicine (1964)
6. Cucchetti, A., Vivarelli, M., Heaton, N., Phillips, S., Piscaglia, F., Bolondi, L., La Barba, G., Foxton, M., Rela, M., O'Grady, J., Pinna, A.: Artificial neural network is superior to meld in predicting mortality of patients with end-stage liver disease. Gut (2007)
7. Delen, D., Oztekin, A., Kong, Z.: A machine learning-based approach to prognostic analysis of thoracic transplantations. Artif. Intell. med. **49**, 33–42 (2010)
8. Freeman, R., Wiesner, R., Edwards, E., Harper, A., Merion, R., Wolfe, R.: Results of the first year of the new liver allocation plan. Liver Transplant. **10**, 7–15 (2004)
9. Kamath, P.S., Wiesner, R.H., Malinchoc, M., Kremers, W.K., Therneau, T.M., Kosberg, C.L., D'Amico, G., Dickson, E.R., Kim, W.R.: A model to predict survival in patients with end-stage liver disease. Hepatology **33**, 464–470 (2001)

10. Lau, L., Kankanige, Y., Rubinstein, B., Jones, R., Christophi, C., Muralidharan, V., Bailey, J.: Machine-learning algorithms predict graft failure following liver transplantation. Transplantation **101**, 125–132 (2016)
11. Leise, M., Kim, W., Kremers, W., Larson, J., Benson, J., Therneau, T.: A revised model for end-stage liver disease optimizes prediction of mortality among patients awaiting liver transplantation. Gastroenterology **140**, 1952–1960 (2011)
12. Malinchoc, M., Gordon, F., Peine, C., Rank, J., Borg, P.: A model to predict poor survival in patients undergoing transjugular intrahepatic portosystemic shunts. Hepatology **31**, 864–871 (2000)
13. Merion, R., Wolfe, R., Dykstra, D., Leichtman, A., Gillespie, B., Held, P.: Longitudinal assessment of mortality risk among candidates for liver transplantation. Liver Transplant. **9**, 12–20 (2003)
14. Myers, R., Shaheen, A., Faris, P., Aspinall, A., Burak, K.: Revision of meld to include serum albumin improves prediction of mortality on the liver transplant waiting list. PLoS One **8**, e51926 (2013)
15. Pérez-Ortiz, M., Gutiérrez, P.A., Ayllón-Terán, M., Heaton, N., Ciria, R., Briceño, J., Martínez, C.: Synthetic semi-supervised learning in imbalanced domains: constructing a model for donor-recipient matching in liver transplantation. Knowl. Based Syst. **123**, 75–87 (2017)
16. Rana, A., Hardy, M., Halazun, K., Woodland, D., Ratner, L., Samstein, B., Guarrera, J., Brown, R., Emond, J.: Survival outcomes following liver transplantation (soft) score: a novel method to predict patient survival following liver transplantation. Am. J. Transplant. **8**, 2537–2546 (2008)
17. Sharma, P., Schaubel, D., Sima, C., Merion, R., Lok, A.: Re-weighting the model for end-stage liver disease score components. Gastroenterology **135**, 1575–1581 (2008)
18. Wiesner, R., Edwards, E., Freeman, R., Harper, A., Kim, R., Kamath, P., Kremers, W., Lake, J., Howard, T., Merion, R., Wolfe, R., Krom, R., Colombani, P., Cottingham, P., Dunn, S., Fung, J., Hanto, D., McDiarmid, S., Rabkin, J., Teperman, L., Turcotte, J., Wegman, L.: Model for end-stage liver disease (meld) and allocation of donor livers. Gastroenterology **124**, 91–96 (2003)

Towards Automated Performance Status Assessment: Temporal Alignment of Motion Skeleton Time Series

Tanachat Nilanon, Luciano P. Nocera, Jorge J. Nieva, and Cyrus Shahabi

Abstract Patient Performance Status (PS) is used in cancer medicine to predict prognosis and prescribe treatment. Today, PS assessments rely on assessor's observation, which is susceptible to biases. A motion tracking system can be used to supplement PS assessments, by recording and analyzing patient's movement as they perform a standardized mobility task e.g. getting up from office chair to sit on examination table. A temporal alignment of the extracted motion skeleton time series is then needed to enable comparison of corresponding motions in mobility task across recordings. In this paper, we apply existing state-of-the-art temporal alignment algorithms to the extracted time series and evaluate their performance in aligning the keyframes that separate corresponding motions. We then identify key characteristics of these time series that the existing algorithms are not able to exploit correctly: task left-right invariance and vertical-horizontal relative importance. We thus propose Invariant Weighted Dynamic Time Warping (IW-DTW), which takes advantage of these key characteristics. In an evaluation against state-of-the-art algorithms, IW-DTW outperforms them in aligning the keyframes where these key characteristics are present.

Keywords Motion skeleton time series · Temporal alignment · Performance status assessment

T. Nilanon (✉) · L. P. Nocera · C. Shahabi
University of Southern California, Los Angeles, CA 90089, USA
e-mail: nilanon@usc.edu

L. P. Nocera
e-mail: nocera@usc.edu

C. Shahabi
e-mail: shahabi@usc.edu

J. J. Nieva
USC Norris Comprehensive Cancer Center, Los Angeles, CA 90033, USA
e-mail: jorge.nieva@med.usc.edu

© The Editor(s) (if applicable) and The Author(s), under exclusive license to Springer Nature Switzerland AG 2021

A. Shaban-Nejad et al. (eds.), *Explainable AI in Healthcare and Medicine*, Studies in Computational Intelligence 914,
https://doi.org/10.1007/978-3-030-53352-6_32

1 Background and Motivation

Advances in sensor electronics and computer vision have led to widespread availability of single-unit color+depth (RGB+D) motion tracking systems. While these systems do not have the same level of accuracy as traditional motion capture systems, their portability and ease-of-use have made them an attractive choice for use in various medical applications, including rehabilitation and home monitoring [12]. One such system is the Microsoft Kinect, whose usability has been broadly validated for many computer vision applications, including object detection, human pose, and action recognition [6].

Performance status (PS) assessment has been used in cancer medicine to identify patients with an increased risk of complications and poor outcomes. However, widely-utilized PS scales such as the Eastern Cooperative Oncology Group (ECOG) scale [7] rely on qualitative PS descriptions, making PS assessments susceptible to biases and inter-rater disagreements [9]. These susceptibilities could potentially be addressed by supplementing PS assessments with a motion tracking system. During a clinic visit, patients could be recorded performing a set of standardized mobility tasks, then a PS score could be calculated, potentially utilizing both the observation/interview and the recordings. .

To provide a concrete example, we first describe the ChairToExamTable task used in our study (see Fig. 1). To perform this task, a patient seated in a standard office chair is asked to walk over and use a stepper to get up and sit on an examination table. This standardized task was developed based on consultation with clinicians to include movements informative to PS assessment.

As a standardized mobility task generally has broad instructions and is performed differently by different patients, a temporal alignment between the recordings is needed to avoid comparing disparate motions across recordings. A temporal alignment between two recordings can be visualized as a mapping between each of their time steps, as in Fig. 1. While these temporal alignments could be annotated manually, the process is time-consuming and would again be susceptible to biases; hence, a systematic and automated approach is preferred.

Based on a comprehensive review of our Kinect recordings, we have identified the following key characteristics of standardized-mobility-task motion skeleton time series that adversely affect state-of-the-art temporal alignment algorithms:

- **Task Left-right Invariance**: Different patients can start walking with different foot (left/right) first. This should be taken into account when generating temporal alignment. An algorithm that is not invariant to this task characteristic could generate an incorrect temporal alignment (see Fig. 1).
- **Vertical-horizontal Relative Importance**: In a typical 3-dimensional Cartesian coordinate system, two axes represent the horizontal plane and one axis represents the vertical direction. However, for a standardized mobility task, the vertical position is more relevant to PS assessments and should be given more weight than

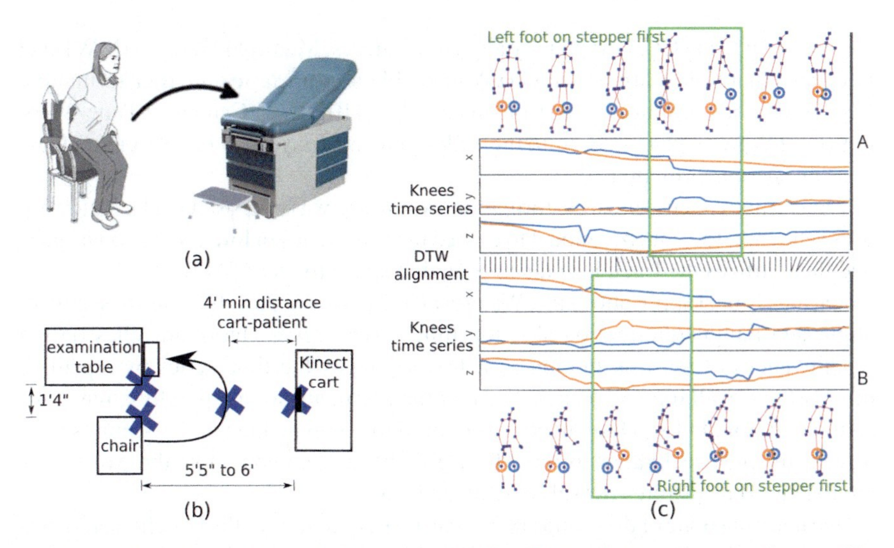

Fig. 1 **(a)** ChairToExamTable task illustration. **(b)** ChairToExamTable task set-up diagram. The small rectangle next to the examination table depicts the stepper. **(c)** Patients A and B stepping on the stepper to get up on the examination table; A steps with their left foot first while B steps with their right foot first. DTW is unable to correctly align the corresponding motions

the horizontal position, as movements in vertical direction involve changes in the subject's potential energy and require larger energy expenditure than movements in the horizontal plane.

In this paper, we propose an unsupervised temporal alignment algorithm for motion skeleton time series, designed to exploit the key characteristics above. Our method is built on DTW due to its robustness and strong baseline results.

2 Related Works

Temporal alignment of time series has rich literature, starting with Dynamic Time Warping (DTW) [8], which computes an optimal warping path whose cost between time steps is their Euclidean distance. While DTW has wide applications in various domains, its ability is limited in handling time series of different view or high dimensionality since it operates directly in the observation space.

In [14], Canonical Time Warping (CTW) was proposed for temporal alignment of human behavior, where Canonical Component Analysis (CCA) was integrated with DTW to enable linear spatial transformation of features alongside the computation of the warping path. In Generalized Canonical Time Warping (GCTW) [13], warping path basis functions were further introduced to reduce the search space for the optimal warping path, significantly reducing its time complexity.

Instead of using CCA to spatially align the features, Manifold Warping (MW) [11] integrates manifold learning with DTW to enable non-linear feature transformation. Recent methods continue to carry on this trend. In [10], Deep Canonical Time Warping (DCTW) was proposed, utilizing fully-connected layers to enable hierarchical non-linear transformation.

Departing from feature transformation, Autowarp was proposed in [1]. Autowarp uses sequence-to-sequence model to embed the time series in low-dimensional space and then utilizes the Euclidean distance in that space to guide DTW.

On the clinical application side, Wang et al. [12] proposed an algorithm to segment, align, and summarize motion skeleton time series. Their algorithm relies on the repetitiveness of the task motion as it first segments the time series by repetitive motion (e.g., walking two steps). Then all the segments are temporally-aligned for summarization. In [3], Hasnain et al. proposed to compute kinematics features such as velocity and acceleration as input for DTW distance computation; these distances are then used as features for subsequent analysis.

We note that none of these approaches can take advantage of the key characteristics of motion skeleton time series described in Sect. 1.

3 Methods

3.1 Multivariate Time Series and Temporal Alignment

The extracted motion skeletons can be represented as time series. We define a d-dimensional *time series* T of length N, $T \in \mathbb{R}^{d \times N}$: $T = (t_1, t_2, \ldots, t_N)$ where each $t_i \in \mathbb{R}^d$ is a d-dimensional vector. We then define a temporal alignment $\mathcal{A}$ between two time series $T_x \in \mathbb{R}^{d \times N_x}$ and $T_y \in \mathbb{R}^{d \times N_y}$ as a sequence of ordered pairs: $\mathcal{A}(T_x, T_y) = ((a_{x1}, a_{y1}), (a_{x2}, a_{y2}), \ldots, (a_{xL}, a_{yL}))$ where L is the length of the alignment, $a_{xi} \in \{1, 2, \ldots, N_x\}$, $a_{yi} \in \{1, 2, \ldots, N_y\}$, and the pair (a_{xi}, a_{yi}) denotes that $t_{xa_{xi}}$ is aligned with $t_{ya_{yi}}$. The alignment path is constrained to satisfy the boundary, monotonicity, and continuity conditions: $(a_{x1}, a_{y1}) = (1, 1)$, $(a_{xL}, a_{yL}) = (N_x, N_y)$, $a_{xi} \leq a_{x(i+1)} \leq a_{xi} + 1$, and $a_{yi} \leq a_{y(i+1)} \leq a_{yi} + 1$.

3.2 Dynamic Time Warping

Let $\mathcal{D}(T_x, T_y)_{i,j}$ be the distance between t_{xi} and t_{yj}. For Euclidean distance, $\mathcal{D}(T_x, T_y)_{i,j} = \sqrt{\sum_{k=1}^{d} (t_{xik} - t_{yjk})^2}$. DTW optimizes for a temporal alignment (warping) path $\mathcal{A}(T_x, T_y)$ where the total cost $\sum_{i=1}^{L} \mathcal{D}(T_x, T_y)_{a_{xi}, a_{yi}}$ is minimized. This cost can be described as a recurrence relation:

$$C(i,j) = \begin{cases} \mathcal{D}(T_x, T_y)_{i,j} & \text{if } i \text{ or } j = 1 \\ \mathcal{D}(T_x, T_y)_{i,j} + \min \begin{pmatrix} C(i-1,j) \\ C(i,j-1) \\ C(i-1,j-1) \end{pmatrix} & \text{otherwise} \end{cases}$$

where $C \in \mathbb{R}^{N_x \times N_y}$ is the accumulated cost array and the total cost is $C(N_x, N_y)$.

Our method is built on DTW due to its robustness and strong baselines. In the following sections, we explain how the key characteristics can be exploited.

3.3 Handling Task Left-Right Invariance

In a mobility task, patients are not usually instructed on which side of their body they have to move first. For example, they can start walking with their left or right foot first; or when instructed to step on a stepper, they can put their left or right foot on the stepper first. To construct a temporal alignment algorithm that is indifferent to this left vs. right variations, we extend the DTW recurrence relation with a separate cost array for each task-equivalent *feature mapping* and allow the alignment path to switch between these mappings.

In our application, the mobility task suggests that we should be indifferent to any switching between the left/right side of the lower/upper body. Let $S = \{O, A, L, AL\}$ contains these feature mappings to which our algorithm should be indifferent (see Fig. 2b for illustration). The recurrence relation $C_s, \forall s \in S$ can then be written as:

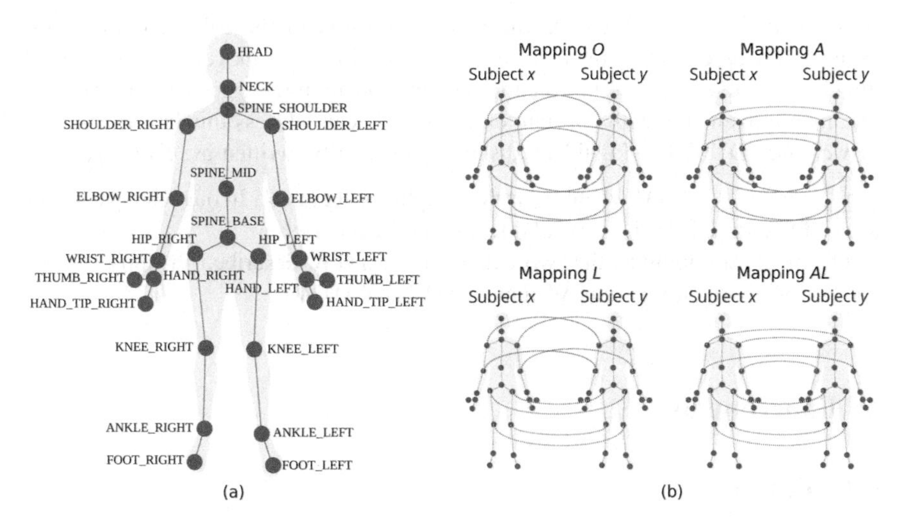

Fig. 2 **(a)** Diagram of skeleton anatomical node (n = 25) extracted from recordings by Kinect SDK v2.0. **(b)** Task-equivalent mappings between the left/right side of the lower/upper body. Some lines are omitted for clarity

$$C_s(i,j) = \begin{cases} \mathcal{D}(\mathcal{F}_s(T_x), T_y)_{i,j} & \text{if } i \text{ or } j = 1 \\ \mathcal{D}(\mathcal{F}_s(T_x), T_y)_{i,j} + \min_{t \in S} \begin{pmatrix} \mathcal{D}(\mathcal{F}_s(T_x), \mathcal{F}_t(T_x))_{i,i} \\ + \\ \min \begin{pmatrix} C_t(i-1,j) \\ C_t(i,j-1) \\ C_t(i-1,j-1) \end{pmatrix} \end{pmatrix} & \text{otherwise} \end{cases}$$

where $\mathcal{F}_s$ denotes a transformation that maps according to s, and C_s denotes the accumulated cost array for mapping s. Note that $\mathcal{D}(\mathcal{F}_s(T_x), \mathcal{F}_t(T_x))_{i,i}$ is the cost of switching the warping path between task-equivalent mappings s and t.

To approximate $\mathcal{F}_s$, we can *reflect* the nodes between the left and the right side of the body. We define the *SPINE_BASE reference frame* to be centered at SPINE_BASE (see Fig. 2a for diagram), where its x-axis is the horizontal component of the vector pointing from HIP_RIGHT to HIP_LEFT, its y-axis is the upward vertical vector, and its z-axis is generated by the right-hand rule. After a transformation into this reference frame, a node position on one side of the body (x, y, z) can be reflected to another side of the body as $(-x, y, z)$. Then the node position can be transformed back into the global reference frame. To minimize errors, this SPINE_BASE reference frame is used for the lower body while the *SPINE_SHOULDER reference frame* (defined similarly) is used for the upper body.

3.4 Handling Vertical-Horizontal Relative Importance

In a standardized mobility task, the vertical position of the subject is more relevant to PS assessments than the horizontal position, as movements in vertical direction involve changes in potential energy and require larger energy expenditure than movements with constant elevation. We can address this in a framework of weighted DTW. For Euclidean distance, this can be written as: $\mathcal{D}(T_x, T_y)_{i,j} = \sqrt{\sum_{k=1}^{d} w_k (t_{xik} - t_{yjk})^2}$ where w_k is the weight for feature k. In our case, we set w_k to w_y if feature k is in the vertical direction and 1 otherwise.

Our method combines the two extensions to DTW described in Subsects. 3.3 and 3.4. We name it Invariant Weighted DTW (IW-DTW).

4 Experiments

4.1 Dataset

Our dataset was derived from the Analytical Tools to Objectively Measure Human Performance (ATOM-HP) clinical project, which will be partially described here.

To examine the feasibility of using a motion tracking system, a wearable activity tracker, and a set of patient-reported outcome questionnaires in PS assessment, a multi-center observational clinical study was performed with a population of cancer patients. Participants were scheduled to come in for a clinic visit twice during the study period, during which ECOG scores and Kinect recordings were collected for the ChairToExamTable task [2–5].

After exclusion of partial, noisy, and nonconforming-to-instruction recordings, 64 recordings remain. The recordings have the minimum, median, and maximum length of 110, 275, and 665 frames, respectively. The motion skeleton time series are then extracted from the recordings using Kinect SDK v2.0 and have a sampling rate of 30 Hz. Each frame originally contains 75 features ((x, y, z) of the 25 skeleton nodes); however, the extracted `WRIST`, `HAND`, `THUMB`, `HAND_TIP`, `ANKLE`, and `FOOT` nodes are found to be unreliable and therefore excluded from analysis. Thus, each frame is left with 13 skeleton nodes.

4.2 Evaluation Metrics

All Kinect recordings were annotated by annotators reviewing a visualization of the extracted motion skeleton time series. We annotated 4 keyframes important for mobility task movement analysis. The first keyframe is defined to be when the subject starts building momentum to stand. The second keyframe is defined to be when the subject starts moving one foot to walk. The third keyframe is define to be when the subject has one foot on the stepper and starts an upward motion towards the examination table. Finally, the fourth keyframe is defined to be when the subject is fully seated. Note that these 4 keyframes divide the task into 3 subtasks. Kinematics features such as accelerations and timings of these subtasks are of interest from a clinical standpoint e.g. the first subtask contains the subject's motions standing up from seated.

We posit that for any analysis performed on motion skeleton time series to be justified, it should never compare a motion in one subtask to another motion in a different subtask. Consequently, the evaluation metrics we use will reflect this. For a pair of motion skeleton time series T_x and T_y, let a computed temporal alignment between them be $\mathcal{A}(T_x, T_y)$ and let their annotated keyframes be S_{xj} and S_{yj} for $j \in \{1, 2, 3, 4\}$. We define keyframes estimated using the computed temporal alignment as $\widehat{S}_{xj} = \text{mean}(\{a \mid (a, S_{yj}) \in \mathcal{A}(T_x, T_y)\})$ and $\widehat{S}_{yj} = \text{mean}(\{a \mid (S_{xj}, a) \in \mathcal{A}(T_x, T_y)\})$. We then define *Keyframe Mean Absolute Error (MAE)* as $\text{mean}\left(|\widehat{S}_{xj} - S_{xj}|, |\widehat{S}_{yj} - S_{yj}|\right)$.

Overall MAE is then calculated as the average over all possible pairs of time series. To further observe how task left-right invariance affects the performance of a temporal alignment algorithm, we also compute the MAE averaged over pairs where subjects moved different foot first in keyframes 2 and 3.

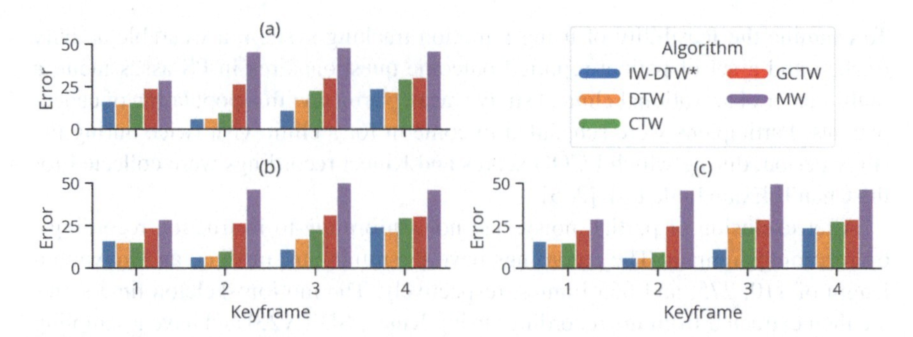

Fig. 3 Evaluation results (MAE) averaged over **(a)** all pairs of recordings and pairs where different foot was moved first in **(b)** keyframe 2 and **(c)** 3 . Task left-right invariance manifests in keyframes 2 (starts walking) and 3 (first foot on stepper)

5 Results

We evaluated IW-DTW against DTW, CTW, GCTW, and MW (see Fig. 3). Autowarp and DCTW were excluded due to poor convergence during model training.

Averaged over all possible pairs of motion skeleton time series, IW-DTW performed better than all baselines in aligning keyframes 2 and 3, where the alignment needs to handle task left-right invariance, but performed slightly worse in aligning keyframes 1 and 4, where task left-right invariance does not manifest. For keyframe 1 (starts standing up), MAE of IW-DTW (16.5) is slightly worse than the best baseline DTW (15.3). For keyframe 2 (starts walking), MAE of IW-DTW (6.4) is similar to the best baseline DTW (6.5). For keyframe 3 (uses stepper to push self up), MAE of IW-DTW (11.2) is distinctly better than the best baseline DTW (17.6). For keyframe 4 (seated on examination table), MAE of IW-DTW (24.2) is slightly worse than the best baseline DTW (22.1).

Restricting our comparison to the pairs of time series where the subjects move different foot first in keyframe 2, keyframe 2 MAE of IW-DTW (6.8) is slightly better than the best baseline DTW (7.2). As for keyframe 3, keyframe 3 MAE of IW-DTW (11.6) is better than the best baseline DTW (24.4).

6 Discussion

IW-DTW outperformed state-of-the-art algorithms in aligning keyframes when task left-right invariance clearly manifests, such as starting to walk or putting one foot on a stepper. However, when task left-right invariance does not manifest, such as getting up or being fully seated, IW-DTW performed slightly worse. We hypothesize that this is because the weighting mechanism is not expressive enough to capture local changes in relative axes importance. For example, when the subject starts getting up, the most importance axes for alignment would be the horizontal axes (note how the body needs to bend forward before any significant vertical movement can occur). Our learned w_y (4.25) weight for the vertical direction impedes IW-DTW in correctly aligning these keyframes.

Deep learning methods hold promise in being able to extract relevant representations and learn directly from the data. However, we observed that for our use case, long sequence length still proved to be a big impediment for gradient-based optimization used in deep learning. In Autowarp [1], the authors have successfully employed their model for 4-dimensional sequences of median length 53; however, it failed to converge when training on our 39-dimensional sequences of median length 275. In DCTW [10], their experiment that is most similar to ours also has median length <100. Downsampling could potentially make Autowarp work for our case, but key movements such as standing up can last less than 15 frames and could be inadvertently filtered out by downsampling.

Online learning also holds promise in learning a classification task as new data comes in. However, the small number of samples ($n = 64$) make our task a poor fit for online learning. Furthermore, as our goal is to supplement traditional PS assessment, we prefer an algorithm that does not change its output with new data, so that its efficacy can be studied and validated in a randomized controlled trial.

Acknowledgements This research has been funded in part by US National Cancer Institute under award number P30CA014089, USC Integrated Media Systems Center (IMSC), and unrestricted cash gifts from Oracle, Microsoft, and Google. Any opinions, findings, and conclusions or recommendations expressed in this material are those of the authors and do not necessarily reflect the views of any of the sponsors. T. Nilanon was also supported in part by DPST, IPST, Thailand.

References

1. Abid, A., Zou, J.: Autowarp: learning a warping distance from unlabeled time series using sequence autoencoders. In: Advances in Neural Information Processing Systems (NeurIPS), October 2018
2. Broderick, J.E., May, M., Schwartz, J.E., Li, M., Mejia, A., Nocera, L., Kolatkar, A., Ueno, N.T., Yennu, S., Lee, J.S.H., Hanlon, S.E., Cozzens Philips, F.A., Shahabi, C., Kuhn, P., Nieva, J.: Patient reported outcomes can improve performance status assessment: a pilot study. J. Patient-Reported Outcomes **3**(1), 41 (2019)
3. Hasnain, Z., Li, M., Dorff, T., Quinn, D., Ueno, N.T., Yennu, S., Kolatkar, A., Shahabi, C., Nocera, L., Nieva, J., Kuhn, P., Newton, P.K.: Low-dimensional dynamical characterization of human performance of cancer patients using motion data. Clinical Biomech. **56**(December 2017), 61–69 (2018)
4. Kao, J.Y., Nguyen, M., Nocera, L., Shahabi, C., Ortega, A., Winstein, C., Sorkhoh, I., Chung, Y.C., Chen, Y.A., Bacon, H.: Validation of Automated Mobility Assessment Using a Single 3D Sensor. In: Hua, G., Jégou, H. (eds.) European Conference on Computer Vision (ECCV) Workshops, vol. 3, pp. 162–177. Springer, Cham (2016)
5. Nguyen, M.N.B., Hasnain, Z., Li, M., Dorff, T., Quinn, D., Purushotham, S., Nocera, L., Newton, P.K., Kuhn, P., Nieva, J., Shahabi, C.: Mining Human Mobility to Quantify Performance Status. In: IEEE International Conference on Data Mining (ICDM) Workshop (2017)

6. Ofli, F., Chaudhry, R., Kurillo, G.: Berkeley Multimodal Human Action Database (MHAD). In: IEEE Workshop on Applications of Computer Vision (WACV), pp. 53–60 (2013)
7. Oken, M.M., Creech, R.H., Tormey, D.C., Horton, J., Davis, T.E., McFadden, E.T., Carbone, P.P.: Toxicity and response criteria of the Eastern Cooperative Oncology Group. Am. J. Clin. Oncol. **5**(6), 649–656 (1982)
8. Rabiner, L., Juang, B.H.: Fundamentals of Speech Recognition. Prentice-Hall Inc., Upper Saddle River (1993)
9. Schnadig, I.D., Fromme, E.K., Loprinzi, C.L., Sloan, J.A., Mori, M., Li, H., Beer, T.M.: Patient-physician disagreement regarding performance status is associated with worse survivorship in patients with advanced cancer. Cancer **113**(8), 2205–2214 (2008)
10. Trigeorgis, G., Nicolaou, M.A., Zafeiriou, S., Schuller, B.W.: Deep canonical time warping. In: IEEE Conference on Computer Vision and Pattern Recognition (CVPR), pp. 5110–5118 (2016)
11. Vu, H.T., Carey, C.J., Mahadevan, S.: Manifold warping: Manifold alignment over time. Proceedings of the National Conference on Artificial Intelligence **2**, 1155–1161 (2012)
12. Wang, R., Medioni, G., Winstein, C.J., Blanco, C.: Home monitoring musculo-skeletal disorders with a single 3D sensor. In: IEEE Computer Society Conference on Computer Vision and Pattern Recognition Workshops, pp. 521–528 (2013)
13. Zhou, F., De La Torre, F.: Generalized canonical time warping. IEEE Trans. Pattern Anal. Mach. Intell. **38**(2), 279–294 (2016)
14. Zhou, F., de la Torre, F.: Canonical time warping for alignment of human behavior. In: Advances in Neural Information Processing Systems (NeurIPS), pp. 1–9 (2009)